P9-ARG-787

TECHNICAL COLLEGE OF THE LOWCOUNTRY
LEARNING RESOURCES CENTER
POST OFFICE BOX 1288
BEAUFORT, SOUTH CAROLINA 29901-1288

GREAT EVENTS

1900-2001

GREAT EVENTS

1900-2001
REVISED EDITION

Volume 8
1998-2001
Indexes

From

The Editors of Salem Press

SALEM PRESS, INC.

Pasadena, California   Hackensack, New Jersey

TECHNICAL COLLEGE OF THE LOWCOUNTRY
LEARNING RESOURCES CENTER
POST OFFICE BOX 1288
BEAUFORT, SOUTH CAROLINA 29901-1288

Editor in Chief: Dawn P. Dawson

Managing Editor: R. Kent Rasmussen *Research Supervisor:* Jeffry Jensen
Manuscript Editor: Rowena Wildin *Acquisitions Editor:* Mark Rehn
Production Editor: Joyce I. Buchea *Page Design and Graphics:* James Hutson
Photograph Editor: Philip Bader *Layout:* William Zimmerman
Assistant Editor: Andrea E. Miller Eddie Murillo

Cover Design: Moritz Design, Los Angeles, Calif.

Cover photos: Center image—Corbis, Remaining images—AP/Wide World Photos
Half title photos: Library of Congress, Digital Stock, AP/Wide World Photos

Copyright © 1992, 1994, 1996, 1997, 2002, by Salem Press, Inc.
All rights in this book are reserved. No part of this work may be used or reproduced in any manner whatsoever or transmitted in any form or by any means, electronic or mechanical, including photocopy, recording, or any information storage and retrieval system, without written permission from the copyright owner except in the case of brief quotations embodied in critical articles and reviews. For information address the publisher, Salem Press, Inc., P.O. Box 50062, Pasadena, California 91115.

© 2002 *Great Events: 1900-2001, Revised Edition*
© 1997 *The Twentieth Century: Great Scientific Achievements, Supplement* (3 volumes)
© 1996 *The Twentieth Century: Great Events, Supplement* (3 volumes)
© 1994 *The Twentieth Century: Great Scientific Achievements* (10 volumes)
© 1992 *The Twentieth Century: Great Events* (10 volumes)

∞ The paper used in these volumes conforms to the American National Standard for Permanence of Paper for Printed Library Materials, Z39.48-1992 (R1997).

Library of Congress Cataloging-in-Publication Data
Great events : 1900-2001 / editors of Salem Press.— Rev. ed.
      v. cm.
Includes index.
    ISBN 1-58765-053-3 (set : alk. paper) — ISBN 1-58765-061-4 (vol. 8 : alk. paper)
    1. History, Modern—20th century—Chronology. 2. Twentieth century. 3. Science—History—20th century—Chronology. 4. Technology—History—20th century—Chronology.
D421 .G627 2002
909.82—dc21

2002002008

First Printing

PRINTED IN THE UNITED STATES OF AMERICA

CONTENTS

CONTENTS

COMPLETE LIST OF CONTENTS

VOLUME 1

VOLUME 2

VOLUME 3

VOLUME 4

VOLUME 5

VOLUME 6

xxiii

VOLUME 7

VOLUME 8

GREAT EVENTS

1900-2001

Hun Sen and Norodom Ranariddh Form Cambodian Coalition

The newly restored Cambodian monarchy avoided renewed civil war when the nation's two most powerful political leaders agreed to a new power-sharing arrangement.

What: International relations; Government

When: November 30, 1998

Where: Phnom Penh, Cambodia

Who:

NORODOM SIHANOUK (1922-   ), king from 1993 and former chief of state of Cambodia

NORODOM RANARIDDH (1944-   ), prince of Cambodia

HUN SEN (1951-   ), prime minister of Cambodia from 1998

Civil War

The twentieth century history of Cambodia was dominated by Norodom Sihanouk, prince of the Norodom line of Cambodia's royal family, who was chosen by the French government in 1941 to rule Cambodia as a puppet king. Revered as a living god by Cambodia's people, King Norodom Sihanouk successfully maneuvered the French out of Cambodia in 1954; abdicated the throne in favor of his father, Norodom Suramarit, in 1955; and assumed the position of prime minister until a 1970 U.S.-backed coup replaced him with a pro-Western government willing to side with the United States against North Vietnam.

Prince Sihanouk sought refuge in China, where he organized a government in exile, the Royal Government of National Union, allied with Cambodia's communist Khmer Rouge party. Sihanouk and the Khmer Rouge successfully defeated the U.S.-backed Cambodian government in 1975. With Khmer Rouge support, Prince Sihanouk served as a figurehead chief of state until 1976, when his former communist allies placed him under house arrest. The Khmer Rouge slaughter of an estimated one million Cambodians over a three-year period beginning in 1976 led some Khmer Rouge to defect to communist Vietnam.

In 1979, Vietnam invaded Cambodia and drove the Khmer Rouge out of the population centers and into the mountains, where they continued to wage limited guerrilla warfare. Vietnam installed a new Cambodian government made up of Khmer Rouge defectors, among them Hun Sen, who was appointed foreign minister, then prime minister in 1985. Prince Sihanouk, released from house arrest, did an about-face and returned to China, where he sided with the Khmer Rouge in opposing the Vietnamese "occupation" of Cambodia.

Cambodian politics has always been a series of rapidly shifting alliances. From 1979 to 1991, Prince Sihanouk continuously negotiated and rejected peace agreements with Hun Sen's government, threatened to break with the Khmer Rouge, and considered retirement to China. On August 27, 1991, Sihanouk deserted his Khmer Rouge allies and signed a peace agreement with Prime Minister Hun Sen, which preserved a place for him in a new government that would seek a United Nations-sponsored election in Cambodia.

Restoration of the Monarchy

The United Nations supervised and administered Cambodia until elections could be held. The Khmer Rouge were initially permitted to participate in the new government, but when Sihanouk returned to Cambodia, he recommended that the Khmer Rouge be denied a governmental role and be prosecuted for their crimes against the Cambodian people. As president of a transition government, Prince Sihanouk turned over the leadership of his royalist political party, the National United Front for an Independent, Neutral, Peaceful and Cooperative Cambodia (FUNCINPEC), to his son, Prince Norodom Ranariddh.

The United Nations sent 22,000 soldiers, police, and officials into Cambodia to disarm the country's rival factions and organize elections. The May, 1993, elections gave fifty-eight seats to Prince Ranariddh's royalist party and fifty-one seats to Prime Minister Hun Sen's Cambodia's People's Party (CPP) in the nation's 120-seat national assembly. The remaining eleven seats were distributed between two other parties. The national assembly approved a new constitution for Cambodia, making the country a constitutional monarchy with Norodom Sihanouk restored as king. Cambodia's two major politicians, Prince Ranariddh and Prime Minister Hun Sen, accepted a power-sharing arrangement in which Prince Ranariddh and Hun Sen were made first and second co-prime ministers, respectively.

The alliance between FUNCINPEC and the CPP was always shaky. Each party appointed its own ministers to each government agency and department, creating political stalemate. Hun Sen led a July, 1997, coup against Prince Ranariddh and the FUNCINPEC party on the pretext that they were forming an alliance with the hated Khmer Rouge, stockpiling weapons to kill Hun Sen, and planning to eliminate the CPP. The royalists' headquarters in the capital was seized, FUNCINPEC legislators and government officials were killed, arrested, or fled, and Prince Ranariddh, fearing for his life, fled to neighboring Thailand.

King Sihanouk refused to bless Cambodia's new government with Hun Sen as first prime minister but offered, unsuccessfully, to mediate the dispute between Hun Sen and Ranariddh. Hun Sen's coup against a democratically elected government angered the United Nations and the Association of South East Asian Nations (ASEAN). Bomb attacks against other anti-Hun Sen politicians within Cambodia; Hun Sen's overruling of King Sihanouk's stated wishes; banishment of the king's half-brother, Prince Norodom Sirivuddh; and the plan to try Prince Ranariddh for treason

caused the United Nations to deny Hun Sen's government the right to Cambodia's seat at the organization and resulted in various nations denying Cambodia further international financial aid.

Bowing to international pressure, Hun Sen agreed to a peace agreement with Prince Ranariddh. Prince Ranariddh acknowledged illegal activities for which he was convicted but then pardoned by King Norodom Sihanouk. New elections were held in July, 1998, giving Hun Sen a majority of the national assembly seats. On November 30, the national assembly approved the creation of a new coalition government pairing the CPP and FUNCINPEC. Hun Sen was elected prime minister of Cambodia, and Prince Ranariddh was elected president of the national assembly. A new legislative chamber, a senate, was created as an upper house headed by CPP leader Chea Sim.

Consequences

Years of war and conflict destroyed much of Cambodia's infrastructure and impoverished many of its people. In addition, the human rights of many of its citizens were violated during the Khmer Rouge regime. The new coalition government, recognized and supported by the United States, brought some measure of peace and stability to the region. Cambodia was able to regain its seat in the United Nations and soon was attempting to gain full membership in ASEAN. The stated goals of the new coalition were to strengthen democracy and the rule of law, create a transparent and accountable government, improve the economy, and alleviate poverty.

King Norodom Sihanouk's advancing years and ill health raised the issue of succession to the throne. In Cambodia, royal succession is by election not by birth order. As long as Prince Ranariddh had remained in exile and with treason charges against him, he would have been ineligible to be king. The peace agreement renewed his eligibility. After the agreement, Hun Sen became recognized as the power broker in Cambodia, but the monarchy remained a valuable moral check on the prime minister's authority and a focus for national reconciliation.

William A. Paquette

Scandal Erupts over 2002 Winter Olympic Games in Utah

> *The biggest scandal in Olympic history occurred when it was revealed that International Olympic Committee members accepted bribes to vote for city sponsors of Olympic Games.*

What: Sports; Crime
When: December, 1998
Where: Salt Lake City, Utah
Who:
MARC HODLER, International Olympics
Committee member coordinating Salt
Lake City preparations
DAVID JOHNSON, senior vice president and
bid leader for the Salt Lake
Organizing Committee
THOMAS WELCH, bid leader for the Salt
Lake Organizing Committee
CHRIS VANOCUR, reporter at KTVX-TV in
Salt Lake City

The Olympic Scandal Breaks

On November 24, 1998, Chris Vanocur, an investigative reporter at KTVX-TV in Salt Lake City, Utah, announced that a letter from the Salt Lake Organizing Committee (SLOC) indicated that inappropriate personal benefits were extended to family members of the International Olympic Committee (IOC) to secure the required bids to host the 2002 Winter Olympic Games. The letter revealed that the SLOC had provided a scholarship to Sonia Essomba, daughter of late IOC member Rene Essomba of Cameroon. The next day, the SLOC denied charges of bribery by describing its activities as "humanitarian assistance." On December 8, the SLOC reported that thirteen individuals had received more than $393,871 in tuition assistance.

National Public Radio reporter Howard Berkes followed up bribery allegations by contacting Marc Hodler, the IOC member who coordinated the Salt Lake Olympic preparations. Hodler emphatically stated that the scholarships were im-

proper. Hodler believed this information would give him the leverage he needed to reform the IOC by limiting the gift giving that takes place between bidding cities and committee members. On December 12, Hodler alleged corruption in determining the location of the Olympics not only in Salt Lake City but also in Atlanta, Georgia; Nagano, Japan; and Sidney, Australia.

Documenting the Corruption

After the IOC bidding system became an international scandal, investigations began at all levels of the Olympics organization. The Utah attorney general's office began a preliminary criminal inquiry that ended when the Department of Justice, Federal Bureau of Investigation, and U.S. Internal Revenue Service initiated an investigation into whether bribery, fraud, public corruption, and tax laws were broken. Former senator George Mitchell was appointed to lead an investigating commission to look into bribery and corruption allegations for the U.S. Olympics Committee. In Switzerland, IOC president Juan Antonio Samaranch appointed an ethics commission to establish future bidding guidelines for potential cities.

The SLOC established an ethics panel headed by Brigham Young University professor Barbara Day Lockhart. This panel conducted an internal investigation, concluding that SLOC bid bosses David Johnson and Thomas Welch hid their IOC lobbying scheme from their board of directors. Frank Joklik, president and chief executive officer of the SLOC, and Johnson, SLOC vice president, were forced to resign, and the committee was reorganized. Investigations revealed that Johnson and Welch hired several consultants to identify IOC members whose votes could be secured by "specific personal benefits." One such consultant, Mahmoud Elfarnawani, claimed that he could ensure the Arab vote for Salt Lake City and was paid $148,260 between 1992 and 1996. Johnson and Welch developed a document listing IOC members whose votes they believed could be bought with "personal benefits." It was revealed that bribes included cash, medical treatment, scholarships, firearms, job offers, shopping sprees, and vacations for IOC family members. In the end, Johnson and Welch were indicted on fifteen federal counts of conspiracy, fraud, and racketeering.

Other investigations alleged that the Nagano bid committee spent an average of $22,000 on each of sixty-two visiting IOC members in addition to providing the millions of dollars necessary to build the Olympic Museum in Lausanne, Switzerland. The Australian Olympic Committee offered IOC members monetary inducements to vote for Sydney the night before winning the bid over Beijing by two votes. Ultimately, four IOC committee members resigned, and six were expelled for accepting bribes and inducements.

Consequences

The Salt Lake City bidding scandal caused immediate internal and external investigations into Olympic committees at all levels. The first criminal charges brought against an individual resulted in Utah businessman David Simmons pleading guilty to a federal misdemeanor tax charge for creating a sham job for the son of IOC member Kim Un-yong. U.S. Olympic official Alfredo La Mont also pleaded guilty to two counts of tax fraud. Johnson and Welch were scheduled to stand trial during the summer of 2001. In July of that year, U.S. district juge David Sam dis-

missed four the fifteen felony counts against the defendants and indefinitely postponed their trial on the remaining counts.

Meanwhile, on March 1, 1999, the IOC announced changes designed to eliminate corruption in the bidding process. Under the new rules, IOC members were to be banned from travel to Olympic bid cities; a small panel was to select the final two candidate cities, which would then be voted on by the full IOC membership; and strict limits would be placed on IOC members' contact with bidders. The IOC established an ethics commission with the majority of its members from outside the committee. Finally, an audit was released showing that the Olympics were in good financial standing, with bank deposits and trust funds totaling $237 million at the end of 1998.

This scandal changed the way in which the Olympics were investigated and reported. The Olympics committees, which once were considered above reproach, came to be regarded much the same as any other powerful international organizations. In the past, most coverage was provided by sports reporters; however, major news organizations began assigning political and investigative reporters to cover all but the actual sporting events.

Although the bidding scandal did not cause any city to be stripped of hosting the Olympic Games, in the first vote on an Olympic site since the scandal broke, Turin, Italy, was selected over Sion, Switzerland. This was seen as backlash against the IOC and Hodler, the IOC member who made corruption allegations in the site selection process.

Gerald S. Argetsinger

Exxon and Mobil Merge to Form World's Largest Oil Company

> *In 1909, antitrust action divided John D. Rockefeller's Standard Oil Trust. The two largest parts became Exxon and Mobil, giant global corporations. Their reintegration in 1998 created both the largest U.S. corporation and the world's largest petroleum company.*

What: Business; Energy
When: December 1, 1998
Where: Exxon corporate headquarters, Irving, Texas
Who:
JOHN D. ROCKEFELLER (1839-1937), founder of Standard Oil Trust in 1882
LEE R. RAYMOND (1938-   ), chairman and chief executive officer of Exxon and the merged Exxon Mobil Corporation
LUCIO "LOU" NOTO (1939-   ), chairman and chief executive officer of Mobil Oil, vice chairman of Exxon Mobil Corporation

Black Gold

The U.S. petroleum industry began in the nineteenth century in the oil fields of Pennsylvania, grew on the production of kerosene, and consolidated under John D. Rockefeller's leadership. While others focused on exploration and production, Rockefeller concentrated on controlling the refining industry. By 1900, his Standard Oil Trust had diversified into the production and distribution of the lubricants demanded by expanding U.S. industry, and Rockefeller had become the richest man in the world.

The power and wealth that Rockefeller amassed through his monopolistic control over the U.S. petroleum industry inevitably attracted the attention of turn-of-the-century government regulators armed with the Sherman Antitrust Act (1890). In 1911, following a two-year battle with the Department of Justice, the Standard Oil Trust was split into thirty-four companies. The larg-est of these were Jersey Standard (the Standard Oil Company of New Jersey), which eventually evolved into Exxon (1972), and Socony (Standard Oil Company of New York), which eventually became Mobil Oil (1968).

Within the Standard Oil Trust, Jersey Oil and Socony's strengths were in refining and marketing. Split apart from their regular sources of petroleum with the breakup of the trust, both companies found themselves short of crude oil and had to search for oil at home and abroad. Within two decades, each had become a fully integrated company with production, refining, and marketing departments, and both had become fully integrated multinationals, with extensive overseas operations in upstream (exploration, drilling) and downstream (refining, marketing, technology research) activities.

These companies first began to collaborate again in their international ventures. As before, their collaboration occurred within a monopolistic framework: the Seven Sisters, the international oil cartel composed of Exxon, Royal Dutch/Shell, Mobil, Texaco, Gulf Oil, Standard Oil of California, and British Petroleum (BP), which as late as 1950 controlled 90 percent of all oil production outside of the United States and the Soviet Union. Inside the United States, however, the threat of antitrust action continued to discourage collaboration between the principal petroleum companies.

During Ronald Reagan's presidency (1981-1989), the U.S. government's opposition to the merger of giant U.S. corporations waned significantly. One of the fallouts of the oil crisis of 1973, when the Arab oil embargo of countries supporting Israel in the October Arab-Israeli war triggered a quadrupling of the price of oil, was the

AP/Wide World Photos

Exxon chairman Lee Raymond (left) and Mobil chairman Lucio Noto, at a press conference announcing their companies' merger.

emergence of a new cartel controlling the price and availability of oil being exported to the developed, importing world. This time, the members were the oil-producing states themselves, organized within the Organization of Petroleum Exporting Countries (OPEC). By the time Reagan took office, the supply disruptions caused by the January, 1979, revolution in Iran and the Iraq-Iranian war in September of that year had caused another energy crisis, and the federal government had become much less concerned with maintaining competition inside the U.S. petroleum industry than with reducing the United States' 50 percent dependency on overseas oil suppliers.

The Merger

Most of the oil industry mergers during the Reagan years involved larger petroleum corporations acquiring smaller oil companies. The motivation for the mergers, however, differed considerably during the earlier and later part of the

1980's. During the decade's early years, profits resulting from the surge in oil prices from under two dollars per barrel in 1970 to nearly forty dollars per barrel by 1980 were used by many companies to acquire smaller petroleum firms. Then came the collapse of oil prices in the middle 1980's, and cost-cutting considerations became the driving force for petroleum industry mergers until the rebounding price of oil at decade's end removed that incentive.

A decade later, cost-cutting considerations again encouraged oil industry mergers. For several years after the United States-led 1991 eviction of Iraqi forces from Kuwait (the Persian Gulf War), the price of oil was either flat or falling on the world market. Indeed, in the months preceding the Exxon-Mobil merger, a glut of oil had driven the price of oil to below twelve dollars per barrel, its lowest point in twelve yeas. In this setting, an Exxon-Mobil merger held substantial financial advantages. Analysts estimated that as many as nine thousand jobs could be eliminated, and overall savings would be between $3 billion and $7 billion.

The merger also produced other financial benefits. With a combined market capitalization of more than $236 billion, Exxon-Mobil became the world's third largest company and the dominant actor in the petroleum field, nearly twice the size of its largest competitor, BP/Amoco. It was therefore better positioned to take advantage of the economies of scale necessary to prosper in the capital-intensive petroleum industry, in which refineries had to be ten times larger than those of the previous generation to be profitable. Similarly, in the uncertain economic and political worlds of oil exploration and development, the merger spread Exxon-Mobil's operations—and risks—over a wider area of the globe.

Finally, though less easily quantified, the merger was facilitated by the similar corporate cultures of these two descendants of Rockefeller's Standard Oil Trust. For two years before the merger, Mobil had discussed joint ventures with such prominent oil firms as Amoco, BP, Conoco, and Chevron. Frequently, however, corporate incompatibility as much as financial considerations blocked the transaction

Consequences

The agreement between Lee Raymond, Exxon's chairman and his counterpart at Mobil, Lou Noto, was completed November 30, 1999. To acquire Mobil, Exxon paid $77.2 billion, a record in the world of corporate mergers. By any standard, the resultant Exxon Mobil Corporation represented the petroleum industry's flagship, with operations in more than two hundred countries, a fleet of more than forty tankers, 13,000 miles of pipeline, 40,000 service stations, and more than 20 billion barrels of oil and gas reserves worldwide. The merger also led to another record almost immediatcly. Only slightly more than a year later, riding the crest of rising oil prices, Exxon Mobil recorded an annual profit of $16.9 billion, a corporate world record.

Joseph R. Rudolph, Jr.

Puerto Ricans Vote Against Statehood

A majority of Puerto Rican voters turned down statehood in a plebiscite on the future political relationship between Puerto Rico and the United States.

What: International relations; National politics

When: December 13, 1998

Where: Puerto Rico

Who:

PEDRO J. ROSELLÓ (1944-   ), former governor of Puerto Rico, New Progressive Party

ANÍBAL ACEVEDO VILA (1962-   ), former senator of Puerto Rico, president of the Popular Democratic Party

RUBÉN BERRÍOS MARTÍNEZ (1939-   ), former senator of Puerto Rico, Puerto Rican Independentist Party

Colonialism and Commonwealth

On July 25, 1898, during the Spanish-American War, the United States invaded Puerto Rico. After hostilities ended under the Treaty of Paris signed on December 12, 1898, the United States occupied and subsequently colonized Puerto Rico. Following World War II, the United Nations placed Puerto Rico on its protected territories list, after which the United States was required to submit an annual report on the overall situation in Puerto Rico. On July 25, 1952, Puerto Rico's new status of Estado Libre Asociado (free associated state), popularly known as commonwealth, was ratified, and this status was interpreted as enabling Puerto Rico to be self-governing.

In 1953, in response to a U.S. petition, the U.N. General Assembly removed Puerto Rico from its protected territories list. However, in the 1970's and in the decades that followed, the U.N. Special Committee on Decolonization proclaimed that Puerto Rico was a colony and the United States must begin the decolonization process immediately. The committee considered Puerto Rico to be a colony because the United

States reserves the right to be the final authority regarding Puerto Rico's future and can overturn any decisions made in Puerto Rico's legislature. For example, in 1996, the U.S. Congress voted to eliminate Section 936 of the Internal Revenue Service code, which gave tax-exemption incentives to attract companies to the island, despite opposition from Puerto Rico and its supporters.

Politics and the Plebiscite

In March of 1998, the U.S. House of Representatives passed the United States-Puerto Rico Political Status Act by a vote of 209-208, allowing Puerto Rico to hold a referendum on its political status. This referendum would be binding on the U.S. Congress, which meant that the United States would have to abide by the decision favored by the majority of the voters in Puerto Rico. Taking advantage of this situation, the New Progressive Party (PNP) prostatehood governor Pedro J. Roselló pushed for Puerto Rico to hold its own plebiscite by the year's end, hoping to prove to Congress that statehood was the favored status among Puerto Ricans.

Controversy quickly arose about the wording of the ballot. The mostly PNP legislature, which favored statehood, had written the description for all of the alternative options in such an unfavorable way that those who supported these options could no longer vote for them. The Popular Democratic Party (PPD), headed by Aníbal Acevedo Vila, which supported commonwealth status, objected because the description for the commonwealth option included the statement that U.S. citizenship—granted in 1917—was not guaranteed under commonwealth status. The PPD claimed that the addition of this phrase to the option was designed as a fear tactic that would skew the votes in favor of statehood. The PPD launched a protest campaign in favor of the option "none of the above." Some who believed

in independence for Puerto Rico joined the campaign as a political maneuver to voice their opposition and to help ensure that statehood would not receive the majority vote. However, because this plebiscite was not sanctioned by and not binding on the United States, many who favored independence refused to participate in what they considered to be a farce.

The plebiscite was held on December 13, 1998, on the heels of devastation caused by two major hurricanes, Georges in September and Mitch in October. The plebiscite coincided with the distribution of U.S. relief checks to help rebuild homes and businesses in Puerto Rico. Many Puerto Ricans believed the timing of the plebiscite was part of the prostatehood government's strategy and meant to demonstrate the benefits of good relations with the United States and how economic assistance would be guaranteed and augmented under statehood. Despite these efforts, a slim majority of Puerto Ricans voted against statehood. When the votes were tallied, 50.2 percent voted for the option "none of the above," followed by 46.5 percent for statehood, 2.5 percent for independence, 0.1 percent for commonwealth, and 0.2 percent for free association.

Consequences

The successful protest campaign demonstrated that Puerto Ricans were strongly divided on the issue of Puerto Rico's political status. The U.S. Senate did not ratify the House measure, and the issue of the political status of Puerto Rico

has not been addressed by the United States since the plebiscite.

In April of 1999, massive civil disobedience broke out across political party lines in Vieques, an island municipality southeast of the Puerto Rican mainland, when the U.S. Navy accidentally killed one civilian and wounded several others during routine bombing exercises. This activity was spearheaded by Rubén Berríos Martínez, former senator from the Puerto Rican Independentist Party, in protest against the U.S. Naval occupation of two-thirds of the island for guerrilla warfare training, bomb testings, and air-land-sea operations. Although the people of Puerto Rico demanded the Navy evacuate the island for health, ecological, and moral reasons, the United States did not concede.

In the November, 2000, elections, the residents of Puerto Rico cast their votes largely in favor of a procommonwealth legislature and elected the first female governor, Sila M. Calderón. During her campaign, Calderón vowed to secure the removal of the U.S. Navy from Vieques. In choosing her, the people of Puerto Rico showed their dissatisfaction with the prostatehood government, which was shrouded in scandal and corruption. However, because the new government leaders were in favor of Puerto Rico remaining under commonwealth status, experts doubted that a plebiscite would be held soon and could only guess whether it would be sanctioned by and binding on the United States.

María Elizabeth Pérez y González

United States and Great Britain Bomb Military Sites in Iraq

> *U.S. president Bill Clinton and British prime minister Tony Blair authorized the bombing of five Iraqi radar sites on the eve of the impeachment vote against Clinton.*

What: War; National politics
When: December 16, 1998
Where: Iraq and Washington, D.C.
Who:

BILL CLINTON (1946-   ), president of the United States from 1993 to 2001

TONY BLAIR (1953-   ), prime minister of Great Britain from 1997

SADDAM HUSSEIN (1937-   ), president of Iraq from 1979

Violation of Treaty Obligations

At the conclusion of the Persian Gulf War in early 1991, restrictions imposed on Iraq by the United Nations included an agreement that required United Nations (U.N.) inspectors to visit sites within the country for the purpose of verifying compliance with the treaty terms governing the production of biological, chemical, and nuclear weapons. Over the next several years, Iraq repeatedly violated the agreement and attempted to renegotiate new terms, backing down only when the threat of retaliation appeared imminent.

On October 31, 1998, Iraqi officials suspended all cooperation with the U.N. inspectors, prompting U.S. president Bill Clinton and British prime minister Tony Blair to order additional troops and ships into the region for a possible military response. While the military coordinated the movement of men and material into the Persian Gulf area, U.N. secretary-general Kofi Annan contacted Iraqi president Saddam Hussein in an effort to persuade him to reconsider his position. As negotiations continued, Clinton authorized National Security Adviser Sandy Berger to inform the Joint Chiefs of Staff that the strike should commence on the following day. At 8:00 A.M.

Eastern standard time on November 14, the Cable News Network (CNN) reported that an Iraqi government announcement would be made shortly, prompting Clinton to temporarily place the strike on hold. Later that morning, a letter arrived from Baghdad in which Iraq agreed to permit the return of U.N. inspectors, but since the dispatch contained what appeared to be conditions, Clinton rejected the offer. The following day, November 14, Clinton received a second letter from Iraq, clarifying its position and informing him that the "conditions" were meant as preferences only. A third letter that officially rescinded Iraq's decision to end cooperation with U.N. inspectors arrived later that afternoon. Clinton suspended the U.S. military operation and contacted Blair, who then ordered British troops to follow the lead of the American forces.

On November 17, U.N. inspectors returned to Iraq, but within one week, Iraqi leaders denied their requests for access to specific documents, claiming that the demands were provocative and designed to justify a military attack against their country. For the next three weeks, Iraq continually denied U.N. inspectors access to the reports, leading Richard Butler, the chief U. N. inspector, to issue a statement on December 15 stating that Hussein had failed to fulfill the earlier promises and instead had placed additional restrictions on U.N. investigators.

Military Action Begins

On December 16, as the U.N. team left Baghdad, Clinton conferred with his national security advisers regarding the situation. Domestically an attack against Iraq on the eve of the House of Representatives' vote on impeachment articles might be suspicious, but international concerns over initiating an attack during the Muslim holy month of Ramadan, then just days away, required an immediate response. Clinton authorized the deployment of more than two hundred aircraft and twenty warships, including the USS *Enterprise* along with fifteen B-52 bombers carrying more than four hundred cruise missiles and a variety of other arms. Targets included military installations and possible chemical, biological, and nuclear weapons sites as well as the barracks of the Republican Guards. A second attack on the following day knocked out additional military air defense sites for the protection of U.S. and British pilots. Although the U.N. Security Council held an emergency session throughout the day, no course of action was decided upon.

Meanwhile, the U.S. Navy commenced Operation Desert Fox, attacking at 3:10 P.M. Eastern

standard time, deploying two hundred Tomahawk missiles from ships located in the Persian Gulf. One and one-half hours later, the president informed Congress of the attack. Shortly after 5:00 P.M. Eastern standard time, the first missiles landed in Iraq as the White House announced that a "substantial" attack was underway. During prime time, Clinton informed the American public about the strike against Iraq and warned Saddam Hussein that any violation of treaty obligations would be dealt with swiftly and severely. An hour later, Republicans voted to delay the impeachment hearings.

Consequences

Critics of President Clinton questioned the timing of the air strike, raising concerns that Op-

The Pentagon released these photographs of the military intelligence headquarters in Baghdad before (left) and after (right) the U.S. air strike.

eration Desert Fox was initiated to push the impeachment hearings from the front page of the news. Congressional members, including Representatives Joe L. Barton and Ron Paul of Texas, Senator Paul Coverdell of Georgia, Senator Bill Frist of Tennessee, and Senator Richard Shelby, supported military action against Hussein but argued that Clinton had already paused the strike before and should not have initiated hostilities while facing the impeachment vote. Others, including both Republicans and Democrats such as Senate Minority Leader Tom Daschle, Senator John Chafee, and Senator Sam Brownback, supported the president's action, arguing that military action was necessary immediately because the Muslim holy month prevented strikes for the next thirty days, during which time, Hussein could have prepared his forces for a confrontation.

During the next year, Iraq rebuilt most of the military sites destroyed in the bombing. The United Nations, concerned with the increasing threat from Iraq, proposed a resolution designed to persuade Iraq into once again accepting U.N. inspectors in exchange for a loosening of restrictions on the country's economy. On December 19, 1999, the United Nations passed a resolution, by a vote of 11-0, with Russia, China, France, and Malaysia abstaining, calling for the removal of the $5.26 billion limit imposed on the sale of oil for food and the lifting of sanctions against the country for renewable 120-day periods once inspectors verified disarmament. Iraq rejected the resolution and prepared to accept the consequences. Although official sanctions remained in place for most items, the U.N. agreed to gradually increase the total amount of oil that could be sold to pay for food and other humanitarian items such as medical supplies.

Cynthia Clark Northrup

House of Representatives Impeaches Clinton

> *The U.S. House of Representatives approved two articles of impeachment alleging that President Bill Clinton committed perjury and obstructed justice in concealing his affair with former White House intern Monica Lewinsky. It rejected charges that Clinton committed perjury in the Paula Jones case and abuse of power.*

What: Government; Law
When: December 19, 1998
Where: Washington, D.C.
Who:

BILL CLINTON (1946-   ), president of the United States from 1993 to 2001
MONICA LEWINSKY (1973-   ), former White House intern
KENNETH STARR (1946-   ), independent counsel
HENRY HYDE (1924-   ), House Judiciary Committee chairman
PAULA JONES (1967-   ), former Arkansas state employee

The Path to Impeachment

In 1998, Americans learned about President Bill Clinton's extramarital affair with Monica Lewinsky and his attempts to hide the nature of the relationship from lawyers of Paula Jones, who was pursuing a sexual harassment lawsuit against him, and from his family, friends, and the American public. Independent counsel Kenneth Starr in September, 1998, submitted a report to Congress alleging that Clinton had lied under oath and obstructed justice in his cover-up of the affair.

The House on October 8 started proceedings toward possible impeachment. Republicans held 228 of 435 House seats but lost five seats in the November 3 midterm elections. Republican Representatives wanted to vote on impeachment before the 105th Congress adjourned and their narrow majority declined. Polls, however, showed that a majority of Americans supported Clinton.

Articles of Impeachment

During November and early December, 1998, the House Judiciary Committee, consisting of twenty-one Republicans and sixteen Democrats, heard presentations by Starr, constitutional experts, and counsels for the president and the committee. Starr recommended that the president be impeached on several counts. Five former prosecutors argued that it would be difficult to prove that Clinton intentionally lied. Charles Ruff, the White House counsel, termed the president's conduct reprehensible but urged the House not to overturn "the mandate of the American people." The committee did not ask the main figures in the case to testify.

Committee chairman Henry Hyde, a Republican from Illinois, sent a list of eighty-one questions to Clinton, who admitted "wrong" conduct but strongly denied that he had lied under oath or urged anyone to do so. The White House on December 8 presented a 184-page legal brief defending Clinton. It claimed that Clinton had sought to find a job for Lewinsky long before she became a witness in the Jones suit and that Clinton's conversation with his personal secretary Betty Currie had taken place before she was asked to testify in the case.

After bitter debate, the committee on December 11-12 adopted four articles of impeachment. The articles, based on Starr's earlier recommendations, alleged that Clinton had (I) given "perjurious, false, and misleading testimony" to a federal grand jury on August 17, 1998, about his relationship with Lewinsky; (II) lied about his relationship with Lewinsky in a January 17, 1998, deposition in the Jones case; (III) obstructed justice by influencing Lewinsky and Currie to testify

2989

AP/Wide World Photos

President Bill Clinton addresses a group of Democrats after the House of Representatives voted to impeach him.

falsely and conceal evidence; and (IV) abused the powers of his office by lying to Congress about his actions. Articles I, III, and IV were adopted 21-16 along straight party lines. One Republican opposed Article III, which passed 21-17. The Jones case had already been settled out of court, with Clinton consenting to pay her $850,000. An alternative motion to censure Clinton, proposed by committee Democrats and endorsed by the White House, failed 22-14. Several moderate Republicans, representing districts that had supported Clinton in 1992 and 1996, faced primary challenges in 2000 from anti-Clinton conservatives and announced they would vote to impeach.

The House debated the four impeachment articles on December 18 and 19, climaxing a turbulent week in which the United States and Great Britain launched intensive air strikes against Iraq and House speaker-designate Bob Livingston of Louisiana withdrew his candidacy after admitting a past extramarital affair. Hyde argued that "lying under oath . . . is a public act" and constitutes "willful, premeditated, deliberate corrup-

tion of the nation's system of justice." Democrat John Conyers, Jr., of Michigan countered that "impeachment was designed to rid this nation of traitors and tyrants, not attempts to cover up extramarital affairs."

After fourteen hours of emotional debate, the House on December 19 approved two impeachment articles. Article I, accusing Clinton of perjury, passed 228 to 206, with five Republicans dissenting and five Democrats consenting. Article III, charging Clinton with obstruction of justice,was approved 221-212. Twelve Republicans were opposed and the same five Democrats voted affirmatively. The House rejected Articles II and IV, 205-229 and 148-285 respectively. House leaders blocked a censure resolution from reaching the floor.

President Clinton, who appeared on the White House lawn with his wife, Hillary, and about one hundred Democrats after the voting, asked the Senate to punish him without removing him from office. Former presidents Gerald Ford and Jimmy Carter urged the Senate to support a censure resolution.

Consequences

The House had impeached an American president for just the second time. Congress in 1868 had impeached and acquitted Andrew Johnson. The Senate trial, presided over by U.S. Supreme Court Chief Justice William Rehnquist, began January 7, 1999. Thirteen Republican House managers, led by Hyde, prosecuted the case. Republicans controlled fifty-five Senate seats but needed support from at least twelve Democrats to secure the two-thirds majority required to convict and remove Clinton from office.

Senate leaders, hoping to avert partisan bitterness during the House proceedings, consented to a shortened trial. The managers and the White House legal team presented their cases, but no live witnesses were summoned. Senators heard a videotaped deposition from Lewinsky. According to Lewinsky's deposition, no one had asked her to deny having a sexual relationship with Clinton in her affidavit in the Jones case.

The Senate, on February 12, acquitted Clinton on both impeachment articles. Article I, alleging grand jury perjury, was defeated 45-55, and Article II, charging obstruction of justice, fell 17 votes short on a 50-50 tally. No Democrat voted for either article. For just the second time, an American president had been impeached and acquitted. The outcome showed the institutional differences between the highly partisan House and more collegial Senate and the inescapable power of political arithmetic. The nation, meanwhile, learned about constitutional history, the impeachment process, perjury and conspiracy laws, and congressional procedural rules.

Before leaving the presidency, Clinton on January 19, 2001, admitted that he gave false testimony about his relationship with Lewinsky and struck a deal with the independent counsel to avoid indictment. Clinton agreed to pay a $25,000 fine, accept a five-year suspension of his Arkansas law license, and forego seeking federal reimbursement for any legal fees.

David L. Porter

Haiti's President Préval Dissolves Parliament

> *After two years of stalemate between himself and parliament, René Préval, the president of Haiti, dissolved parliament and appointed a prime minister by decree.*

What: National politics
When: January 12, 1999
Where: Port-au-Prince, Haiti
Who:
RENÉ PRÉVAL (1943-   ), president of
    Haiti from 1996 to 2001
JACQUES EDOUARD ALEXIS, prime minister
    of Haiti from 1999 to 2001

Election of Préval

In 1986, a popular uprising ended the twenty-eight-year dictatorship of the Duvalier family. In the ten years that followed, the military staged four coups, making the attempt to develop a democracy difficult. In December of 1990, Jean-Bertrand Aristide, an activist Roman Catholic priest who accepted liberation theology, was elected president of Haiti, winning 67 percent of the vote. The election of Aristide represented a powerful popular uprising that united grassroots workers, peasants, and student organizations. Aristide's program of social and economic reforms threatened the position of the elite. He was removed from office by the military in September, 1991.

The United States opposed the military dictatorship but did not approve of the anti-imperialism policies of Aristide. No effective sanctions were imposed on the military junta by U.S. president George Bush, and President Bill Clinton followed the same policy. The Clinton administration entered negotiations with the Haitian military and reached an agreement that provided for the retirement of the military and the return of President Aristide. The United States sent fifteen thousand troops to Haiti to provide law and order and to retrain a police force.

Aristide was restored in September, 1994, but his ability to put his program into effect was severely compromised by the circumstances of his return to office. The United States, the Haitian military, and business leaders of Haiti forced Aristide to compromise in return for a loan of $700 million, $200 million of which was used to pay the Haitian debt to the United States. In 1996, Aristide supported the electoral campaign of René Préval, whom he had chosen as his successor. Haitians showed little interest in the electoral process. Only 5 percent of eligible voters participated in the local and municipal elections of 1997.

Difficulties for Préval

Préval's government was faced with problems from the beginning. He came to office amid charges of electoral fraud and with a parliament controlled by the opposition. Parliament not only refused to approve the four nominees for prime minister that Préval had submitted but also would not approve a budget for the nation. The stalemate created by the dispute between the executive and legislative branches made it impossible to have a functioning government and blocked Haiti from receiving foreign aid. Haiti had not had an effective government since June, 1997, when Prime Minister Rosny Smarth resigned, protesting that the elections had been rigged with Préval's complicity to favor Aristide loyalists.

On January 11, 1999, President Préval announced on television that he would end the government crisis by appointing a prime minister without parliamentary approval. The next day, he selected the minister of education, Jacques Edouard Alexis, as the prime minister. As soon as the prime minister chose his cabinet, Préval decreed the prime minister and his cabinet to be the new government. At the same time, he pledged to work with the opposition to create a trustworthy and nonpartisan electoral commission by February 3.

The president of the senate, Edgard Le Blanc, said that Préval had staged a coup against the democratic institutions of Haiti. Because the legislature was more unpopular than Préval, it could do little more than protest. Ordinary Haitians were much more concerned with the struggle against grinding poverty than with politics. Because their terms had expired shortly after the new prime minister was appointed, the legislators decided to extend their terms until a new government could conduct elections. Préval prevented the extension by dissolving parliament and calling for new elections. He replaced the mayors of cities whose terms had expired with interim management teams.

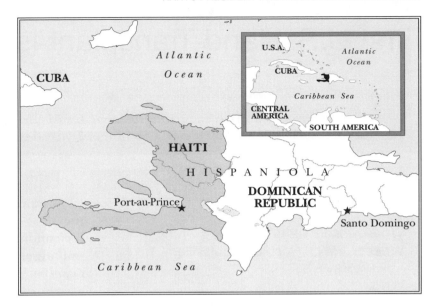

Street riots prevented the appointment by the February 3 deadline of the electoral council, which had the responsibility of organizing and conducting elections and counting ballots. The naming of the cabinet by the February 2 deadline was also delayed. On the next to last day of February, the Haitian supreme court refused to settle the dispute between the legislature and executive as requested by the parliament.

On March 26, Préval appointed a fifteen-member cabinet to end the almost two years of political impasse. The first order of business for the new government was the organization of elections, but most of the political parties said they would boycott the elections. Political unrest and failure to create the provisional electoral council to supervise the voting caused the elections to be postponed three times.

Because the reforms he wanted to put into effect were not advocated by Préval, former president Aristide broke with Préval and formed a new political party, the Organization of People in Struggle, to oppose the official Lavalas Party. Aristide said that Préval was too much under the influence of the United States and the economic elite of Haiti.

On May 2, 2000, elections for member of parliament and local officials were finally held. Despite charges of irregularities, the observers from the Organization of American States said that the elections were creditable. Runoff elections were delayed until July and August. The results gave the party of former president Aristide eighteen of the nineteen seats in the senate and about one-third of the seats in the lower house. Despite criticism within Haiti and from abroad, the government said that it would stand by the election results.

Consequences

The failure of the legislature and the executive to agree on any political procedure or program and the opposition to the long delayed elections by most political parties called into question the possibility of developing democracy in Haiti. Further political unrest could easily bring about more rioting and lead to economic and political chaos. Although the elections of 2000 returned Aristide to power, the military was not disarmed nor purged of the leaders of past coups. The military leaders remained in a position to use the political situation to justify another coup, which would halt the development of democracy in Haiti.

Robert D. Talbott

First U.S. Hand Transplant Is Performed

A team of surgeons performed a successful hand transplant operation in Louisville, Kentucky, enabling the recipient to perform twisting and gripping functions and to feel sensation in the hand.

What: Medicine
When: January 24, 1999
Where: Jewish Hospital, Louisville, Kentucky
Who:
MATTHEW SCOTT (1961?-   ), hand transplant recipient
WARREN C. BREIDENBACH, lead hand surgeon, head of surgical team
JON W. JONES, JR., lead transplant surgeon

A Complex and Risky Procedure

Hand transplant surgery is an extremely complex procedure. Unlike a solid organ transplant, a hand transplant involves multiple tissues: bones, tendons, cartilage, muscle, fat, nerves, blood vessels, and skin. After the difficult surgery is performed, a major possible complication is rejection, the natural response of the patient's immune system not to accept the alien hand.

An early hand transplant was attempted in 1964 in Ecuador in South America. The transplant was rejected within two weeks. By the 1980's, scientists had refined therapies using drugs called immunosuppressants, which prevent rejection of transplanted tissue or organs by suppressing the body's immune system. These drugs make organ transplants more practical, but carry risks for the recipient, including possibly fatal infections and an increased risk of developing cancer among other conditions. Because of the side effects of immunosuppressive drugs, transplants at first were limited to life-saving procedures, such as heart or liver transplants. Because a hand transplant is not a life-saving procedure, the risks were considered to outweigh the potential benefits.

Advances in immunosuppressive therapies in the 1990's allowed transplants to be performed for conditions that were not life-threatening. A multidisciplinary team of researchers and physicians, including transplant immunologists, hand surgeons, and micro-surgeons from Jewish Hospital in Louisville, Kentucky, and the University of Louisville developed a pioneering hand transplant program that was granted the first approval for a program of its kind in the United States in July, 1998.

The hand transplant procedure was primarily intended for healthy individuals who had experienced the loss of a hand or forearm because of trauma such as accident or amputation, rather than for congenital abnormalities (individuals born without a hand). Replacement hands would come from donors who met the criteria of total and irreversible brain damage, with the consent of their families.

A hand transplant operation was performed on an Australian man in Lyons, France, in September, 1998, by a group of doctors flown in from around the world. This was widely acknowledged to be the world's first hand transplant.

The First U.S. Hand Transplant

The first hand transplant procedure in the United States was performed at Jewish Hospital in Louisville, Kentucky, on January 24-25, 1999. The recipient was Matthew Scott, a thirty-seven-year-old paramedic from New Jersey. Scott lost his dominant left hand in a fireworks accident at the age of twenty-four. He had been using a prosthesis (artificial hand) and had relearned many activities necessary for daily living. However, an artificial hand does not allow twisting motions, has limited gripping ability, and has no feeling. The transplant included about two inches of forearm, to allow a functioning

2994

wrist. The donor was a brain-dead fifty-eight-year-old man.

The surgery lasted for fourteen hours, and was performed by a team that included Warren C. Breidenbach, lead hand surgeon, and Jon W. Jones, lead transplant surgeon. After surgery, Scott was placed on a combination of immunosuppressive drugs at a reduced dosage to lower the risks of cancer and infection associated with antirejection medication. This lower dosage of antirejection drugs was innovative. Because the hand transplant was not a life-saving procedure, the drug treatment could be less aggressive than with organ transplant patients.

Scott remained in the Louisville area for three months, for follow-up biopsies and laboratory evaluations to monitor immunosuppressive drug therapy, to watch for possible episodes of rejection, and to begin intensive physical therapy. A year after surgery, he had good hand function, range of motion, grip, and sensation, allowing him to tie his shoes, turn the pages of a newspa-

per, and throw a baseball, among other actions. He could sense hot and cold and feel pain. He would continue watching for rejection and monitoring of the effects of immunosuppressive drugs for the rest of his life.

Consequences

The most significant impact of Scott's hand transplant surgery in January, 1999 is the advance in immunosuppressive drug therapy that made its success possible. Many people felt that the hand would be rejected almost immediately, or that Scott would not be able to live with the risk of infection and side effects of the antirejection drugs.

Clint Hallam, the Australian man who received the hand transplant in France in 1998, did not fare as well. He began having difficulties with rejection and stated dramatically that he had no feeling in his "dead man's hand." Eventually, the new hand was amputated by one of the surgeons on the transplant team. The transplant team

From left to right, Warren C. Breidenbach, Jon W. Jones, Jr., and Gordon R. Tobin answer questions regarding the first transplant of a hand in the United States.

2995

stated that Hallam had not followed through with the antirejection drug therapy nor did he remain under the regular care of his physicians after the transplant.

Subsequent hand transplants have been considered successful, including double hand transplants for individuals who have lost both hands in accidents. However, the hand transplant procedure is controversial. Discussion centers on whether hand transplants are ethical or wise, because limb transplants, unlike heart or liver transplants, are not essential to life, and the drugs that must be taken to prevent rejection by the immune system carry the risk of serious side effects. A transplant recipient must take these immunosuppressive drugs for the rest of his or her life.

Do the benefits of a hand transplant outweigh the risks? Advocates of the transplant say that the improvement in daily living, the increased motion, function and feeling in a transplanted hand as opposed to a prosthesis, and the psychological benefits of a warm human part make the risk of immune suppression worth it. Opponents of the procedure question the ethics of performing a procedure with potentially life-threatening risks in a case where a life is not threatened. James Herndon of Harvard Medical School, who questions the ethics of the hand transplant operation, suggests that the procedure be limited to patients who are already taking immunosuppressive drugs for a life-threatening problem or those who have lost both hands, while the medical community waits for further advances in immunosuppressive therapy.

Susan Butterworth

Scientists Trace HIV to Chimpanzees

Careful scientific research found evidence showing that HIV-I, the virus behind the lethal human ailment known as AIDS, originated among chimpanzees of west-central Africa.

What: Biology; Genetics; Health; Medicine
When: January 31, 1999
Where: Chicago, Illinois
Who:
BEATRICE H. HAHN (1955-   ), a biologist at the University of Alabama at Birmingham

From Chimpanzees to Humans

Scientists long suspected that the virus causing the lethal disease AIDS (acquired immunodeficiency syndrome) that infects humans might be traced to other primates—the nearest living animal relations to human beings. Genetic studies offered good evidence that HIV-I, the strain of HIV responsible for the global human AIDS epidemic, can be traced directly to chimpanzees of west-central Africa.

These findings were announced on January 31, 1999 by Dr. Beatrice H. Hahn of the University of Alabama at Birmingham at the sixth annual conference on Retroviruses and Opportunistic Infections Meeting in Chicago. A new simian immunodeficiency virus (SIV) in a subspecies of the chimpanzee, *Pan troglodytes*, had been discovered. This virus is similar to HIV-I and also the troglodytes subspecies lives in the area where earlier researchers had determined that the HIV epidemic began.

It is believed that although the virus may have infected chimpanzees for thousands of years, they, unlike humans, showed no symptoms of the infection. Apparently some unfortunate person of the region first contracted the disease as a result of close contact with a chimpanzee. People in the region both hunted chimpanzees in the wild for food and keep them as pets, thereby creating opportunities for frequent exposure to SIV strains.

Viruses and Retroviruses

The first disease known to be caused by a virus was the one that causes an infection of tobacco plants called tobacco mosaic disease. Russian scientist Dimitri Iwanowski recognized in the 1890's that the agent responsible is infectious like bacteria, but much smaller and therefore not visible under a light microscope.

Later, in the 1950's, when the electron microscope was first utilized, viruses could be seen for the first time. Also, their basic chemical nature was determined. Despite considerable diversity among the many kinds, all have similar basic structures: cores of nucleic acids (DNA or RNA) surrounded by protein shells.

All viruses are infectious for some particular species of plant, animal, bacterium, or protist (algae, protozoans), or fungus. Generally, each kind of virus is specific for a particular host species and is not capable of infecting other species. For example, a tomato plant is not infected by the tobacco mosaic virus, and humans are not infected by the dog distemper virus. However, exceptions do exist: The infection of humans by the HIV virus, previously known only in chimpanzees, is an example. It should be noted that humans and chimpanzees, both primates, are closely related; approximately 98 percent of their genes are common to both.

Knowledge of how viruses reproduce has been gained primarily from studies of a special group of viruses called bacteriophages or phages. These are viruses that infect and reproduce within bacterial cells. First, the phage attaches itself by means of its tail fibers to the wall of the bacterium. Next, the DNA (or RNA) of the phage is injected into the host cell where it replicates (duplicates) itself several times. These molecules stimulate the machinery of the host cell to synthesize protein molecules used to make the coats of new phages. The next step is the assembly of the shells and DNA molecules to form new phages. At this stage, the host bacterium bursts open, releasing the numerous, newly formed viruses, any one of which can infect a new, previously uninfected, bacterium. Thus, within short time, a few viruses (or only one) may infect and lyse (break down) a large number of host cells. In much the same way, human cells are infected by cold and flu viruses. Viruses that reproduce in this manner are said to have a lytic life cycle.

In contrast, certain other viruses infect and reproduce by means of a lysogenic life cycle. Examples are viruses that cause herpes and AIDS in humans. Lysogenic viruses, upon entering host cells are less aggressive, often requiring months before they are detected. Rather than almost immediately lysing the host cell, the DNA of the invading virus becomes incorporated into the chromosomes of the host cell. At this stage, the host cell is not damaged. In fact, as the host cell undergoes its usual division, the viral DNA is replicated along with the host chromosomal DNA. At some later time, the virus may switch to the

AP/Wide World Photos

Luc Montagnier, co-discoverer of HIV, warns governments of the need for further research into an AIDS vaccine.

lytic cycle, destroy host cells, producing symptoms of the disease.

The HIV virus, which causes AIDS in humans, has a lysogenic cycle, but is also given the designation of a retrovirus, indicating that information within host cells flows in reverse of the usual direction. On entering a human host cell, the RNA of the HIV virus is used to make complementary DNA (cDNA). The cDNA is then integrated into the host chromosome. At a later time, after months or years, the viral DNA of the host chromosomes is used to make viral RNA around which protein shells are added to make new AIDS viruses. Thus, the major feature of the retrovirus life cycle is the flow of information from RNA to DNA, the reverse of the usual direction. Due to the complexity and many stages of this life cycle, anti-AIDS drugs have the potential for controlling the disease by interfering with some part of the chain of events necessary to produce and release new HIV viruses.

Consequences

The recent evidence indicates that the HIV virus hopped the "species barrier" which usually separates species. Also, that the hop occurred only once, and somewhat earlier, than previously thought. The first unfortunate, and anonymous, African victim probably lived in west-central Africa during the 1950's. Apparently, the human epidemic that began with him (or her) was confined to Africa for two or three decades before becoming known there, and, subsequently, throughout the world.

There are several implications of these recent findings. First of all, they lay to rest several erroneous notions about the origin of HIV. One of these was that it had been spread accidentally by way of an early polio vaccine.

At the beginning of the twenty-first century, researchers hoped that further studies of native chimpanzee populations would lead to an understanding of why the SIV does not cause chimpanzees to become sick. This information may be used to better protect humans from the lethal AIDS virus. Meanwhile, other scientists were taking other approaches. Included were the development of more effective anti-AIDS drugs and an AIDS vaccine.

Thomas E. Hemmerly

Researchers Generate Nuclear Energy on Tabletop

In an effort to produce alternative sources of clean energy, researchers generated nuclear energy on a tabletop by both fusion and fission.

What: Physics; Energy
When: February, 1999, to January, 2000
Where: Lawrence Livermore National Laboratory, Livermore, California
Who:

TODD DITMIRE (c. 1969-   ), associate professor of physics, University of Texas at Austin

THOMAS E. COWAN, associate professor of physics, University of California, Berkeley

KENNETH B. WHARTON and
VICTOR P. YANOVSKY, physicists

Generating Energy

Between February, 1999, and January, 2000, physicists at the Lawrence Livermore National Laboratory (LLNL) succeeded in producing energy for the first time from nuclear fusion, as well as from nuclear fission, on a tabletop. The ultimate goal of these tabletop nuclear experiments was to produce alternative sources of clean energy to generate electricity.

Nuclear energy research is focused on developing a reliable energy alternative to the present burning of fossil fuels. Nuclear energy can be released by the processes of fission and fusion. Nuclear fusion combines, or fuses, small nuclei into larger nuclei and releases energy that can be used to generate electricity. Controlled nuclear fusion offers a number of advantages over other energy sources in that first, the fuel is widely available and virtually inexhaustible; second, the process is environmentally favorable, producing no greenhouse gases and no long-lived radioactive by-products; and third, it is inherently safe. The main technical difficulty in producing con-trolled fusion is that the reacting nuclei initially repel each other because of the electrical interaction between like charges. The charged nuclei must approach each other fast enough so that the strong nuclear force is greater than the electrical repulsion. The interacting nuclei must then be kept together long enough at high enough temperatures for fusion to occur.

When hot enough, the fusion fuel, typically consisting of hydrogen isotopes, is separated into its positively charged nuclei and negatively charged electrons, forming a plasma. One approach for containing the burning fusion fuel is inertial confinement. In inertial confinement fusion, a small pellet of fusion fuel is compressed and confined by its own inertia for a long enough time at a high enough temperature and density that fusion reactions occur.

Nuclear fission releases energy when heavy nuclei split into lighter nuclei. Controlled fission is initiated when a slow neutron strikes the heavy nucleus, typically uranium-235. Commercial fission reactors depend on a chain reaction in which each fission event releases neutrons that induce additional fission events. Fission has two major weaknesses as an alternative source of energy: Some by-products are intensely radioactive and long-lived, and the fuel is not plentiful.

Tabletop Fusion and Fission

In February, 1999, Todd Ditmire, Victor Yanovsky, Kenneth Wharton, Thomas Cowan, and other researchers at the Lawrence Livermore National Laboratory used the most powerful lasers available to produce energy for the first time from nuclear fusion on a tabletop. The top of the table measured 4 feet (1.2 meters) by 11 feet (3.4 meters). Clusters of deuterium, or heavy hydrogen, molecules were bombarded with high-

2999

powered laser pulses. When the laser beam was focused on a very small volume of deuterium, the molecular clusters were heated to tens of millions of degrees Celsius, stripping electrons from the deuterium atoms to form a plasma. As the superheated clusters exploded, some of the fast-moving deuterium ions fused together with high enough velocity to produce helium-3 nuclei, energetic neutrons, and heat energy. In order for the nuclear fusion process to produce commercial power, the fuel must be confined long enough to ignite a self-sustaining thermonuclear reaction. However, the plasma generated in tabletop fusion disperses too quickly.

To enable the generation of sustained fusion, the lab at Livermore planned to add the National Ignition Facility. This facility was to be as large as a football stadium and to contain 192 extremely powerful lasers. In this facility, an inertial confinement fusion reaction is to be generated by focusing intense beams from the lasers onto a pea-sized pellet consisting of two hydrogen isotopes, deuterium and tritium. Plans were to detonate five pellets every second. The heat generated from these interactions would be used to drive turbines that generate about ten billion watts of electrical energy. In order for this prototype model to become a practical power plant, the costly lasers that initiate the fusion process must be reduced in size, cost, and complexity.

In addition to the production of nuclear fusion reactions on a tabletop, Cowan and a team of researchers from the University of California, Berkeley, produced nuclear fission reactions on a tabletop at the Lawrence Livermore National Laboratory in January, 2000. A solid-gold target mounted on a sample holder containing uranium was first hit by a lower-energy laser pulse to produce a swarm of electrons at the target's surface. These electrons were then accelerated to high energies by blasting the gold target with a high-power beam from the world's first thousand-trillion watt laser.

The accelerating electrons produced gamma rays that liberated high-energy neutrons from the gold. The neutrons in turn bombarded uranium-238 samples, causing them to undergo nuclear fission. Later analysis of the target revealed the presence of radioactive isotopes of gold and other elements, which are all products of the nuclear reactions initiated by the high-energy neutrons.

Consequences

Tabletop nuclear reactions initiate a new class of nuclear physics experiments. Tabletop fusion research is directed toward creating a new, safe source of clean energy with an abundant fuel supply. Although it is not promising as a commercial energy source, tabletop fusion promises a scientific payoff. This research is providing new information about the behavior of fusion plasmas and insights about the future design and development of commercial fusion power plants that can generate economical, reliable electricity.

Tabletop fission research has generated new insights into the fundamental interplay between light and solids, the interactions of gamma ray photons with one another, the physics of plasmas, and the generation of energy from sources other than fossil fuels. The results of tabletop fusion and fission show that lasers can be used to perform desktop physics experiment to study nuclear physics and astrophysical processes without the necessity of scheduling a major multimillion-dollar particle accelerator facility for such work.

Alvin K. Benson

King Hussein of Jordan Dies

King Hussein ruled for almost forty-six years, keeping disaster at bay in a region almost continuously torn by armed conflict. Upon his death, his son, Abdullah, acceded to the throne, espousing his father's politics of moderation, peace, and stability.

What: Government; International relations
When: February 7, 1999
Where: Amman, Jordan
Who:
HUSSEIN IBN TALAL (1935-1999), king of Jordan from 1952 to 1999, also known as Hussein I
ABDULLAH IBN HUSSEIN, (1962-   ), son of King Hussein, also known as King Abdullah II from 1999
HASSAN IBN TALAL (1947-   ), brother of King Hussein

King Hussein's Legacy

King Hussein ruled Jordan for almost forty-six years, the longest reigning monarch in the Middle East. He provided stability in Jordan in a turbulent era, maintained working relations with his embattled neighbors, and contributed to the difficult peace process. He was often viewed as a symbol of moderation, prudence, and reason in an area in which those qualities were in short supply. Becoming king at age eighteen (in 1953), he soon faced political unrest and narrowly survived several assassination attempts, but he eventually won the respect and support of his people. He belonged to the Hashimite Dynasty, which traces its lineage to the Prophet Muḥammad (the founder of Islam). This gave Hussein a distinct advantage in a Muslim country, and he had the trust and support of the Jordanian army, another asset when his rule was in jeopardy.

Hussein endeavored to calm the bitter Israeli-Palestinian confrontations, a delicate task as 70 percent of the Jordanian population are Palestinians who sought refuge in Jordan in the wake of five Arab-Israeli wars and recurring civil strife under Israeli rule. Most of these Palestinians (who became Jordanian citizens) were violently opposed to Israel and very supportive of the governments advocating its destruction. However, under Hussein, Jordan and Israel made peace. He made a concerted effort to integrate Palestinians into the mainstream of Jordanian society, thus fostering stability.

Hussein moved the country toward increased democracy, legalizing political parties and opening the way to free political expression (by 1999, there were twenty parties in Jordan, ranging in ideology from communist to Islamist). He supported the writing of a national charter, formally ratified and implemented in 1991, which now provides the constitutional basis for future governments. Hussein was thus a factor for stability at home and an element of moderation in this divided region.

Hussein's Illness and Death

In June, 1998, Hussein was diagnosed with lymphatic cancer (he had undergone successful surgery for prostate cancer in 1992). For six months, he was treated at the Mayo Clinic in Rochester, Minnesota. In October, 1998, though frail as a result of his cancer treatment, Hussein left the clinic at the invitation of President Bill Clinton to help break the deadlock in the Israeli-Palestinian negotiations at the Wye River Plantation near Washington, D.C. He returned home to Amman on January 19, 1999, and received a hero's welcome.

During Hussein's treatment at the Mayo Clinic, Crown Prince Hassan, his brother, had served as regent. However, Hussein took exception to Hassan's stewardship and political maneuvering by Hassan's supporters. Hussein re-

AP/Wide World Photos

A Jordanian grieves as another holds a portrait of the deceased King Hussein.

minded the country that, although gravely ill, he was still in charge and, a week after his return, demoted Hassan, naming his eldest son, Abdullah, the new crown prince (Abdullah had been crown prince until the age of three; but the dangerous conditions at home and in the Middle East in the mid-1960's had led Hussein to name his brother Hassan heir to the throne).

The ceremony making Abdullah crown prince took place at the military airport in Amman as Hussein was returning to the Mayo Clinic (his deteriorating condition required urgent treatment). However, the end was near, and Hussein returned home on February 4, 1999, and died three days later on February 7. The country witnessed an outpouring of national grief and an international expression of sympathy: More than fifty world leaders attended the funeral, including President Bill Clinton, three former U.S. presidents, and every major Arab and Israeli official.

Consequences

Crown Prince Abdullah was crowned as Abdullah II on June 9, 1999, at the age of thirty-seven (Abdullah I, his great-grandfather, was the first king of Jordan). The political transition was remarkably smooth. The sudden and unforeseen casting aside of Prince Hassan after decades of prominence in Jordan's ruling circles did not cause any backlash. The new king was reported to have the support of the royal family, including Prince Hassan. Abdullah was trained as a military commander. Like his father, he graduated (1980) from the Royal Military Academy at Sandhurst, England, and served in Jordan's armed forces. In 1993, he was appointed deputy commander of Jordan's elite Special Forces, and retained the post until he became king. His popularity among the armed forces, long a foundation of the monarchy's power, was an asset in his new role; however, experts speculated as to whether he could lead the nation as well as his father had.

Abdullah had demonstrated a desire to be close to the people of Jordan. He was known to have disguised himself as a taxi driver and a television reporter to get to know the working people and understand their outlook. He pledged to carry on the domestic and international policies of his father and was determined to continue his cautious efforts toward peace and to preserve Jordan's ties to Western powers. The latter have an interest in Jordan's stability and in its support of the delicate Arab-Israeli peace process. Western powers were likely to provide assistance in Jordan's economic struggle (a foreign debt of $8.4 billion, a population expected to double in less than twenty-three years, 1.3 million refugees to support, and a rate of unemployment approaching 20 percent). For example, President Clinton promised to support debt relief and to provide as much as $450 million in economic assistance. Although the task ahead would not be easy; King Abdullah's commitment and leadership were reassuring.

Jean-Robert Leguey-Feilleux

Scientists Slow Speed of Light

A Danish physicist and her collaborators reduced light's speed from 186,000 miles (299,274 kilometers) per second to 38 miles (61 kilometers) per hour.

What: Physics
When: February 18, 1999
Where: Rowland Institute for Science, Cambridge, Massachusetts
Who:
LENE VESTERGAARD HAU (C. 1960-   ), teacher, Department of Physics, Harvard University, and experimental physicist, Rowland Institute for Science
STEPHEN E. HARRIS (1936-   ), professor of electrical engineering and applied physics, Edward L. Ginzton Laboratory, Stanford University

Slowing Light's Speed

On February 18, 1999, a long letter appeared in *Nature*, a preeminent scientific journal, explaining how Lene Vestergaard Hau and her colleagues had spectacularly slowed down the speed of light from 186,000 miles (299,274 kilometers) per second to 38 miles (61 kilometers) per hour. According to physicist Albert Einstein, nothing can travel faster than light—it is the universe's final speed limit. However, scientists had long known that a light beam moves more slowly through such materials as air, water, and glass. What was remarkable about the results reported in *Nature* is that Hau got light to travel 20 million times more slowly than its speed in a vacuum.

Instead of glass, the material that Hau used was a tiny cloud of sodium atoms cooled to a temperature close to absolute zero. Like light's speed, absolute zero is another of the universe's ultimate limits because it is the temperature—459.67 degrees below zero Fahrenheit (273.15 degrees below zero Celsius)—at which nothing can be colder. When sodium atoms are cooled to these extremely low temperatures, they behave

in very strange ways. In this ultracold world, the motion characterizing atoms at ordinary temperatures nearly stops. Quantum mechanics is the science that deals with this world of the very small and very cold, and one of its foundational insights is the uncertainty principle, which states that if one does not know the speed of an atom very accurately, one cannot be sure where it is.

In this quantum world, individual atoms cannot be pinpointed, and they lose their identities. Consequently, they merge into a superatom consisting of millions of atoms. This dense glob is called a Bose-Einstein condensate, because, in 1924, Indian physicist Satyenda Nath Bose and Einstein predicted this unusual phenomenon. In 1995, scientists actually made Bose-Einstein condensates in the laboratory, and these coherent atomic clusters became the subjects of many studies in the United States and Europe.

How Light Was Slowed

To get light to move very, very slowly, Hau and her collaborators needed both supercooled sodium atoms and some modern optical equipment. They began their project in the spring of 1998. To reach extremely low temperatures, Hau's group used a combination of lasers and evaporative cooling. These low temperatures reduced thermal motions of the sodium atoms and increased their density so that a Bose-Einstein condensate was formed.

Ordinarily, this Bose-Einstein condensate, because of its high sodium-ion density, would block light as effectively as a brick wall, but in the early 1990's, Stanford physicist Stephen E. Harris discovered a way to make this thick cloud of sodium atoms transparent. His technique, called electromagnetically induced transparency, changes the way light moves through a material by modifying how this material absorbs light. Specific materials such as sodium absorb, or block, light of a particular wavelength, but a laser can change the quantum properties of the sodium atoms so that they no longer absorb light at that wavelength. Hau directed such a laser beam, called a coupling beam, through a very tiny cloud of nearly motionless sodium atoms.

She then aimed a second laser beam, called the probe pulse, which was at right angles to the coupling beam, through the now-transparent supercooled cluster of sodium atoms. The light speed of this probe beam became astonishingly slow. Furthermore, the refractive index of the sodium-atom condensate became enormous. The refractive index is a number that measures the ratio of the speed of light in free space to its velocity in a particular material. In other words, the refractive index gives scientists a way to determine the degree to which a material bends light. The sodium condensate created by Hau's group had a refractive index that was a million times greater than any previously measured value (and 100 trillion times that of glass fiber).

Consequences

In 2001, Hau and other groups of scientists were able to bring light essentially to a stop and then get it moving again. Superslow light has become a subject of intense study in various laboratories around the world, and many of the scientists involved in these investigations predict that their discoveries will have practical applications. For example, the techniques developed by Hau and others could allow scientists to make highly precise measurements of the properties of atoms. Because the sodium condensate's large refractive index can be precisely controlled, it may be possible to make devices that easily change light frequencies from the infrared through the visible to the ultraviolet.

Hau's discovery, which made insightful use of optical apparatus, might lead to the development of new optical equipment. Hau herself mentioned the possibility of optical instruments that function at very low light levels. Her ways of controlling light's speed might also give scientists new ways of designing low-power lasers and optical memory devices. Other possible applications include highly sensitive night-vision glasses and television screens with very bright projected images.

Perhaps the greatest potential impact of slow light will be in the area of optical computers, which would operate with photons instead of electrons. Laser-condensate combinations could lead to ultrafast switches that could be turned on or off by a single photon. Quantum computers that relied on light to transfer information could process data at much greater speeds than electronic computers.

None of these applications has become a reality, but the gain in understanding and control of light and Bose-Einstein condensates achieved by Hau and Harris have led some commentators to predict that they will eventually win a Nobel Prize for their discoveries.

Robert J. Paradowski

White Supremacist Is Convicted in Racially Motivated Murder

John William King, a self-proclaimed white supremacist, was convicted of murder in the death of James Byrd, Jr., an African American who was dragged on a chain behind a pickup truck in Jasper, Texas, in June, 1990.

What: Civil rights and liberties; Crime
When: February 23, 1999
Where: Jasper, Texas
Who:

JOHN WILLIAM KING (1975?-   ), self-
proclaimed white supremacist
RUSSELL BYRD, JR. (1950?-1998), African
American murder victim

The Crime

On February 23, 1999, a jury convicted John William King, a twenty-four-year-old former convict and self-proclaimed white supremacist, of kidnapping and murder in Jasper, Texas. The victim, Russell Byrd, Jr., a forty-nine-year-old unemployed African American from Jasper, had been beaten and dragged behind a pickup truck for about three miles (nearly five kilometers). Following the verdict, King was sentenced to death.

In the early morning hours on Sunday, June 7, 1998, three men, twenty-three-year-old John William King, thirty-one-year-old Lawrence Russell Brewer, Jr., and twenty-three-year-old Shawn Berry offered James Byrd, Jr., a ride in the back of Berry's pickup truck in Jasper. With Byrd riding in the truck bed, the three men drove to a local convenience store. There King took the wheel and drove to a dirt road east of town. At this point, the three men got out of the truck and began beating their passenger until he appeared to be unconscious. At this point, Berry apparently ran away while King and Brewer chained Byrd to the pickup truck. Byrd, who was alive and conscious, was then dragged about three miles (nearly five kilometers) down the road before the men unchained him. Investigators found

Byrd's tennis shoes, shirt, wallet, keys, and dentures along the route. His head and right arm had been severed when the body rolled into a roadside ditch and slammed into a concrete culvert.

According to an autopsy report, much of Byrd's skin was stripped and shredded; he had numerous broken bones, and his elbows, kneecaps, lower back, and both heels were ground to the bone. At the site where Byrd was beaten, investigators found a cigarette lighter with the word "Possum" inscribed on it. Possum was King's nickname in prison. They also found a torque wrench set with the name "Berry" inscribed on it. By nine o'clock Sunday night, Berry had been arrested for several traffic violations, and his truck was impounded. Investigators found blood splattered on the undercarriage of the truck as well as red clay and vegetation on the truck similar to the clay and vegetation the killer's truck had driven through. All three men were subsequently arrested. At King's apartment, police found items indicating he was sympathetic with white supremacist groups, including a Ku Klux Klan drawing of three horsemen galloping from a fire.

The Trials

The dragging death of Byrd was seen as a throwback to lynchings that had occurred decades earlier in the American South. It drew international attention to Jasper, a town of eight thousand. The first man to stand trial was King. Prosecutors argued that King was planning on forming his own white supremacist group, the Texas Rebel Soldier Division of the Confederate Knights of America, in Jasper and had hoped to use the death of Byrd to help recruit members. King, whose arms and torso were covered with

Nazi and white supremacist tattoos, was linked to the death scene by prosecution witnesses, his cigarette lighter, and his DNA (deoxyribonucleic acid) on a cigarette butt.

Prosecutors also placed in evidence a letter that King had tried to smuggle out of prison to Brewer in which he asserted that the three men had made history regardless of the verdict and would die proudly if need be. After two and one-half hours of deliberation, the jury, which had chosen its only African American member as foreman, found King guilty of murder on February 23, 1999. The jury subsequently sentenced King to death. King was the second white sent to death row in Texas for killing a black person since capital punishment had been reinstated in the 1970's. Texas had executed only one white for killing a black—in 1854.

In September, 1999, Brewer was tried in Bryan, Texas, for the same crime. Brewer, who had met King in prison, had also adorned himself with racist tattoos and had joined the Confederate Knights of America. Brewer was also found guilty of capital murder and sentenced to death.

The trial of Berry took place in Jasper in the fall of 1999. Like his two codefendants, Berry was found guilty of capital murder. Berry's lawyers had argued that Berry had felt threatened by his racist companions when they dragged Byrd to death, and Berry blamed King and Brewer with devising and carrying out the plan to kill Byrd. Nine people testified at the sentencing phase that Berry did not meet the death penalty test of being a future threat to society. Among Berry's witnesses was a psychiatrist who had testified against Brewer and King in the first two trials. The jury accepted the argument and sentenced Berry to life. The sentence meant that Berry had to spend a minimum of forty years in prison. Although Byrd's family was satisfied with the conviction, they felt Berry's sentence should have been the death penalty.

Consequences

The Byrd murder reminded Americans of the racial hatred that still existed in the United States. Although lynchings seldom occurred, hate crimes were more common. For example, according to the U.S. Justice Department, 5,396 racial hate crimes were reported in 1996.

The brutality of the murder drew national and international attention. The fact that the crime was undertaken by members of a white supremacist group added to the infamous nature of the crime. Although all three men were convicted of capital murder, the crime was a reminder to the United States that racial violence remains a part of American society.

William V. Moore

Plan to Save California Redwoods Is Approved

> *A last-minute agreement reached by a lumber company, the state of California, and the U.S. government saved the famous Headwaters Grove of old-growth redwoods from certain destruction.*

What: Environment; Business
When: March 2, 1999
Where: Humboldt County, California
Who:

CHARLES HURWITZ (1940-   ), Pacific Lumber owner

DIANNE FEINSTEIN (1933-   ), U.S. senator from California

PETE WILSON (1933-   ), California governor from 1991 to 1999

GRAY DAVIS (1942-   ), California governor from 1999

BRUCE BABBITT (1938-   ), U.S. secretary of the interior from 1999

A Debate Over Redwoods

In March, 1999, the Pacific Lumber Company, the U.S. government, and the state of California reached a last-minute agreement that put a grove of old-growth redwoods under public protection and placed logging restrictions on other tracts of forest. For years, environmentalists had been trying to put an end to the clear-cutting of California's famous redwood groves. The redwoods were considered important because of their extreme age (some may be one thousand years old) and their historic significance. Moreover, the redwoods provided a habitat for such endangered species as the marbled murrelet, a seabird. Also, studies indicated that the redwoods create the ecosystem around them, turning the fog and mist common in this part of California into rain that falls on the surrounding forest. This rain can measure as much as 4 inches (10 centimeters) a night.

In the months before the agreement was finally reached, relationships between the loggers and environmentalists had grown increasingly tense. This tension reached a peak with the death of David Chain, an activist belonging to the Earth First! organization. While protesting the logging operations of Pacific Lumber, Chain was crushed to death under a tree felled by a logger. This event, as well as a rapidly approaching deadline regarding the use of federal funds, put pressure on all sides of the issue to reach an agreement.

The Agreement

The plan to save the California redwoods had begun some three years earlier. U.S. senator Dianne Feinstein, a Democrat from California, won a crucial piece of legislation to provide federal money for the purchase of land. Feinstein was further credited with bringing the negotiators to the table and keeping them there when it seemed that the entire deal would fall apart. California governor Pete Wilson helped hammer out both the state legislation and regulations that would govern not only the sale of the land but also logging by Pacific Lumber on the acres it retained. These regulations were a key issue for environmentalists but almost broke the deal with the lumber company. Gray Davis, California governor at the time of the signing of the agreement, continued Wilson's policies with a few modifications. Nevertheless, the deal nearly fell apart in the last moments. The offer of federal money allocated for the purchase of land was scheduled to expire at midnight on February 28, 1999. Had the deal not been reached, the money would have returned to the federal treasury. Pacific Lumber owner Charles Hurwitz balked at restrictions the state wanted to place on his logging operations in other parts of the state. Only last-minute reinterpretations of how much lumber the company would be allowed to cut moved Hurwitz to sign the agreement just before midnight.

Under the terms of the agreement, Pacific Lumber and its parent firm Maxxam, received $380 million from the state of California and the U.S. government for the purchase of Headwaters Forest. The forest immediately became a protected area, open to the public for hiking and viewing. The purchase included 3,000 acres (1,215 hectares) of forest that had never been logged and another 4,500-acre (1,823-hectare) area that had experienced some lumbering. Another $100 million went to Maxxam over the next five years for the purchase of additional stands of historic redwoods. The U.S. Bureau of Land Management and the state of California agreed to comanage the land as a preserve.

In addition, the agreement banned logging in another 8,000 acres (3,240 hectares) of redwood forest as part of the Habitat Conservation Plan, a plan designed to protect endangered species. Pacific Lumber also agreed to the terms of the Habitat Conservation Plan through the year 2049 on another 211,000 acres (85,455 hectares) that it owned. Further, logging was banned in buffer zones around rivers and streams where the Coho salmon thrived. Finally, the total amount of lumber that could be harvested by Pacific Lumber was restricted by the agreement.

Consequences

Many hailed the Headwaters deal as an important and historic moment for the environment and business. The agreement served as an example of how preservation and progress can accommodate each other, and there were those who suggested that the deal could serve as a model for future cooperation between environmentalists and big business.

Interior Secretary Bruce Babbitt announces the plan to set aside an ancient redwood grove in California. At far right is California senator Dianne Feinstein.

3009

Most environmentalists agreed that the pact was not perfect. The Sierra Club, for example, worried that the last-minute changes in the interpretation of the Habitat Conservation Plan gave the lumber company a lot of power to change the terms of the agreement. Likewise, the Environmental Protection Information Center pointed to gaps in the legislation that might lead to further destruction of salmon habitats. Regardless, the director of the Headwaters Sanctuary Project in Oakland, California, expressed satisfaction with the deal, suggesting that it was a huge step toward the protection of the redwoods.

In a statement made while announcing the terms of the agreement, Interior Secretary Bruce Babbitt said that it is up to the government and private industry to demonstrate that it is possible to preserve old-growth forests, protect endangered species, and maintain the ability to have sustainable harvesting of lumber.

Most participants agreed that the final impact of the Headwater agreement might not be evident for many years. Nevertheless, the Headwaters agreement was one of the first attempts to balance both environmental and economic needs. The consequences of the pact may be far-reaching as other communities attempt to protect their natural surroundings while maintaining their economic base.

Diane Andrews Henningfeld

New Fault Is Discovered Under Los Angeles

> *Researchers announced that they had found a fault directly under the Los Angeles metropolitan area that could cause massive damage if an earthquake occurred along it.*

What: Disasters; Earth science

When: March 5, 1999

Where: Los Angeles, Whittier, Pasadena, and Northridge, California

Who:

JOHN SHAW, a Harvard University professor

PETER SHEARER, a Scripps Institution of Oceanography professor

CHARLES FRANCIS RICHTER (1900-1985), an American seismologist

CHARLES RUBIN, a Central Washington University geologist

Finding Faults

A number of faults—large, active cracks in the earth, capable of causing destructive earthquakes—cut through the Los Angeles metropolitan area. The most famous is the San Andreas fault, which runs from east of Los Angeles north to San Francisco and beyond. The San Andreas fault represents the boundary between the North American and the Pacific plates. Plates are very large pieces of Earth's crust, some larger than continents, that make up Earth's surface. Most of the motion between these two plates occurs along the San Andreas fault, east of Los Angeles. Slippage along the San Andreas caused the 1906 San Francisco earthquake. Among the other major faults found in the Los Angeles metropolitan area are the Sierra Madre, Elysian Park, and the Santa Monica faults.

Using data gathered by oil companies, researchers John Shaw of Harvard University and Peter Shearer of the Scripps Institution of Oceanography discovered the presence of yet another fault in the area, which they called Puente Hills fault system. The results of their research were first published in *Science* on March 5, 1999. This newly identified fault lies under several hundred square miles of densely populated urban areas. Its lowest known point is more than 9 miles (14.5 kilometers) below Earth's surface, and its highest point is 1.8 miles (2.9 kilometers) below Dodger Stadium and the skyscrapers of downtown Bunker Hill. The Puente Hills system is a blind thrust fault, so called because the fracture in Earth's crust does not reach the surface; therefore, it cannot be seen. Such faults release energy by suddenly rising, a motion that is particularly destructive to buildings on the surface. Thrust faults have been recognized all over the world, with two in the Los Angeles area, the Elysian Park fault, which runs underneath downtown Los Angeles, and the Northridge fault, which ruptured during the 1994 Northridge quake. Many still-unknown blind thrust faults may exist in southern California. The Puente Hills fault extends from Coyote Hills in northern Orange County, near the town of Brea, and runs for about 25 miles (40 kilometers) to the west, beneath downtown Los Angeles.

The Danger Below

The energy released by an earthquake is expressed in terms of its magnitude, using a scale developed in 1935 by Charles Francis Richter of the California Institute of Technology to measure the relative size and strength of earthquakes. If an earthquake has a magnitude of seven, it is ten times more powerful than an earthquake that has a magnitude of six; a six is one hundred times more powerful than a magnitude of four. According to researchers Shaw and Shearer, the Puente Hills fault could potentially generate what seismologists classify as a "major" earthquake (magnitude 7.0 or higher on the Richter scale). Sudden ruptures along any segment of the Puente Hills thrust fault could produce strong earthquakes. Even stronger temblors could result if all three segments ruptured

3011

simultaneously. The land around the Puente Hills fault has been moving at rates of between 0.5 and 0.9 millimeters (0.02 and 0.03 inches) per year. Researchers believe the fault probably caused the 1987 Whittier Narrows earthquake, a magnitude 5.9 tremor on October 1, 1987, that resulted in eight deaths, two hundred injuries, and $358 million in property damage.

To predict the future behavior of the Puente Hills fault, geologist Charles Rubin of Central Washington University and his colleague, Scott Lindvall, excavated a trench across the fault to determine its previous behavior. They examined buried evidence (layers of soil, gravel, sand, and charcoal) at a site near Pasadena. Rubin and Lindvall then dated the charcoal found in the various layers and estimated that at least two prehistoric earthquakes had shifted one edge of the fault upward more than 30 feet (9 meters).

Faults that move the ground vertically are called reverse faults; the Puente Hills fault system is thus a blind thrust reverse fault. These types of faults are potentially very damaging. For example, the Northridge blind thrust fault generated a magnitude 6.7 earthquake in 1994. This earthquake killed fifty-seven people, left twenty thousand homeless, and resulted in $40 billion in damage. It was one of the most expensive natural disasters in U.S. history.

Consequences

The Los Angeles region has been dramatically affected by four large earthquakes over the past thirty years. The Puente Hills fault is capable of large surface displacements as well as strong ground motions. These ground motions can disrupt roads, railways, phone lines, power lines, and water systems. They can also produce significant damage to modern buildings. "Los Angeles is caught in a vise," said Harvard's Shaw. "It is locked between converging sections, or plates, of Earth's crust, carrying North America and part of the Pacific Ocean floor." As the plates collide, rocks beneath the city are shattered, folded, and cut by faults of many types. The Los Angeles Basin is contracting at a rate of 7 to 8 millimeters (0.28 to 0.31 inches) per year.

Stress building up along parts of the newly discovered fault, then rupturing, could cause a series of large quakes every 250 to 1,000 years. Furthermore, a violent breach along the entire fault at once could produce a convulsion three times the intensity of the Northridge earthquake. Such a cataclysm could happen every 500 to 2,000 years and result in an estimated three thousand to eight thousand deaths and damages of about $200 billion.

Denyse Lemaire

Czech Republic, Poland, and Hungary Join NATO

When the North American Treaty Organization admitted the first former Soviet bloc countries—the Czech Republic, Poland, and Hungary—into its organization, it created a historic alliance.

What: International relations
When: March 12, 1999
Where: Truman Presidential Library, Independence, Missouri
Who:
BILL CLINTON (1946-   ), president of the United States from 1993 to 2001
MADELEINE ALBRIGHT (1937-   ), U.S. secretary of state from 1997 to 2001

A Decade Debate

On March 12, 1999, three formerly Communist countries—the Czech Republic, Poland, and Hungary—joined the North Atlantic Treaty Organization (NATO), creating an alliance between countries that had once been enemies. In 1949, twelve North American and Western European countries formed the original alliance. Those twelve NATO member were joined by four others: Greece and Turkey in 1952, West Germany in 1955, and Spain in 1982.

With the fall of the Berlin Wall in 1989 came the collapse of communist governments in Eastern Europe, the fall of the Soviet Union in 1991, and the end of the Cold War. The new Eastern European countries sought to renew their pre-communist alliances with Western Europe, and one way was to seek admission into NATO, which continued to exist despite the Cold War's end. NATO established a partnership relationship with non-NATO European countries, including those in Eastern Europe and even Russia.

This was the first step that enabled NATO to consider admitting the new members from the region. After joint exercises with East European forces, NATO military leaders were ready to admit the countries, but political issues still needed to be settled. Although a number of Eastern European countries, including Slovenia, Romania, Lithuania, Latvia, and Estonia, sought admission, the first countries the alliance considered were the Czech Republic (the western part of Czechoslovakia; the eastern portion became Slovakia in 1993), Poland, and Hungary.

The decision to accept the former Warsaw Pact members was controversial for both the East and West. In 1999, NATO was debating its role in the post-Cold War age, particularly in regard to its position in the Balkans, where war had broken out in Kosovo. Some in the West feared admission of former communist satellites might lead to espionage involving NATO secrets. Furthermore, Russia saw the admission of its former satellites as a possible threat.

A Historical Alliance

U.S. president Bill Clinton, took the lead in sponsoring NATO expansion into Eastern Europe. Clinton believed that the admission of former Soviet bloc countries would solidify the European-American alliance and make war in Europe less likely. As early as 1994, the U.S. government persuaded Poland to join in NATO military exercises as an opening for cooperation. Resistance in Poland caused the foreign minister Andreze Olechowski to resign in January, 1995, but most Poles favored their country's joining the alliance. In campaigning for his second term, President Clinton emphasized his support for the admission of some Eastern European countries to NATO by the spring of 1999.

All sixteen of the NATO members needed to ratify admission of the new entries. The countries petitioning for admission saw NATO membership not simply as a security arrangement but also as a means of improving their economies,

3013

Czech foreign minister Jan Kavan signs papers as U.S. secretary of state Madeleine Albright watches at the Harry S. Truman Library in Independence, Missouri.

stabilizing their region, and guaranteeing their new democratic politics. In 1996, NATO agreed to admit Poland Hungary, and the Czech Republic, although some members supported different countries for the first round of expansion. Many people thought Slovakia would enter at the same time, but because of domestic Slovakian concerns and the desire to go slow on NATO expansion, the country was dropped from the initial plans of expansion.

The admission of the three countries took place at the Harry S. Truman Presidential Library in Independence, Missouri. The site was selected partly because Truman had been president when NATO was created. U.S. secretary of state Madeleine Albright received the official

documents ratifying the acceptance of the three countries from their foreign ministers. In deference to Russia, the ceremony was not accompanied by much fanfare.

Bureaucratic procrastination and questions regarding expenditures for upgrading the Eastern European forces to NATO standards and military policies that had to be changed to follow the alliance's guidelines delayed the final requirements for admission. Particularly at issue was the NATO requirement for civilian control over the military. Old-line communist generals refused to accept this new subordination. However, the advance of democracy in Eastern Europe accompanied by the retirement or firing of the old guard and the willingness of the new politicians to accept their responsibility in this area caused the objection to civilian control to disappear.

Consequences

The three countries added more than 200,000 new troops to the alliance. One of the immediate tasks of the new members was support for the war in nearby Serbia, for which they were not completely prepared. They contributed soldiers for peacekeeping and auxiliary service in the Balkans. The alliance required the three to bring their armed forces up to NATO standards, and the other members agreed to help finance the $1.5 billion required for the expansion. Nevertheless, the expenses necessary for upgrading seriously strained the budgets of the impoverished Eastern European governments, and the improvements required many years. Initial efforts stressed training, professional development, restructuring, and communications rather than

weapons and armament. Although the new members had already made some efforts to upgrade their forces, they fell short of their goals, which they hoped to attain by March, 1999.

The admission of the three former Warsaw Pact members opened the way for more Eastern European countries from the Baltic (Latvia, Lithuania, and Estonia) to the Balkans (Bulgaria, Romania, Albania, and Slovenia) to join NATO. The relationship of NATO to the former republics of the Soviet Union such as the Ukraine remained a stumbling block for further expansion, which was put off until at least 2002. In the meantime, the immediate effects on the Czech Republic, Poland, and Hungary were both positive and negative. Their goal of becoming closer to the West was realized, but they also found themselves unexpectedly in the midst of the war in the former Yugoslavia and having to use their limited resources to upgrade their militaries.

Frederick B. Chary

First Nonstop Around-the-World Balloon Trip Is Completed

After twenty years of intense competition, adventurers Bertrand Piccard and Brian Jones conquered one of aviation's last challenges, traveling around the world in a balloon.

What: Space and aviation; Transportation
When: March 20, 1999
Where: Château d'Oex, Switzerland, to Mauritania, Africa
Who:
BERTRAND PICCARD (1958-   ), Swiss medical doctor
BRIAN JONES (1947-   ), British engineer and flying instructor

The Great Balloon Race

When Bertrand Piccard and Brian Jones crossed the 9.27 degrees west latitude over Mauritania in the Breitling Orbiter 3 after a flight of 25,361 miles (40,806 kilometers) on March 20, 1999, they became the first balloonists to circumnavigate the globe. Their flight ended an intense competition, closely followed in the press, which had produced thirteen expeditions during the previous sixteen months.

Modern sport ballooning experienced a revival in 1978 when Ben Abruzzo, Maxie Anderson, and Larry Anderson successfully crossed the Atlantic Ocean. A number of records were set during the following three years, but no one came close to completing an around-the-world trip, the ultimate challenge for balloonists. Between 1993 and 1999, twenty-one attempts were made; thirteen lasted less than a day. From 1997, new technologies heightened both the chance of success and the level of competition. In January, 1997, Steve Fossett flew from St. Louis, Missouri, to India, setting records for consecutive miles (11,265; 18,125 kilometers) and hours (146.44). Altogether Fossett made five attempts, including a

1998 expedition that established a new world distance record (14,236 miles; 22,905 kilometers).

The Breitling watch company became a major corporate player in the balloon race, sponsoring three attempts to circle the globe. The first Breitling Orbiter, piloted by Piccard and Wim Verstraeten, came down in the Mediterranean in January, 1997, after only a few hours of flight. The second Orbiter, flown by Piccard, Verstraeten, and Andy Elson, put down in Myanmar (Burma) in February, 1998, after setting a new duration record (233.55 hours).

Piccard and Jones, pilots of the third and eventually successful flight, were both ballooning veterans who began flying at an early age. Piccard's grandfather, Auguste Piccard, was the first man to reach the stratosphere in a balloon (1931), using a pressurized capsule that he had invented. Bertrand was an early hang-gliding enthusiast and had established a world altitude record. The Breitling Orbiter 3 mission was his fourth attempt to circumnavigate the earth in a balloon. Jones, who began flying at the age of sixteen, spent thirteen years in the Royal Air Force. In 1986, he found his passion in ballooning.

Away At Last

Many doubted that the Orbiter 3 would be successful when it was launched on March 1, 1999. This giant silver balloon, manufactured by Don Cameron, was 170 feet (52 meters) tall and weighed 9 tons (8 metric tons). This hybrid, experimental balloon had separate compartments for helium, which provided lift, and hot air, which helped control altitude.

The envelope was so fragile that it could not be inflated for tests, but there was no time for further modifications. The cable and wireless team

3016

of Andy Elson and Colin Prescott had been aloft since February 18. As former expedition partners with the Orbiter 2, Piccard and Elson remained friends and were in frequent contact by fax while aloft. In addition, in various parts of the world, five other teams, including that of their toughest competitor, British tycoon Richard Branson, were nearing their launch dates.

Piccard and Jones, both tense and hopeful, lifted off from the Swiss Alpine village of Château d'Oex on March 1. For the first three days, the balloon traveled south and west to catch jet streams over northern Africa. In northern Mali, the pilots turned eastward, flying across Algeria, Libya, Egypt, and Saudi Arabia to avoid Iraq, where they were not guaranteed airborne security. On March 7, Piccard and Jones were surprised to learn that Elson and Prescott had been forced to ditch their balloon in the Pacific Ocean off the Japanese coast after a flight of more than 11,000 miles (17,699 kilometers).

The Orbiter 3 crossed China in 14 hours on March 9-10. The following day over the Pacific

Ocean, they faced a choice between two jet streams: one heading toward the United States and the other toward Mexico and Cuba. Following the advice of meteorologists, they adopted the longer, southern route to take advantage of more stable weather conditions. On March 15, the balloon entered the jet stream southeast of Hawaii at the highest altitude of the trip, 36,300 feet (11,064 meters), and at a speed of 81 miles per hour (130 kilometers per hour). At its strongest, the jet stream pushed the balloon forward at 144 miles per hour (232 kilometers per hour), the fastest speed reached during the twenty days. The expedition was threatened on March 18 over the Gulf of Mexico when Orbiter 3 was ejected from the jet stream and began to head for Venezuela instead of Africa. Piccard and Jones ascended to 35,000 feet (10,668 meters) to catch an easterly wind, escaping almost certain failure had they drifted farther south.

Catching another jet stream, the Orbiter 3 shot across the Atlantic Ocean, approaching speeds of 100 miles per hour (161 kilometers per

The Breitling Orbiter 3 balloon passes over the Swiss Alps.

hour). Piccard and Jones officially circumnavigated the globe after 19 days, 1 hour, and 49 minutes as they reached 9 degrees, 27 minutes west latitude, 36,000 feet (10,973 meters) above Mauritania. Almost twenty hours later, they landed gently in the Egyptian desert, 300 miles (482 kilometers) southwest of Cairo, where they were picked up by an Egyptian army helicopter. Piccard and Jones had officially covered 25,361 miles (40,806 kilometers), breaking the distance record set previously by Fossett in 1998 by 11,000 miles (nearly 18 kilometers). The Orbiter 3 also set a duration record, staying aloft for 477.47 hours.

Consequences

On March 20, 1999, Jones and Piccard earned the respect of the world's aviation community by conquering one of the last great challenges of flight. They received a prize of $1 million but had flown more for the challenge than for the money, setting an example for adventurers around the world. When Piccard had doubted the wisdom of risking so much for an aviation record, he had been encouraged by Jean-Christophe Jeauffre, founder of the Jules Verne Adventure Association. "It's not a question of whether or not you have a *right* to fly," Jeauffre insisted. "You have a *duty*. Mankind needs people to do things like this. People are going to dream with you." Piccard and Jones were also awarded the Olympic Order in recognition of their "exploit for peace," exciting the imagination of people worldwide as they flew high above more than thirty countries in their quest.

Ellen Powell

North Atlantic Treaty Organization Wars on Yugoslavia

> *The North Atlantic Treaty Organization (NATO) took military action against Yugoslavia after that country began a campaign of terror against the ethnic Albanian majority of the Serbian province of Kosovo. NATO feared that this campaign would destabilize neighboring countries and might lead to a broader Balkan war.*

What: War

When: March 24-June 10, 1999

Where: Province of Kosovo, Republic of Serbia, Yugoslavia

Who:

SLOBODAN MILOŠEVIĆ (1941-   ), president of Yugoslavia from 1997 to 2000

IBRAHIM RUGOVA (1944-   ), leader of the Kosovar Albanians, elected president of the unrecognized Republic of Kosovo

The Causes

Yugoslav leader Slobodan Milošević rose to power in the Serbian Republic of Yugoslavia by playing on the fears and resentments of the majority Serbs. Serbs were particularly worried about the growing majority of ethnic Albanians in Kosovo, a province of Serbia. Milošević gained support from Serbian nationalists by withdrawing Kosovo's autonomous status. In 1989, he imposed direct Serbian control over the province and its 90 percent Albanian majority. In 1990, the ethnic Albanians of the province proclaimed a Republic of Kosovo and, in 1992, elected a Kosovar Albanian intellectual, Ibrahim Rugova, president. The Serbs, who declared the election meaningless, removed ethnic Albanians from positions in civil administration, the police, education, and the economy. Rugova and the moderates counseled passive resistance and noncooperation. They argued that this would win the Kosovar Albanians the support of the West.

The West, however, was slow to intervene in the wars that accompanied the breakup of the former Yugoslavia in 1991 and 1992. It was only in August, 1995, that the North Atlantic Treaty Organization (NATO) intervened decisively in Bosnia to put an end to the conflict there. However, to the dismay of the Kosovar Albanians, the peace accords formulated at Dayton, Ohio, in November 1995, did not address their plight. The United States supported the sanctity of existing frontiers and hesitated to favor the Kosovar Albanians, fearing that to do so would promote destabilizing separatism. Militant Kosovar Albanians, arguing that Rugova's moderation had failed to produce results, organized the Kosovar Liberation Army (KLA), which began a guerrilla campaign against Serbian police and officials. Milosevic responded with unrestrained repression.

In September, 1998, after a series of massacres of Albanians by Serb forces, the United Nations called for a cease-fire as well as talks between the Yugoslav government and representatives of the Kosovars. When Serb attacks continued, NATO authorized air strikes in October, 1998. The attack was aborted at the last minute when Milošević agreed to withdraw Yugoslav troops, allow Kosovar refugees to return to their homes, and permit two thousand unarmed monitors from the Organization for Security and Cooperation in Europe (OSCE) to verify compliance. Nevertheless, violence continued.

Following massacres of Kosovar civilians by Serbian forces in January, 1999, NATO demanded that the Yugoslav government participate in peace talks with Kosovar representatives

AP/Wide World Photos

This U.S. Navy photograph shows a cruise missile being launched from the USS Philippine Sea *during an attack on Yugoslavia from the Adriatic Sea.*

or face air strikes. At first, both sides, refused to agree to terms put forth by NATO at Rambouillet, France. After intense pressure the Kosovar representatives agreed on March 18 to autonomy for the province under NATO protection, even though it would continue to remain a constituent part of Serbia. Milošević not only rejected the agreement but also moved more troops into the province. As more Albanians were killed and a massive flight of Albanians began toward Albania and Macedonia, the United States attempted to force Milošević to back down. When threats failed, NATO credibility was on the line. The OSCE observers were withdrawn on March 20, and NATO air attacks began before dawn on March 24.

The War

NATO's leadership had mistakenly believed that a few days of bombing would persuade Milošević to accede to their demands. President Bill Clinton's promise that the United States would not commit ground troops to the conflict and the reluctance of the United States and the rest of NATO to expose their pilots to antiaircraft defenses prevented low-altitude attacks and impeded NATO's effectiveness against Serb ground forces. NATO also did not take into consideration the importance of the Kosovo issue to Milošević and the Serbs.

Despite weeks of bombing, including strategic and military targets in Serbia proper, the Serbs drove approximately 850,000 ethnic Albanians from the province of Kosovo and another 500,000 into the hills. NATO then decided to put more pressure on the people of Serbia by knocking out their electricity and urban water systems. The NATO alliance remained committed to the conflict in spite of a growing number of Serbian civilian casualties and open expressions of Russian opposition. Damage to Serbia's infrastructure, growing Serbian military casualties, and the failure of Russian displeasure to break the resolve of NATO finally persuaded the Serbian regime that it would have to accept NATO's demands.

An acceptable neutral leader, President Martii Ahtisaari of Finland, brought to Milošević NATO's demands that the Serbian forces leave Kosovo, that all Albanian Kosovars be allowed to return, and that a NATO force occupy Kosovo to guarantee their security. Milošević gave way on June 2, 1999. According to the agreement, the KLA would be disarmed, and Kosovo, although granted substantial autonomy, would still be a part of Serbia.

On June 10, after confirming that the Serbians had begun to withdraw their forces, NATO suspended its bombing campaign. The United Nations Security Council, with China abstaining, approved a NATO-led international force to enforce peace in the province and a U.N. interim administration to guarantee substantial autonomy and to prepare for free elections. On June 12, the first NATO troops entered the province.

Consequences

NATO and the United Nations were unable to prevent acts of vengeance against Serbian civilians by returning ethnic Albanians, and many fled the province. In addition, the idea that Kosovo would remain part of Serbia has been rejected not only by the KLA but also by the moderates led by Rugova.

Within Serbia, the war finally undermined the authoritarian rule of Milošević. He was defeated in the Yugoslavian presidential election on September 24. Although he fraudulently attempted to force his opponent, Vojislav Kostunica, into a runoff, a popular uprising forced Milošević to admit defeat. Kostunica was sworn in as president on October 7, 2000, and proceeded to dismantle Milošević's apparatus of power. Finally in Serbian parliamentary elections on December 23, Kostunica's eighteen-party reform coalition, the Democratic Opposition, humiliatingly defeated Milošević's Socialist Party.

Bernard A. Cook

Assisted-Suicide Specialist Kevorkian Is Found Guilty of Murder

> *After gaining notoriety as an assisted-suicide practitioner, Jack Kevorkian was convicted of second degree murder for the videotaped assisted-suicide of Thomas Youk in Michigan.*

What: Crime; Human rights
When: March 26, 1999
Where: Michigan
Who:
JACK KEVORKIAN (1928-   ), physician
JUDGE JESSICA COOPER (1946-   ),
    Oakland County circuit court judge
THOMAS YOUK (1945-1997), Kevorkian's
    patient

Death by Delivery

On March 26, 1999, Jack Kevorkian, the physician who prominently assisted suicides, was found guilty of second-degree murder in the death of his last patient, Thomas Youk. He was also convicted of using a controlled substance (lethal drug) after his medical license had been revoked. On April 13, 1999, Judge Jessica Cooper sentenced Kevorkian to ten to twenty-five years for second-degree murder and three to seven years for the use of a controlled substance. His sentences were to be served concurrently.

Kevorkian believes that terminally ill or severely handicapped people should have the right to end their suffering and die with dignity. He was devoted to helping them achieve this end. Over the course of the years, he built several machines that would help with this process. All the patient had to do to begin the suicide process was press a button. This easy-to-use design was very important because many of these patients were severely crippled or paralyzed. He used lethal drugs or carbon monoxide to bring about the death of his patients.

Kevorkian had assisted in the suicides of at least 130 people before Youk's death. His first patient was a fifty-four-year-old woman with Alzheimer's disease, who died on June 4, 1990. Over the next seven years, he helped people with a variety of diseases and painful conditions end their lives. Not all were terminally ill. Some were in unbearable pain from chronic (nonfatal) conditions, and several of his patients were found to have no physical ailments at all.

A Public Challenge

Kevorkian wanted to challenge the legal system to make assisted suicide not only legal but also acceptable. Therefore, he was extremely selective in choosing his first client. He videotaped his interview with her, discussing her decision. To get many of his early patients, Kevorkian advertised in the newspapers. In his advertisement, he sought those who were "oppressed by a fatal disease, severe handicap, or a crippling deformity." He offered to help them kill themselves, free of charge. He proclaimed himself a practitioner of "obitiatry," literally a doctor of death. He interviewed the potential patient, often videotaping the interview as he had done with his first patient, and many were helped to die within a few days of meeting Kevorkian.

Kevorkian sought out media attention and became bolder and more flamboyant with each patient. Questions were raised as to his real motives because he sought media attention so avidly. He was ordered to stand trial for his part in several of the deaths, but charges were always dismissed, or he was acquitted. He carried on with his work despite court orders to stop and eventually lost his medical license. After Kevorkian's conviction, his videotapes, including that of his first patient, were reviewed, and numerous people questioned whether many of his patients had the mental capacity or understanding to make the decision to die.

For a death to be ruled an assisted suicide, the patient had to take an active role in the process. Legally speaking, Kevorkian went too far in the death of Thomas Youk on September 17, 1998. Youk suffered from amyotrophic lateral sclerosis (ALS), a degenerative nerve disease. Youk was so severely affected by the disease that he was no longer able to even push a simple button to contribute to his death. Kevorkian injected him with drugs intravenously. This crossed the line from assisted suicide to euthanasia (mercy killing) because the patient could not take an active role in the process. Kervorkian videotaped this death, made it known to the media what he did, and dared prosecutors to come after him.

To further push the point, two months after Youk's death, Kevorkian took the tape to the news show *60 Minutes*. The videotape was shown on television. It explicitly showed Kevorkian's actions in Youk's death. Once again, he challenged the government of the state of Michigan to prosecute him, for if it did not, he said, the state was condoning this type of death and accepting its legality. At this point, the state stepped in, arrested Kevorkian, and charged him with murder. The jury agreed, convicting him of second-degree murder. In sentencing him, Oakland County circuit court judge Cooper said, "You had the audacity to go on national television, show the world what you did, and dare the legal system to stop you. Well, sir, consider yourself stopped."

Consequences

The resolution of this case affected both the supporters and opponents of euthanasia. Those who oppose euthanasia, such as the International Anti-Euthanasia Task Force, began using Kevorkian as an example of how the concept of euthanasia can be abused. In examining the work of Kevorkian, they were quick to point out that many of Kevorkian's patients were not terminally ill, and several were not even physically ill. Knowing this, Kevorkian still used them to further his own goal. They claim that Kevorkian always had a morbid curiosity with death, which he displayed throughout his career, making him unable to hold a position for long and eventually forcing him into early retirement.

Jack Kevorkian makes his closing statement before the jurors.

Many of those who support euthanasia, such as members of the Hemlock Society and the Euthanasia Research and Guidance Organization (ERGO), agreed that the conviction of Kevorkian might be for the best. They believed that Kevorkian's early work was well intentioned. However, he became too greedy for media attention, always flaunting his activities in order to be in the spotlight. He became more daring, and after a while, his attention seeking was actually beginning to harm the cause of euthanasia. Therefore, many euthanasia supporters felt it was best if Kevorkian ceased activity before more damage was done to their cause. As Judge Cooper said in the sentencing of Kevorkian, his conviction was not about euthanasia. "It was all about you, sir. It was about lawlessness. It was about disrespect for a society that exists and flourishes because of the strength of the legal system."

Mary Ellen Campion

Dow Jones Industrial Average Tops 10,000

> *The Dow Jones Industrial Average, the most widely used benchmark for stock market performance, surpassed an important plateau when it passed 10,000. It took the average less than four years to double in value, bringing new enthusiasm and optimism to the investing public.*

What: Business; Economics
When: March 29, 1999
Where: New York Stock Exchange, New York City
Who:
CHARLES HENRY DOW (1851-1902), creator of the Dow Jones Industrial Average
ALAN GREENSPAN (1926-), chair of the Federal Reserve

Pulse of the Market

When Charles Henry Dow created the Dow Jones Industrial Average (DJIA, Dow) on May 26, 1896, he probably never dreamed that it would become the standard measure of the stock market's performance for more than a century. At the time, little or no reliable information was available about companies, and crooked brokers and unscrupulous corporate insiders often manipulated stock prices. Prudent investors stayed away from stocks and invested in bonds that paid predictable interest. Because not all stocks moved in the same direction at the same time, it was difficult to know whether stocks were generally rising or falling.

Dow created the DJIA by taking an average of the prices of twelve stocks. In 1916, the DJIA was expanded to include twenty stocks. The number was raised again in 1928 to thirty, where it remains. When the DJIA's peaks and troughs increased progressively, a bull market prevailed. When both the peaks and troughs progressively declined, it was considered a bear market. Although Dow used the word "industrial" in naming the average, the DJIA has represented a cross-section of the economy. As the economy changed from agrarian to industrial to service and technology, the companies represented in the average also changed. In 1999, the thirty stocks that made up the DJIA were of major companies in their industries and represented about one-fifth of the market value of all U.S. stocks.

From a value of 40.94 in 1896, the DJIA steadily rose to 100 in 1906. Rising stock dividends, increased savings, and easy access to borrowed money continued to push stocks higher. The DJIA peaked at 381 before its crash in October, 1929. Because of excessive borrowing and widespread fraud and price manipulation, people lost not only money but also their faith in the financial markets, and the resulting Great Depression lasted for several years.

Businesses raise money in the securities markets to finance new projects and implement new production technologies that are vital for continued economic growth. If businesses experience difficulty in raising money, it has serious consequences for the economy. Recognizing the importance of the securities market, Congress created the Securities and Exchange Commission (SEC) in 1934 to regulate securities markets to ensure that they were fair and free of fraud and manipulation and that investors had reliable information about companies and securities.

Dow at Giddying Heights

The increased transparency slowly rebuilt the faith of investors in the U.S. stock market. The Dow came out of its decline and hit a 1,000-point mark in 1972. As confidence in the stock market grew, so did the number of investors and the trading volume. By 1999, about half of all Americans owned equities, compared with 14 percent in 1980. From less than 1 billion shares trading in 1960, the trading volume had surpassed 200 billion by 1999, and the dollar trading volume increased from about $45 billion to about $10 trillion. On March 29, 1999, the Dow passed the 10,000 mark. At 10,000, the Dow stood at 300 percent above its level of October, 1990, and 1,200 percent above its level in 1982. The value of stocks held by individuals was also up by 381 percent since 1990. In 1999, the value of stocks held by households represented more than 25 percent of total household assets, up from 8 percent in 1983.

Over the years, the stock market and investors matured, and the days of widespread manipulation and fraud and the resulting wild price swings became a part of history. Eight out of the ten largest yearly percentage increases occurred before 1940, as did ten out of ten of the largest percentage declines. Nine of the ten largest one-day increases took place before 1940, as did eight out of ten of the largest declines.

The Federal Reserve Bank has played a vital role in the stock market increase. Using its monetary policy, the Federal Reserve influences interest rates, inflation, employment, and output and thereby corporate profits and stock prices. Alan Greenspan, the chair of the Federal Reserve, was credited for being the architect of the unprecedented economic as well as the stock market expansion.

Consequences

The Dow at 10,000 was what it reflected: The unprecedented strength of the U.S. economy and the dominance of U.S. corporations in the global marketplace. In 1990, the U.S. share of the world stock market was 36 percent. By 1998, it had increased to 44 percent. Although the stock market continued upward, some stock watchers were concerned about Greenspan's use of the words "irrational exuberance" on December 5, 1996, to express his concern about the unreasonably high level of stock prices. Although stocks declined following his speech, they did not stay down for long. Other market watchers countered Greenspan's remarks by saying that stocks had been undervalued for decades and if they were correctly valued, the DJIA would be at 36,000 or even higher.

On January 14, 2000, the DJIA closed above 11,700, a record high. However, the seemingly unstoppable rise in the DJIA had come to an end. That March, it dipped below 10,000. It soon recovered but fell below 10,000 again in April, 2001. It rallied, then fell again, closing below 9,000 on September, 17, 2001, in response to the terrorist attacks on New York and the Pentagon.

Rajiv Kalra

New York Stock Exchange traders applaud after the Dow closes above 10,000 for the first time.

AP/Wide World Photos

3025

Jury Awards Smoker's Family $81 Million

The family of a long-term smoker who died of lung disease is awarded a record settlement against a tobacco company.

What: Law; Health
When: March 30, 1999
Where: Multnomah County Circuit Court, Portland, Oregon
Who:
Jesse Williams (died 1997), deceased long-term smoker of Marlboro cigarettes
Mayola Williams, widow of Jesse Williams
Philip Morris, cigarette manufacturing company and maker of Marlboro cigarettes

The Verdict

On March 30, 1999, in Multnomah County Circuit Court in Oregon, the jury in the case of *Williams v. Philip Morris* awarded a record $81 million to the estate of Jesse Williams. The plaintiff's attorneys were William A. Gaylord, Raymond F. Thomas, James S. Coon, and Charles S. Tauman. The defense attorneys were Walter L. Cofer, Billy R. Randles, James L. Dumas, and Jay W. Beattie.

Mayola Williams, the widow of Jesse Williams, sought $101 million from Philip Morris, the maker of Marlboro, her late husband's cigarette of choice, on behalf of his estate. The trial began with jury selection on February 22. The panel consisted of six men and six women, including three smokers and four former smokers. After a five-week trial, the jury decided by an 11-1 vote that each party was negligent, and each was 50 percent responsible for Jesse Williams's injury. The panel awarded Mayola Williams $21,485 for economic damages and $800,000 for noneconomic damages (pain and suffering). The jury also voted 9-3 on a separate verdict form that false representations by Philip Morris had contributed to Jesse Williams's injury. The panel

awarded his widow an additional $79.5 million in punitive damages, plus $21,485 in economic damages and $800,000 in noneconomic damages, for a total of $81 million.

The case centered on the health of Jesse Williams, who had been a janitor for the Portland Public Schools. Williams began smoking in his early twenties, while serving in the U.S. Army during the Korean War. Eventually, he was smoking three packs a day, and after 1955, he smoked only Marlboro cigarettes, made by Philip Morris. In October, 1996, he was diagnosed with a small-cell carcinoma of the lungs. He died of the disease five months later, on March 17, 1997, at age sixty-seven. He left a widow, Mayola Williams, and six adult children. Williams had smoked Marlboro cigarettes for more than forty years.

The Issues and Testimony

Mayola Williams defined four specific acts for which Philip Morris should be held liable: for selling a dangerous and defective product, for negligence in failing to warn of the addictive nature of its cigarettes, for fraudulently misrepresenting the health hazards of smoking, and for conspiring with other tobacco companies to conceal those hazards. Philip Morris defended itself by focusing on Jessie Williams's personal history, claiming that it was not responsible for his death because he chose to smoke, failed to heed warning labels on Marlboro packages, and ignored news articles about the hazards of smoking as well as his wife's pleas to stop smoking.

As the trial began, the testimony and arguments focused on two specific questions. The first was that of Philip Morris's possible negligence in the manufacture of its cigarettes and whether that negligence was the cause of Williams's death. The second question was whether Philip Morris made false representations about the causal link between smoking and cancer and whether those representations contributed to Williams's death. A feature of Oregon product liability laws required the jury to decide what percentage of the negligence to assign to Williams or to Philip Morris, because in Oregon, a smoker is barred from receiving an award if a jury determines that he or she bore more than 50 percent liability for the problem over which a suit is brought.

During the testimony, Williams was portrayed as someone who was heavily addicted to nicotine but who trusted the company not to sell a harmful product. Attorneys for Philip Morris claimed that Williams's forty years of smoking were a matter of free choice and willpower and that if he had really wanted to quit, he could have done so at any point. They pointed to statistics from the National Centers for Disease Control and Prevention indicating that millions of people were able to quit smoking. The Williams family attorneys and expert witnesses countered that nicotine dependence is not just psychological but chemical in nature. Neal Benowitz, who had conducted nicotine research at the University of California at San Francisco, testified that smoking is not a free choice if the person is addicted. During the course of the trial, Williams's attorneys made use of internal Philip Morris documents, made public through an earlier settlement. These documents showed that the company had already been aware for quite some time that the nature of the dependency was chemical and that there was a link between smoking and cancer.

Other issues in the trial related to whether the specific form of cancer contracted by Williams was in fact the same one associated with cigarette smoking and whether his cancer actually began before the eight-year period specified by Oregon product liability law. If the cancer was determined to have started earlier, Phillip Morris would not have been liable. The jury's $81 million decision against Phillip Morris included $79.5 million in punitive damages. The punitive damage amount was later reduced to $32 million by a post-trial ruling on May 13, 1999, by Circuit Court Judge Anna J. Brown, who felt any award greater than $32 million violated constitutional guarantees against excessive punishment. On June 15, 1999, Judge Brown denied a motion for a new trial by Philip Morris, which had argued that one of Williams's attorneys inflamed the jury in his closing arguments and did not provide sufficient evidence for a punitive award.

Consequences

Immediately after the Williams case, the value of stock in Phillip Morris and other tobacco companies declined on Wall Street. Investors feared that because Oregon had especially strict liability laws, other cases pending against tobacco companies would be successful. The jury's decision in this case underscored the possibility that large industries could be held responsible for the health of consumers.

Alice Myers

Canada Creates Nunavut Territory

In 1999, the map of Canada changed with the formation of Nunavut Territory, an area in the eastern and central Arctic regions of Canada where the majority of the population was Inuit.

What: Government
When: April 1, 1999
Where: Northwest Canada
Who:
PAUL OKALIK, premier of Nunavut

A New Territory

Nunavut, an Inuit (Eskimo) word meaning "our land," is the name of the territory in Canada that was officially designated on April 1, 1999. Nunavut was carved from the central and eastern part of Canada's Northwest Territories; the western part of the Northwest Territories is still known by that name. Roughly one-fifth the size of Canada, the territory covers about 818,962 square miles (2.1 million square kilometers). In 1999, the total population of Nunavut was 27,100, about 85 percent of which was Inuit.

History

The Nunavut Territory was formed by the Inuit people and the Canadian government to better administer the people of the region. As far back as 1966, the Northwest Territory was seen by the Canadian government as too large and vast a land area to be governed efficiently. The Inuit people began calling for the creation of their own territory in 1976, shortly after the Canadian government decided to negotiate settlements with native peoples who filed land claims.

In 1992, the population of the Northwest Territories voted to form another territory from lands within its boundaries. The boundaries of Nunavut were drawn corresponding to the land area the Inuit people claimed as their traditional territory as well as to the lands they inhabited in the past and present. The territory's boundaries began in the east around Hudson Bay and reached what would become the borders of the new Northwest Territories in the west. It also included many of the islands opposite Greenland and in the Beaufort Sea. In 1993, the Canadian government and the Inuit reached the Inuit Land Claims Agreement, in which the government agreed to create a new territory for the Inuit. The same year, the government passed the Nunavut Act, establishing an implementation commission and setting the basis for the creation of the new territory, which was to be called Nunavut. The formal creation of the Nunavut Territory took place on April 1, 1999.

Nunavut is made up of three regions: Qikiqtaaluk, Kivalliq, and Kitikmeot; these are the Inuit names for regions formerly called Baffin, Keewatin, and Kitikmeot. The Territory of Nunavut is sparsely populated, with twenty-eight communities. Iqaluit (formerly known as Frobisher Bay) on Baffin Island is the largest community with about 3,600 inhabitants and is the new capital of the territory. About one-sixth of the entire population of Nunavut lives in, or near, the capital community. The territory is run as a public government and has the same institutions as any other Canadian territorial government with a commissioner, a cabinet, a legislative assembly, and a public service and territorial court. Paul Okalik, a lawyer, became the territory's elected premier.

The newly formed Nunavut had a young population, with about 40 percent being teenagers. Territorial officials were working hard to better educate this sector of the population. In the past, many Inuit children missed a lot of school because their culture did not place much importance on formal learning but focused more on traditional and practical learning, values, and experiences. It was really more important for the

young Inuits to become familiar with the land and environment for survival and to become efficient in traditional activities such as hunting and trapping. In addition, traveling in and out of the region was expensive, so seeking a higher education at one of Canada's colleges or universities was not economically feasible for many Inuit students and their families. In 1999, the main source of income remained trapping and hunting, so traditional ways of life were still greatly valued and essential.

Consequences

After formally becoming a territory, the Inuit and the other residents of Nunavut Territory still faced many problems. Roughly one-third of the population lived on welfare and other government programs. Those that did have jobs made an average yearly income of around $11,000 or less, although cost for goods and services remained high. The age of the workforce was young, and the educational level was relatively low.

In response, the Nunavut government developed long-term employment plans designed to eventually provide jobs for about 85 percent of the Nunavut population. Money from the Nunavut Land Claims agreement with the Canadian government was expected to improve the economic situation for the Inuits and the other people of Nunavut. The money was to be used for economic growth and development in such activities as natural resource exploration for minerals such as zinc, gold, copper, silver, lead, and even diamonds. Businesses such as hotels and motels to provide services to environmental tourists,

3029

trucking, and shrimp fishing were seen as viable economic outlets to employ the people of Nunavut. Also, under the land claim agreement, three new national parks were to be created within the territory. These parks were expected to increase tourism and thereby raise the amount of tourist dollars brought into the newly formed territory.

Under the 1993 land claims agreement, the people of Nunavut were given the opportunity and right to self-govern and to determine their own destiny. Their territorial government incorporated traditional Inuit values into its departments in order to develop and implement governing policies that reflect Inuit culture. Because its governing body is based on traditional Inuit values, the Nunavut Territory offers the Inuit the opportunity to mold and shape their future and their destiny.

Douglas Heffington

Astronomers Discover Solar System with Three Planets

Careful analysis of light from Upsilon Andromedae revealed the first known multiple-planet system orbiting a normal star.

What: Astronomy; Physics
When: April 15, 1999
Where: San Francisco, California
Who:
R. PAUL BUTLER, astronomer, Anglo-Australian Observatory, Epping, Australia
GEOFFREY W. MARCY, astronomer, San Francisco State University and the University of California, Berkeley
PETER NISENSON, astronomer and member of team at Whipple Observatory, Mount Hopkins, Arizona

Discovery of a New Planetary System

On April 15, 1999, two teams of astronomers, represented by R. Paul Butler, Geoffrey W. Marcy, and Peter Nisenson, independently obtained evidence that three planets orbit the star Upsilon Andromedae, a star not too different from the Sun. Planets are detected about stars other than the Sun by carefully monitoring the spectra of those stars to look for evidence of periodic motion. Starlight is collected with a large telescope and focused onto a slit. Light passing through the slit falls on a diffraction grating that spreads the light out into a rainbow, or spectrum, as it was named by Isaac Newton. The spectrum is actually a series of colored images of the slit. Patterns of dark lines in the spectrum reveal which elements are present in the star. If the star is moving toward or away from Earth, these patterns are shifted slightly toward the blue or toward the red end of the spectrum. This shifting toward the blue or red because of the star's motion is an example of the Doppler shift. To see the small shifts caused by planets, extreme care must be taken to keep the parts of the instruments at a constant temperature, properly aligned, and calibrated.

A planet does not really orbit the center of a star; instead, both the planet and the star orbit the center of mass of the system. (The center of mass is where a teeter-totter must be supported if it is to balance with the star on one end and the planet on the other.) The greatest motion of the star occurs if it has a massive planet very close to it. Therefore, the detection method used is most likely to discover "hot Jupiters," or massive planets orbiting close to their parent stars. Most of the planets discovered so far have been hot Jupiters.

Using an especially sensitive method, the first extrasolar planetary system may have been discovered in 1992, but unlike Upsilon Andromedae, it is a bizarre system of cinders orbiting a dead star. Periodic Doppler shifts in the radio pulses from the pulsar PSR 1257+12 indicate the presence of three small planets near the pulsar and a Saturn-sized planet much farther away. A pulsar is a rapidly rotating neutron star, a stellar remnant of a supernova explosion. It was previously believed that no planetary-sized bodies could survive such an explosion. Perhaps these planets are the remnants of giant planets, or perhaps they were formed from the supernova debris.

The Upsilon Andromedae System

Marcy and Butler announced their discovery of the planet Upsilon Andromedae B in January, 1997. This planet is at least 0.71 Jupiter masses and whirls around the star Upsilon Andromedae A in 4.6 days at a distance of only 0.06 astronomical units (an astronomical unit is the distance between Earth and the Sun). Upsilon Andromedae A is only slightly hotter than the Sun, but it is 30 percent more massive and three times as luminous as the Sun. Compared with sunlight on the Earth, Upsilon Andromedae's scorching rays are 470 times as intense on B.

Surprisingly, the fit between theory and data grew worse as the astronomers made additional measurements at the Lick Observatory. This suggested that Upsilon Andromedae might have another planet. A second team including Nisenson began their own measurements at the Whipple Observatory. On April 15, 1999, the two teams made a joint announcement of two new planets. Upsilon Andromedae C and D have minimum masses of 2.11 and 4.61 Jupiter masses. The elliptical orbit of C takes it from 0.7 to 1.0 astronomical units from its star so that it gets from 3.7 to 1.8 times the intensity of radiation that Earth receives from the Sun. The orbit of D also is elliptical, taking it from 1.5 to 3.5 astronomical units from Upsilon Andromedae A, making its radiation extremes 0.8 and 0.1 times the intensity that Earth receives.

Because all three planets are gas giants, they do not have solid surfaces and cannot have Earth-type life. They may have habitable satellites, but those of B and C are likely to be too hot. A satellite of D might be habitable if it has a thick atmosphere, but D's elliptical orbit would produce extremely harsh seasons. To have Earth-like conditions, an Earth-type planet would require a nearly circular orbit 1.7 astronomical units from Upsilon Andromedae A, but the large masses and elliptical orbits of C and D would probably make such an orbit unstable.

Consequences

The discovery of the Upsilon Andromedae system has at least three important consequences. First, this was the first multiple-planet system discovered orbiting a normal star. Because the detection method yields only an estimate for the minium mass of the planet, some astronomers wondered if those previously discovered planets were actually small stars instead of planets. It was therefore important that the bodies orbiting Upsilon Andromedae were unlikely to be anything other than planets. Second, the discovery encouraged astronomers to search for more multiple planetary systems, and by August, 2001, more than seventy extrasolar planets had been found, including six double planet systems. Third, theories of planetary formation allow Jupiter-like planets to form no closer than 4 astronomical units from a Sun-like star. If true, the Upsilon Andromedae planets must have migrated inward from their formation sites, yet computer models indicate that their current orbits are stable. It will be a challenge to theoreticians to explain this, and it is hoped that this challenge will lead to a better understanding of planetary formation.

Charles W. Rogers

Former Pakistani Prime Minister Bhutto Is Sentenced for Corruption

> *Pakistan's first woman prime minister and her husband were accused and convicted of corruption in a power struggle in the Islamic Republic of Pakistan.*

What: Politics; Crime
When: April 15, 1999
Where: Rawalpindi, Pakistan
Who:

BENAZIR BHUTTO (1953-   ), prime minister of Pakistan from 1988 to 1990 and 1993 to 1996, and leader of the Pakistan People's Party (PPP)

ASIF ALI ZAHDARI, Bhutto's husband and a senator

NAWAZ SHARIF (1949-   ), Bhutto's political rival and prime minister at the time of the trial

PERVEZ MUSHARRAF (1943-   ), general, chief of staff, and subsequent chief executive

Power Struggles in Pakistan

Since its foundation in 1947, the Islamic Republic of Pakistan has alternated between periods of democratic government and of military rule, during which the activities of the political parties have been severely curbed. Many of the political leaders in Pakistan come from wealthy landowning families and have strong regional bases. The Bhuttos were one such family. Benazir Bhutto's father, Zulfiqar Ali Bhutto, led the Pakistan People's Party (PPP), which had its power base in the Sindh, Pakistan's southern province, and in Karachi, its only port and its largest city.

Zulfiqar Ali Bhutto became the country's leader in 1971, remaining in power till 1977, when a military coup under General Mohammad Zia ul-Haq overthrew his government and imprisoned him. He was eventually hanged in April, 1979, while Benazir, his eldest child, and other family members were detained. Benazir spent six years in prison or detention; however, during this time, she campaigned for democratic freedom. In April, 1986, she returned from exile in Great Britain, becoming co-leader of the PPP with her mother.

After Zia's death in a plane crash, democracy was restored in 1988, and Benazir Bhutto's party won an overwhelming electoral victory. She became the first woman prime minister of any Muslim country. However, in August, 1990, her government was dismissed by the president of Pakistan for mismanagement, inability to control regional violence, and corruption. Her rival, Nawaz Sharif, became prime minister until he also was dismissed on corruption charges.

Subsequent elections in October, 1993, returned her to power, but she was dismissed again on almost identical charges in November, 1996. Shortly after, Sharif again became prime minister in new elections in which the fragmented PPP was routed. Subsequently, Bhutto was accused of corruption along with her husband, Asif Ali Zardari, who was also accused of plotting the murder of Bhutto's estranged brother.

The Trial

Both Benazir and her husband were tried under the Accountability Act, which Sharif had pushed through parliament in 1997. An accountability bureau had been set up to deal with the widespread corruption among politicians and civil servants. Zardari's arrest came as no surprise to many: He was popularly known as "Mr. Ten Percent" for his role in government deals. Bhutto's reputation, however, was much better. A Harvard and Oxford graduate, she had been an idealistic campaigner for democracy, and she vigorously proclaimed her innocence. Zardari had been imprisoned in 1996 for other charges. Bhutto left for England with her children in December, 1998, after an initial effort to prevent her leaving the country had been overturned.

Bhutto's pleas of innocence began to be doubted when, in 1997, Swiss authorities froze nineteen secret bank accounts belonging to the Bhutto family on suspicion of money laundering $20 million. Then, in August, 1998, she and her husband were indicted in Switzerland after an investigating magistrate found evidence of off-shore accounts in which illegal kickbacks had been banked. The Swiss authorities, who could not by any stretch of the imagination be accused of playing power politics, made their evidence available to the accountability bureau.

What emerged in the eighteen-month trial was that Bhutto's government had hired a Swiss goods inspection company to inspect all imports into Pakistan to prevent evasion of customs duties, the country's largest source of revenue. To win the contract, the company had agreed to pay a kickback, possibly 6 percent, to the Bhuttos, funneling the monies into an offshore company owned by Zardari, and from there into a Swiss bank account. The Swiss company had deposited nearly $4.3 million in the account. Similar charges against the couple were not considered at the trial, which, by Pakistani standards, was fairly speedy.

The two judges hearing the case imposed a five-year prison sentence on both Bhutto and her husband and fined then $8.6 million, banning them from ever again holding political office. From London, Bhutto pledged to return to Pakistan to appeal her case to the supreme court. The accountability bureau said the two "have finally been proved world-class thieves."

AP/Wide World Photos

Former prime minister Benazir Bhutto, with her son Bilawal Bhutto and daughter Asifa Bhutto.

Consequences

Bhutto duly lodged an appeal in the supreme court, which was due to reach a finding by November, 1999. In the meanwhile, she protested that she had been framed by Sharif and other politicians from the Zia era. She claimed she had not had a Swiss account since 1984. However, the Swiss evidence caused most Western countries, hitherto supportive of Bhutto's championing of democracy, to withdraw their support. Pakistan's firing of nuclear missiles at this time also diverted Western attention away from Bhutto's cause to the threat of nuclear rivalry between Pakistan and its neighbor, India. Nevertheless, Bhutto toured the United States seeking support. The kindest response to her was to see her as a too fond wife, unaware of her husband's greed and ambition. Bhutto, however, was usually viewed as too keen-eyed to be deceived.

Sharif's government appeared to be doing no

better than Bhutto's: The economy was collapsing; Islamic militancy was growing; and corruption seemed as rife. On October 12, 1999, the government was overthrown in a bloodless coup led by General Pervez Musharraf, army chief of staff, who designated himself chief executive and suspended the constitution. The supreme court was immediately caught up in deciding the legality of such action, which it did in May, 2000, even if under the control of the Musharraf regime. The regime's hostility was directed to Sharif, who was accused of treason in December, 1999. After conviction and appeal, he was sentenced to ten years exile in Saudi Arabia but banned from holding political office for twenty-one years.

Under Musharraf, Bhutto's prison sentence was commuted, and the fine was reduced. She was able to return to the country. In the absence of party political activity, disbarment from public office was hardly relevant. She remained head of the PPP, although rivalries with family members had weakened her power base, and a prominent opposition figure. Her husband, Zardari, was not freed from prison. By the latter part of 2001, the accountability bureau had not pressed any further charges against Bhutto; however, the question of whether she could make another political comeback remained unanswered.

David Barratt

Vajpayee's Government Collapses in India

> *Only thirteen months after being elected as India's prime minister, Atal Behari Vajpayee lost a vote of confidence in parliament, forcing creation of a new government or new elections.*

What: Government
When: April 17, 1999
Where: New Delhi, India
Who:
ATAL BIHARI VAJPAYEE (1926-   ), prime minister of India from 1998 and leader of the Bharatiya Jana Party (BJP)
JAYARAM JAYALALITHA, political leader from Tamil Nadu, India
SONIA GANDHI (1947-   ), leader of the Congress Party and widow of former prime minister Rajiv Gandhi

Political Instability in India

Since gaining its independence in 1947, India has had a parliamentary style of democracy based on the British system. Because a large percentage of the adult population votes, Indian politicians must be responsive to the people. Nearly forty political parties exist in India, but until the late 1980's, one political party, the Congress Party, dominated the country's politics, thereby providing a certain amount of stability. Over the years, some of the party's members became corrupt, causing people to mistrust the Congress Party and opening the door for challenges from some of the other parties. This caused some instability in the electoral process because no single party could gain a clear majority, and the parties were forced to form coalitions.

In a coalition government, if one of the parties in the coalition withdraws its support, the government can collapse. When this occurs, a vote of confidence is called in the parliament to determine if a new coalition can be formed. If it cannot, then a new election must be called. In the last ten years, unstable coalitions have led to the formation of five different governments in India.

Vajpayee's Rise and Fall

Until the 1990's, the Bharatiya Jana Party (BJP) was a very small political party from the political right. The BJP rose to national prominence partly because of the growing distrust of Congress Party leaders and their electoral success in several heavily populated states in India's northwestern region. In 1998, Atal Bihari Vajpayee, the leader of the BJP became the prime minister of India, although he had to form a coalition government with twenty-three other parties in order to succeed. The coalition was fragile because the many parties agreed to work together based on political opportunism rather than because of any agreement on the issues.

After only one year in office, one of Vajpayee's coalition partners threatened to pull her support. She was Jayaram Jayalalitha, a former actress and the leader of a political party from the southern state of Tamil Nadu. She accused Vajpayee of not supporting his coalition partners adequately because he was not helping her with some of her political issues. On the other hand, Vajpayee accused Jayalalitha of using her coalition position for her own political goals. This came to a head in April, 1999, when Jayalalitha withdrew the support of the eighteen members of parliament in her political party. A vote of confidence was called for April 17, and Vajpayee lost by only one vote.

As often happens in these cases, another party that was not in the original coalition decided to support Vajpayee, but it did not have enough votes to make a majority. This forced India to form a new government. Sonia Gandhi, the leader of the Congress Party, tried to create a new coalition but could not get enough support. As a result, a new election was called for October, 1999. Until then, Vajpayee was to oversee the government but he was not allowed to make any major decisions.

Consequences

When the BJP first came to power in 1998, its leaders said and did many things that angered some communities in India and around the world. Critics said that Vajpayee and his party were using their position to push their own narrow agenda. At first Vajpayee ignored them. When his government collapsed in April, he realized that he had to govern for the whole country and not just for those who voted for his party. When the October elections were held, he was returned to office with more support than the first time, but he still needed to form a coalition to govern.

Although the Indian people wanted more political stability, they apparently did not want one party to dominate policy as in the past. By giving Vajpayee more support, they were creating more stability, but by forcing him to still depend on a coalition, they were telling him to tone down his own party's point of view and take into account the other parties in the country. Vajpayee was the only prime minister to be returned to office in twenty-seven years.

Carolyn V. Prorok

Littleton, Colorado, High School Students Massacre Classmates

Two Columbine High School students, armed with guns and bombs, entered the school and opened fire. Though some students and teachers escaped, others became trapped in the school. Twelve students and one teacher died before the killing spree ended.

What: Crime
When: April 20, 1999
Where: Littleton, Colorado
Who:
ERIC HARRIS (1981-1999), Columbine student
DYLAN KLEBOLD (1982?-1999), Columbine student

A Suburban Setting

The Denver suburb of Littleton, Colorado, seemed like a safe place to raise families. In April, 1999, with only a few weeks left of school, most students were counting down the days until summer arrived.

At Columbine High School, Eric Harris and Dylan Klebold were both finishing their senior year. To many, both Eric and Dylan seemed like typical normal teenagers. They were both intelligent, liked to bowl, enjoyed listening to music, and spent hours each afternoon playing the video game Doom. Though they did not have many friends, Harris and Klebold were fringe members of the "Trench Coat Mafia," a name given to a group of students who wore black trench coats, played Doom, and made their own videos.

During the months preceding the killings, Harris and Klebold began to express an obsession with bloodshed and hatred. They repeatedly watched the gruesome movie *Natural Born Killers*, started to use German words and expressions from the Nazi era, and created a Web site filled with hate. What people did not know was that Harris and Klebold were also building a collection of weapons, learning how to make pipe bombs from the Internet, and determining when the cafeteria would be most full of students for their attack. While other students at Columbine High School were planning a traditional senior prank, Harris and Klebold were planning something much bloodier.

Ultimate Doom

On the morning of Tuesday, April 20, 1999, Harris and Klebold were preparing to destroy

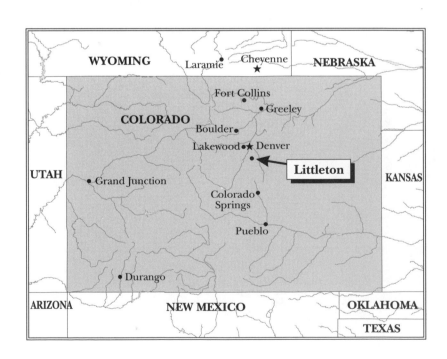

A student and her mother embrace after the shooting at Columbine High School.

hundreds of lives. Just after 11:14 A.M., Harris and Klebold planted two 20-pound (9-kilogram) propane tank bombs under two tables in the cafeteria. The bombs were hidden in duffel bags, with the assumption that they would not be noticed among the hundreds of student backpacks. The bombs were set to detonate at 11:17 A.M. While waiting for the bombs to explode, Harris and Klebold went back to their cars and strapped on their weapons. Though the bombs were expected to kill most of the students, Harris and Klebold were ready to shoot any fleeing students who survived.

Harris and Klebold waited, but nothing happened. The bombs did not explode. At 11:19 A.M., Harris and Klebold decided to take action and entered the school. Once they started shooting,

chaos ruled. Students and teachers scrambled to escape from the gunfire. Harris and Klebold stalked the halls, firing at anyone they saw. Students frantically hid in any space they could find. Some hid in bathroom stalls, in cabinets, under desks, or in closets. Groups of students barricaded themselves into classrooms. Several students used their cellular phones to call their parents. They told their parents that they loved them and that they thought they were going to die.

Harris and Klebold made their way to the library where fifty-six people were trapped. The two killers taunted many of those hiding under the tables. Then they shot them. Harris and Klebold laughed and seemed to enjoy themselves. Within eight minutes, ten students had been killed and twelve more seriously injured in

3039

the library. Then Harris and Klebold left to wander the halls.

Although a police Special Weapons and Tactics (SWAT) team arrived approximately twenty minutes after the first shots were fired, they had no knowledge of who or how many people were shooting. The police searched each of the fleeing students for weapons, afraid that the killers might be among them. It took precious minutes to determine what was happening inside the school.

While wandering the halls, Harris and Klebold looked through classroom windows at students crouching in fear; however, they did not break into the rooms or try to harm the students. Twenty minutes later, the killers returned to the library. After shooting at police through the library windows, Harris and Klebold ended the rampage by shooting themselves a few minutes after noon.

Consequences

Harris and Klebold's shooting rampage shocked the world. Schools, once thought a safe haven for children, had become a place of violence. Students were killing students. Children were killing children. This extreme violence perpetrated by intelligent, wealthy, white students in a suburban town made violence seem possible in every school. Parents demanded to know how and why this happened so that they could prevent it from happening in their children's schools. Society and the media searched Harris and Klebold's backgrounds for clues to their anger and violence.

Violent video games and movies were blamed for making children used to blood, gore, and death. The Internet was also blamed for providing the young killers with blueprints for bombs. Gun control became a major issue after Columbine. Many people questioned whether stricter gun laws could have prevented the tragedy. Although new laws and major changes were proposed, none were immediately passed.

Many also blamed the parents of the killers for the tragedy. Harris had built a Web site that espoused hatred, and the two boys made the thirty pipe bombs in Harris's garage. A large number of people wondered how Harris and Klebold's parents had not noticed the signs of the two youths' growing anger and violence.

Eric Harris and Dylan Klebold had spent more than a year planning their massacre. Though their propane tank bombs did not explode, they managed to kill twelve students and one teacher as well as wound more than two dozen others during their bloody rampage. Harris and Klebold not only brought violence into classrooms, they brought fear.

Jennifer E. Rosenberg

United States to Retain Smallpox Virus Sample

President Clinton announced that the United States would retain its sample of the smallpox virus, one of only two known remaining samples in the world.

What: Medicine; Biology; Terrorism
When: April 22, 1999
Where: Washington, D.C.
Who:

BILL CLINTON (1946-　　), president of the United States from 1993 to 2001

DONALD A. HENDERSON (1928-　　), director of the Johns Hopkins Center for Civilian Biodefense Studies, Baltimore, Maryland

ALAN ZELICOFF, senior scientist, Sandia National Laboratory Center for National Security and Arms Control, Albuquerque, New Mexico

JOSHUA LEDERBERG (1925-　　), scientist and professor emeritus of the Rockefeller University, New York City

Smallpox Virus Retained

On April 22, 1999, President Bill Clinton announced that the United States would continue to preserve samples of the live smallpox virus at the Centers for Disease Control and Prevention (CDC) in Atlanta, Georgia. As a member of the World Health Organization (WHO), the United States had agreed in 1996 that all remaining stocks of the smallpox virus should be destroyed in June of 1999. President Clinton's decision reversed this agreement, causing mixed reactions and opinions within the scientific community.

Smallpox (variola) is a highly contagious and dangerous virus that typically kills 30 percent of those infected. It has caused millions of deaths over the course of human history. Through extensive medical vaccination programs, especially in developing countries, smallpox was gradually eliminated from the world, with the last known case of the virus occurring in Somalia in 1977. The WHO declared smallpox to be totally gone in 1980 and has several times carried out extensive discussion about whether to keep or destroy laboratory samples of the killer virus.

Scientific Controversy

Scientists and military specialists around the world have disagreed for many years on whether stores of the very dangerous smallpox virus should be kept or destroyed. In 1996, the WHO decided that only two laboratories should keep samples of the virus under highly secure conditions: the CDC in the United States and the Vector Institute in Russia. All other WHO countries destroyed their samples of the smallpox virus. The WHO further agreed to destroy the Russian and U.S. smallpox stocks in June of 1999, with a meeting in May of 1999 to confirm that decision.

Beginning in 1996, the United States debated the wisdom of destroying the last remaining smallpox virus stores. Russia had never supported the idea, and as the May, 1999, confirmation deadline approached, much concern arose in the United States about the public health and national security issues surrounding the decision to destroy the virus. Donald A. Henderson, former White House science adviser and director of the Johns Hopkins Center for Civilian Biodefense Studies in Baltimore, Maryland, argued in favor of destroying the virus, saying that it presented a big risk to public health, even in highly protected laboratories. Research on the virus would be expensive and difficult. The virus cannot be studied using animals (only humans get smallpox) and can only be examined in very expensive laboratories under highly secure conditions. Most

3041

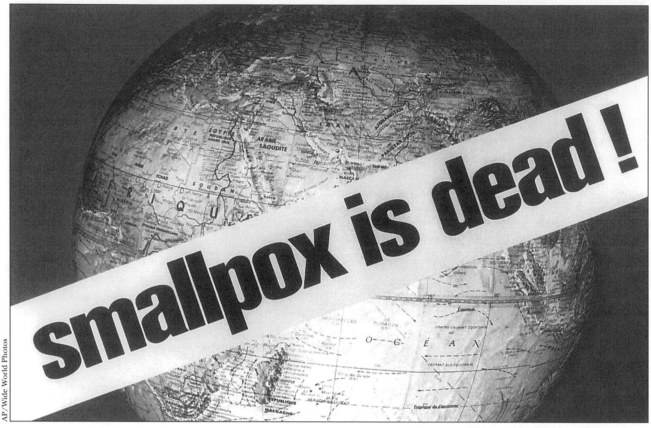

AP/Wide World Photos

The World Health Organization published this poster after declaring smallpox eradicated on May 8, 1980.

scientists do not have access to such labs and would be reluctant to work with such a contagious and dangerous virus.

Other scientists and military leaders argued equally strongly that some samples of the smallpox virus must be retained. Alan Zelicoff, a biodefense expert at the Sandia National Laboratory Center for National Security and Arms Control in Albquerque, New Mexico, objected strongly and publicly to the idea of destroying the smallpox virus samples. Even if Russia and the United States destroyed their samples, it was possible that some other countries might have secret stocks of the virus. Many experts believed that North Korea had such samples. The international threat of bioterrorism—the use of biological weapons—has become stronger in recent years. In the hands of bioterrorists, the smallpox virus could kill millions of people. Stocks of smallpox vaccine still exist but only in small quantities. In addition, a large part of the world's population has never been vaccinated. Scientists and military experts in the United States argued

that the existence of this vulnerable population necessitated the retention of live viral samples, which would function as a potential threat and deterrent to countries that might use the smallpox virus as a weapon.

Some scientists argued that the smallpox virus needed to be kept to study it further and find better drugs for treating it. The virus might be released on purpose by bioterrorists or accidentally from the labs where it is stored, or it might arise naturally from a very similar virus existing in animal populations (for example, monkey pox). A better understanding of the virus and better drugs to fight an outbreak might be developed by study and research of the existing live variola laboratory samples.

Joshua Lederberg, president emeritus of the Rockefeller University in New York City and winner of the Nobel Prize in Medicine in 1958, encouraged the federal government to conduct an external review of the arguments for and against destruction of the variola virus. At the request of various federal agencies, the Institute of Medicine (IOM) prepared a one-hundred-page report with the title *Assessment of Future Scientific Needs for Live Variola Virus.* Although the report did not make a recommendation about whether to destroy the virus, it did conclude that keeping live samples of the virus would be essential "for the development of antiviral agents or novel vaccines to protect against a reemergence of smallpox due to accidental or intentional release of variola virus." President Clinton was persuaded by this argument and signed a memo calling for the preservation of variola in labs.

Consequences

Following President Clinton's announcement, the WHO met on May 24, 1999, and decided to delay the destruction of the smallpox virus stocks in the United States and Russia until at least 2002. Surprisingly, member countries were largely in agreement that more time is needed to study the virus before it is finally destroyed. The WHO resolution also created an international committee of experts who will oversee research on smallpox and recommend a time frame for the final destruction of the virus. Such research might include genome sequencing to understand the structure of the virus better and the development of improved models for testing antiviral drugs and vaccines against the live virus samples. Russia and the United States both expressed a desire to work collaboratively to carry out research on the variola virus samples.

At the national level, President Clinton established an Initiative on Biological and Chemical Weapons Preparedness to strengthen U.S. defenses against bioterrorism. Additional federal funding was set aside to develop vaccines for smallpox and other dangerous viruses that might be used as biological weapons.

Helen Salmon

NATO Bombs Hit Chinese Embassy in Belgrade

> *During air raids over Serbia, U.S. missiles fell on the Chinese embassy in Belgrade, killing three people. Although the U.S. government called the bombing a tragic error, the incident worsened relations between China and the United States.*

What: International relations; War
When: May 7, 1999
Where: Belgrade, Serbia, Yugoslavia
Who:
WESLEY CLARK (1944-), North Atlantic Treaty Organization supreme allied commander in Europe
JAMES SASSER (1936-), U.S. ambassador to China in 1995
JIANG ZEMIN (1926-), president of China from 1993

Air War in Yugoslavia

On May 7, 1999, a U.S. B-2 bomber participating in North Atlantic Treaty Organization (NATO) air raids over Serbia in Yugoslavia hit the Chinese embassy in Belgrade with three missiles, killing three people and injuring twenty. The bombing raids over Belgrade were part of a NATO air campaign in response to Serbian "ethnic cleansing" in its southern province of Kosovo. A province of historical and religious significance for Orthodox Christian Serbs, Kosovo was inhabited mainly by Albanians, most of whom were Muslims. In response to conflicts between the two communities and the growth in Kosovo of the Kosovar Liberation Army (KLA), an Albanian paramilitary force, Serbian president Slobodan Milošević ordered the Serbian army to force Albanians out of the region. Serbian soldiers killed hundreds of Albanian men and boys and forced hundreds of thousands of refugees into neighboring countries.

NATO, led by U.S. forces, began air raids against the Serbian positions in Kosovo and against strategic and military installations in the heart of Serbia, especially the area around the capital, Belgrade. However, in addition to striking military targets, including Serbian railroads and bridges, the bombs often hit civilian buildings. Some countries, including Russia and China, disapproved of NATO's campaign. The bombing began in mid-March, 1999, with seven hundred sorties (bombing flights) a day, and on May 7, bombs hit the Chinese embassy.

A Deadly Mistake

General Wesley Clark, supreme commander of NATO forces in Europe, said the embassy was "mistakenly identified." The Central Intelligence Agency, using incorrect maps, supplied the false information to NATO command. NATO claimed that the pilots believed the embassy was the headquarters of a Yugoslav arms agency. Clark stated in an interview that no pilot or crew error was involved. He said that although the process of selecting targets was at fault, he believed that the incident was a unique occurrence, and he retained his faith in the process. "We're going to continue to intensify this air campaign," he told the American Broadcasting Company. "We're not going to let an incident like this deter us from doing what we think is right and necessary." Russian president Boris Yeltsin, who was trying to negotiate an end to the Kosovo conflict, said the bombing was an outrage and warned of the consequences. Japan and India also protested the bombing, and the following day, NATO suspended night bombing over Belgrade for one night. A NATO spokesperson admitted that its forces had mistakenly bombed a hospital in the Serbian city of Nis.

The Chinese called a meeting of the United Nations Security Council the following day to discuss the bombing and urged the council to take stern measures, but the council only expressed its shock and deep concern and sent condolences to the Chinese victims. Students in Beijing, the Chinese capital, staged the largest anti-American demonstrations since 1972, the year relations between the United States and mainland China resumed after more than twenty years without diplomatic recognition. Demonstrations also occurred in Shanghai, Xian, Nanjing, Macao, Hong Kong, and other Chinese cities.

Over the following days, the demonstrations grew larger as crowds gathered in front of the American embassy burning U.S. flags, overturning cars, and throwing stones, bricks, bottles, and even eggs. The Chinese government said that it urged restraint but backed all "legal" protests. Some observers, however, thought that the Chinese government and Communist Party were behind the demonstrations for political reasons, possibly to bolster their support among the public. Chinese vice president Hu Jintao said the demonstrations reflected the patriotism of the Chinese people and promised the Communist Party's full support. Reports said that the government provided buses for the students to go to the demonstrations.

Furthermore, although the students were demanding an official apology, the government did not inform them that this had already been given or provide information on NATO's explanation of the bombing until three days later. The official Chinese explanation stated that the bombing was retaliation for Chinese support of Serbia. The U.S. ambassador to Beijing, James Sasser, officially apologized and promised American aid to the injured survivors and the families of the deceased. However, some experts believed that the apology occurred too late and that U.S. president Bill Clinton should have immediately called Jiang Zemin, China's president. The State Department claimed, however, that President Jiang refused to take Clinton's calls.

At the time, 20,000 Chinese lived in Yugoslavia (Serbia and Montenegro), and after the bombing, those in Belgrade showed their solidarity with the Serbs in an anti-NATO demonstration. China supported Yugoslavia in part because it feared that ethnic conflict in its own country (with the Tibetan or other minorities), could lead to Western intervention, especially because NATO at this time was expanding by admitting Eastern European countries.

The bombing took place at a time of tension between China and the United States. China hoped to upgrade its trading relation with the United States, achieving Most Favored Nation status and entry into the World Trade Organization, but the U.S. Senate raised objections citing China's poor record with regard to human rights.

Consequences

The bombing hurt NATO's efforts to get wider support for its anti-Serbian campaign. The U.S. State Department urged Americans to postpone visits to China, and the Boston symphony postponed its scheduled tour to the country. Initially, China suspended military talks with the United States and other Western countries, but relations were not permanently ruptured. The bombing also encouraged the Chinese military to ask for increased spending, although economic development remained the country's highest priority.

Frederick B. Chary

Citadel Graduates First Woman Cadet

In May, 1999, Nancy Mace became the first woman to graduate from the Citadel, a state-supported military academy that had been an all-male institution for more than 150 years.

What: Gender issues; Education
When: May 8, 1999
Where: Charleston, South Carolina
Who:
NANCY RUTH MACE (1978-   ), first woman graduate of the Citadel
PETRA LOVETINSKA, Nancy Mace's classmate
SHANNON FAULKNER, first woman admitted to the Citadel
EMORY MACE (1941-   ), Nancy Mace's father

Storming a Male Bastion

On May 8, 1999, Nancy Ruth Mace became the first woman to graduate from the Citadel, a state-supported military college in Charleston, South Carolina. Founded in 1842, the Citadel was well known for its many distinguished alumni, a large percentage of whom went on to leadership roles in civilian life rather than in the military. Many South Carolinians saw graduation from the Citadel as a guarantee of success.

Even though most military academies, including West Point, had begun admitting women in 1976, the Citadel and the Virginia Military Institute (VMI) continued to hold the line. In January, 1993, Shannon Faulkner, a high school senior from Powdersville, South Carolina, received a notice of provisional acceptance to the Citadel. However, as soon as school administrators discovered that Faulkner was a woman, her admission was rescinded. Faulkner filed suit, charging that the all-male policy was unconstitutional.

A three-year legal battle followed. Citadel alumni raised money for their school, wrote outraged letters to newspapers, and distributed bumper stickers urging everyone to "save the males." The Citadel even offered Converse College, a private women's college in Spartanburg, South Carolina, $5 million to start an alternative military program so that South Carolina would be seen as offering women the same opportunities as men.

However, time ran out before that plan could be implemented. On August 11, 1995, the U.S. Supreme Court refused to issue a stay blocking Faulkner from the Citadel. The next day, she moved into the barracks. Faulkner soon became ill, which she attributed to stress. Her detractors insisted that she was not in good enough physical condition to undergo the rigorous first week of training. Faulkner's withdrawal after just six days at the Citadel prompted a wild celebration by the cadets, which was televised throughout the world.

However, less than a year later, the governing board of the Citadel voted to drop the school's all-male admissions policy. Once the U.S. Supreme Court ruled that VMI must admit women, the Citadel had to admit defeat.

Holding the Fort

In August, 1996, Jeanie Mentavlos, Kim Messer, Petra Lovetinska, and Nancy Mace began their first year at the Citadel. There were more than five hundred "knobs," or first-year students, in their class.

Mace later spoke of the hostility they encountered, the glares, the hisses, the taunts, and the expletives. After one semester, Mentavlos and Messer quit. In the lawsuits they filed, one of which was subsequently settled out of court, the women alleged hazing and sexual harassment, even having their clothes set on fire. The Citadel decided to install alarms in the rooms of women students. Eventually fourteen of the cadets named by Mentavlos and Messer were either disciplined or resigned.

After the departure of Mentavlos and Messer, there remained only two women: Lovetinska, a

Czech national, and Mace, whose father was Emory Mace, a highly decorated retired Army general who in 1997 would become the Citadel's commandant of cadets. Both young women were determined not only to stay but also to succeed. Mace's physical exploits soon won her grudging respect from the male cadets. Her first year, she finished fifth out of 150 in a two-mile run. Out of the entire freshman class, she was one of only four to pass the first fitness test. Mace also aimed high in academics. She graduated magna cum laude, winning the Citadel's highest academic honor, the right to wear a gold star on her uniform. This time when her name was called, no one hissed. Among those who congratulated her were some who had led the fight to keep women out of the Citadel.

Nancy Mace (right) shakes hands with commencement speaker Pat Buchanan as she graduates from the Citadel.

Consequences

Because Mace had transferred course credits from a two-year college she had attended, she graduated after only three years at the Citadel. Her classmate and close friend Lovetinska remained at the Citadel for another year. During her senior year, she held the second highest post in her battalion, that of executive officer. In 2000, she made history by becoming the Citadel's first four-year woman graduate.

In the spring of 2001, ten more women were candidates for graduation. One of them, Mandy Garcia, was an outstanding athlete and also held the second highest rank in the corps. The percentage of women in the corps of cadets was still relatively low; there were just 79 out of more than 1,800 students. However, that number represented a 34 percent increase within a two-year period. Moreover, early in March, 81 women applicants had already been accepted for the fall term.

Administrators admitted that the Citadel still had a number of male cadets who resented the presence of women within their ranks. However, every effort was being made to educate these holdouts, for wherever they went after graduation, whether into military service, into political life, or into business, the male graduates of the Citadel would have to deal with women both as equals and as superiors.

Faulkner must be credited for forcing one of the two remaining state-supported all-male colleges in the United States to admit women, and all four of the women who followed her should be honored for helping to pave the way. However, the two who remained, Mace and Lovetinska, deserve the highest praise not only for their courage but also for their wisdom and their persistence. Wisely, they avoided publicity; courageously, they refused to complain. They simply persisted in trying to be the best cadets they could be, thus becoming role models not just for the women who would follow but also for Citadel cadets of both sexes. They won an important victory in an ongoing battle against discrimination.

Rosemary M. Canfield Reisman

U.S. House Report Describes Chinese Spy Activity

A House select committee presented a 700-page report detailing allegations that China engaged in espionage in the United States in an effort to gain U.S. nuclear secrets.

What: International relations
When: May 25, 1999
Where: U.S. Congress, Washington, D.C.
Who:
CHRISTOPHER COX (1952-   ), Republican representative from California, head of the House Select Committee on U.S. National Security and Military/Commercial Concerns with the People's Republic of China
NORMAN DICKS (1940-   ), Democratic representative from Washington
BILL CLINTON (1946-   ), president of the United States from 1993 to 2001

The Cox Report

On May 25, 1999, Christopher Cox, chairman of the House Select Committee on U.S. National Security and Military/Commercial Concerns with the People's Republic of China, and Norman Dicks, Democratic representative on the committee, testified before the House International Relations Subcommittee on Asia and the Pacific on allegations that China spied on the United States' nuclear program and stole secrets to advance its own nuclear program during the past twenty years. Their three-volume, 700-page report, based on the select committee's investigation, became known as the Cox Report (House Report 105-851). The two representatives appeared before the Senate Governmental Affairs Committee a week later.

The Cox Report was part of a congressional effort to implement major policy changes in securing national secrets against alleged Chinese nuclear spying. Nine committees in both the House and Senate engaged in various inquiries into Chinese spying or security lapses at the U.S. Energy Department's nuclear weapons laboratories. Selected to head the House select committee in the summer of 1998, Cox forged a bipartisan group that included four Democrats and four Republican members, to work on this sensitive national security matter. After the six-month investigation, all nine members of the committee agreed that Chinese espionage continued to be a serious threat. On January 4, 1999, the committee submitted to President Bill Clinton its classified report, including thirty-eight recommendations related to export controls and counterintelligence. On May 25, one day before Cox's testimony at the congressional hearing, the committee released an unclassified version of its report.

Alleged Chinese Spying

According to the Cox Report, the nuclear secrets that China obtained from U.S. national weapons laboratories enabled it to design, develop, and successfully test modern strategic weapons sooner than otherwise would have been possible. These thefts of nuclear secrets began as early as the late 1950's and almost certainly continued into the 1990's. The report found that security programs at the national weapons laboratories failed to meet minimal standards. It also found that repeated efforts since the early 1980's had failed to solve the counterintelligence deficiencies at the national labs. Therefore, Chinese operatives working in U.S. nuclear labs and production facilities gathered weapons information on nuclear secrets, including information on the highly controversial neutron bomb and some of the most advanced warheads such as the W-88, also known as the Trident II. The report identified what was called one of the worst security breaches in U.S. history.

The report also alleged that two U.S. defense companies ignored legal requirements and allowed China to obtain information critical to its ballistic missile program. The investigation claimed that China engaged in a concerted campaign to obtain technology through some three thousand commercial ventures and front companies in the United States. China has successfully tested neutron bomb and thermonuclear weapons that were on a par with the United States. These modern thermonuclear weapons took the United States decades of effort, hundreds of millions of dollars, and numerous nuclear tests to achieve. Experts judged China to be capable of designing serial production of such weapons

during the 2000's in connection with the development of its next generation of intercontinental ballistic missiles.

The report targeted the Clinton administration, especially the Energy Department, which is responsible for the U.S. nuclear labs' security. President Clinton provided a written response agreeing with the need to maintain effective measures to prevent the diversion of U.S. technology and prevent unauthorized disclosure of sensitive military information, but not agreeing with all the reports' analysis. A number of defense analysts and some members of the Clinton administration claimed the cumulative effect of the espionage and China's military ambitions has been exaggerated. Some experts and dedicated amateurs around the world have taken a close look at the report and determined that the committee did not have its final report authenticated or examined by real experts, resulting in some errors. Mistakes range from substantial misrepresentation of Chinese aerospace technology to minor errors in dates and hardware designations. The Chinese government denied the allegations and accused the U.S. government of disgracing itself by ceaselessly peddling allegations about China. The foreign ministry of the People's Republic of China denounced the Cox Report's allegations, calling them absurd and groundless.

Consequences

The experts on nuclear proliferation maintained that the United States retains an overwhelming qualitative and quantitative advantage in deployed strategic nuclear forces. Nonetheless, they estimated that if the United States were to confront China's conventional and nuclear forces at the regional level, a modernized Chinese strategic nuclear ballistic missile force would pose a credible direct threat against the United States. The experts agreed that although the Cox Re-

U.S. NATIONAL SECURITY AND MILITARY/COMMERCIAL CONCERNS WITH THE PEOPLE'S REPUBLIC OF CHINA

VOLUME I

SELECT COMMITTEE
UNITED STATES HOUSE OF REPRESENTATIVES

AP/Wide World Photos

This report on Chinese espionage described twenty years of spy activity.

3049

port was full of holes and based on scattered information, it made important recommendations regarding U.S. national security concerns.

After the Cox Report, members of Congress began pledging to pass legislation on everything from tightening restrictions on foreign scientists visiting national weapons labs to giving the Energy Department freer rein to conduct polygraphs to throwing more money into security efforts. Lawmakers demonstrated an eagerness to increase the Energy Department's security and counterintelligence budgets. A Senate subcommittee approved an additional $53 million for counterespionage activities at the department.

Additional resolutions were debated on various measures to tighten security at U.S. weapons labs and to put new restrictions on technology exports to China. The Senate considered legislation that would bar some Chinese officials from visiting the United States.

The Clinton administration implemented some of the House committee's recommendations. The committee, however, cautioned that unilateral U.S. action was not likely to stop the development of the Chinese nuclear weapons program but simply to cause China to get the needed technology elsewhere.

Xiaobing Li

Two New Elements Are Added to Periodic Table

A team of scientists at Lawrence Berkeley Laboratory Nuclear Science Division detected the formation of two new elements, with atomic numbers 116 and 118, as the result of bombarding lead targets with krypton ions in the 88-inch (2.2-meter) cyclotron.

What: Chemistry; Physics
When: June 7, 1999
Where: Lawrence Berkeley National Laboratory (LBNL), California
Who:
VICTOR NINOV, nuclear physicist, LBNL
K. E. GREGORICH, nuclear chemist, LBNL
DARLEANE HOFFMAN (1926-   ) and
ALBERT GHIORSO (1915-   ), nuclear chemists, LBNL

At the Frontier of Element Synthesis

In a statement released on June 7, 1999, nuclear chemists and physicists K. E. Gregorich, Victor Ninov, Darleane Hoffman, and Albert Ghiorso, at Lawrence Berkeley National Laboratory (LBNL) in California announced the discovery of two new, exceptionally heavy chemical elements. Atoms of an element with atomic number 118 in the periodic table were formed in the 88-inch (2.2-meter) cyclotron, in which an intense beam of highly energetic krypton-86 ions had been targeted on a lead-208 sample for more than ten days. Element 118, which had not yet been named, is stable only for 200 microseconds and decays into another new element (atomic number 116) and other elements already known.

The tiny, dense nucleus at the center of each atom is made up of protons (positively charged) and uncharged neutrons. Incredibly strong forces bind the neutrons and protons together, and the positive charge on the nucleus repels the krypton ions that bombard it. Only high-energy ions can overcome this repulsion (the coulomb barrier) and enter the nucleus.

The mass of an atom is accounted for by its total number of neutrons and protons (mass number), and the chemical identity of the atom is related to its number of protons (atomic number). The chemists' periodic table is an arrangement of elements in order of increasing atomic number, organized in rows and columns to bring elements of similar properties together in the same column. Each column of elements forms one of the eighteen groups that make up the periodic table. Elements in the same group (such as xenon and radon) are called congeners of one another.

Elements beyond uranium (atomic number 92) do not occur naturally and must be produced artificially with the aid of particle accelerators such as the cyclotron. In the cyclotron, positively charged atoms (ions) whirl around in a circular path, picking up energy from an alternating magnetic field, and when energetic enough, these ions can hit a target atom and penetrate through its shells of electrons into its nucleus. A direct hit can lead to a compound nucleus, which usually sheds its excess energy by emitting particles or radiation and forming one or more daughter atoms that can be detected. At LBNL, krypton (atomic number 36) ions, each containing 50 neutrons, were produced in very large numbers and used to bombard a target of lead (atomic number 82, 126 neutrons). During ten days, only three new atoms with atomic number 118 (36 plus 82) and with 175 neutrons were produced. Element 118 is of particular interest to chemists because it is expected to be the heaviest element in the noble gas group. Lighter members of this group (such as xenon and radon) have filled extranuclear electron shells, which are particularly stable, causing these elements to resist chemical combination.

The LBNL experiment succeeded by avoiding spontaneous fission, in which the hoped-for heavy nucleus splits instantly into two lighter fragments, without forming a new element. The laboratory's planning was facilitated by the work of theoretical physicists at Berkeley and elsewhere. Theoreticians use mathematical models of nuclei to derive combinations of neutrons and protons that are expected to be relatively resistant to fission. These models predict an island of stability for nuclei with around 114 protons and 184 neutrons. Only the most highly developed accelerators such as the Berkeley 88-inch (2.2-meter) cyclotron are capable of producing high-intensity, high-energy heavy ions such as those used in this work. Equipment of similar capability also exists in Germany at the Society for Heavy Ion Research (GSI) in Darmstadt and in Russia at the Joint Institute for Nuclear Research (JINR) in Dubna, outside Moscow. The accelerators at all these locations are large and complex and require the attention of many scientists and technicians when in operation.

Detection and Identification

The LBNL experiment produced only three atoms of element 118, which had to be separated from a complex mixture of products of the ion bombardment of the lead target. Once separated, the atoms were detected and identified. Because of the short lifetime of the new atoms, special techniques were needed for this crucial phase of the experiment. The Berkeley gas-filled separator (BGS), which enabled scientists to accomplish this complex task, uses three large magnets that sort out the charged particles and allow the detection of selected, relevant masses. Mass is

detected by the time of flight of the particle. Identification of element 118 was confirmed by its pattern of decay: a series of six alpha-particle (helium nuclei) emissions leading first to element 116 and then to other elements as light as atomic number 106 (seaborgium).

Consequences

Transuranium elements have been produced since the 1940's when plutonium (atomic number 94) was synthesized in kilogram quantities for nuclear weapons. Later, elements such as americium and californium found use as radiation sources. Elements beyond atomic number 100 cannot be accumulated and stored but must be generated and studied a few atoms at a time. Nevertheless, information can be obtained about the physical and chemical properties of these elements that improves understanding of the periodic table. For example, fourteen elements following actinium (atomic number 89) are now classified together as an actinide series, similar to the lanthanide series (atomic numbers 58 through 71). This clarifies the chemistry of uranium and thorium, which were previously thought to be congeners of tungsten and hafnium. The lifetimes of new elements are used to check the predictions of theoretical calculations of nuclear stability, and the feedback helps improve theoretical models of the nucleus.

The Berkeley group planned further studies, including the bombardment of bismuth (atomic number 83) with krypton ions in an attempt to synthesize element 119. As studies of the chemical properties of elements 118 and 116 become possible, they should confirm the periodic table groups to which the elements have been assigned.

John R. Phillips

Clinton Denounces Racial Profiling

Racial profiling, a law enforcement policy of using race and ethnicity to identify those most likely to have committed crimes, was declared "morally indefensible" by U.S. president Bill Clinton.

What: Law; Civil rights and liberties; Social reform
When: June 9, 1999
Where: Washington, D.C.
Who:
BILL CLINTON (1943-   ), president of the United States from 1993 to 2001
JOHN CONYERS, JR. (1929-   ), Democratic representative from Michigan.
AMADOU DIALLO (1977-1999), West African immigrant killed by New York City police detectives

Police Profiling

On June 9, 1999, President Bill Clinton ordered federal law enforcement agencies to compile data on the race and ethnicity of people they question, search, or arrest to determine whether people are stopped by law officials merely because of their skin color. "Racial profiling is in fact the opposite of good police work, where actions are based on hard facts, not stereotypes," the president explained. "It is wrong, it is destructive, and it must stop."

He issued a presidential directive requiring the Justice Department to analyze the collected data to determine whether law enforcement agencies engage in racial profiling and what measures could be employed at the national level to eliminate its use. He also challenged state and local police forces, which do the vast bulk of police work in the United States, to follow his lead in determining the extent of the practice. The president's directive came in response to two incidents: the killing of Amadou Diallo, a unarmed West African immigrant, by New York City police detectives, and the beating and torture of Abner Louima, a black Haitian immigrant, in a Brook-

lyn police station. Diallo was shot forty-one times as he entered his apartment building. The four police involved were acquitted after convincing a jury that they had made a mistake. Louima was tortured by police before being released in another case of mistaken identity.

Driving While Black or Brown

Many African American and Hispanic drivers contended that their cars were being stopped and searched more often than those driven by whites, a type of racial profiling popularly known as DWB (driving while black, a play on DUI, or driving under the influence). A report on stops by state troopers on the New Jersey Turnpike confirmed their claim and initiated a controversy. In response, Representative John Conyers, Jr., a Democrat from Michigan, introduced a bill in the House of Representatives calling for the attorney general to examine statistics from local police departments on stops for traffic violations to determine what prompted the stop, the race and ethnicity of the driver, whether the vehicle was searched, and if so, what was found and whether an arrest was made.

The problem of racial profiling of motorists was found to exist far beyond New Jersey, as is illustrated by this story from Oklahoma. In August, 1998, a thirty-seven-year-old U.S. Army sergeant, Rossano V. Gerald, and his young son Gregory were driving across the Oklahoma border when the Oklahoma Highway Patrol stopped them. Gerald was a decorated veteran of the Gulf War and a career soldier. He was an African American man of Panamanian descent. The state police placed both father and son in a closed patrol car with the heat on for two and one-half hours and warned them that a police dog would attack them if they tried to escape. The officers shut off the patrol car's video camera to hide any evidence of misconduct. They searched Gerald's car, rummaged through his belongings, found nothing, and finally released the sergeant and his son. The Geralds apparently were victims of discriminatory racial profiling by police.

Racial profiling is based on the notion that minorities commit most drug offenses. This premise is untrue because five times as many whites use drugs than African Americans or Hispanics. Nevertheless, this false perception results in the harassment and arrest of innocent people because of their skin color. In 1986, the Drug Enforcement Administration (DEA) opened Operation Pipeline, a highway drug interdiction program, as part of the government's war on drugs. The DEA enlisted local police departments to search for narcotics dealers on major highways and told officers that Hispanics and West Indians dominated the drug trade, so they deserved close observation, especially if they wore their hair in dreadlocks and were traveling together in pairs. These racially defined practices became part of police policies across the nation.

Statistics from just one state, Illinois, illustrate the depths of the problem. Although Hispanics made up less than 8 percent of the state's population from 1987 to 1997, they were the subjects of more than 27 percent of the searches made by state troopers. Although African Americans made up less than 15 percent of the Illinois population,

they totaled 23 percent of the searches conducted by police assigned to drug traffic investigations. In one police district in which African Americans made up 24 percent of the local driving-age population, they accounted for 63 percent of the police searches. An American Civil Liberties Union study of the disparities found that while state police asked a much higher percentage of Hispanic motorists than white motorists for consent to search their vehicles, they found illegal goods in a lower percentage of the vehicles driven by Hispanics. Therefore, the author of the study concluded that the searches were made based on race.

Consequences

Racial profiling, according to the U.S. Commission on Civil Rights in its April, 1999, report, has led to the following results: The African American proportion of drug arrests rose from 25 percent in 1980 to 37 percent in 1995. African Americans, who account for 13 percent of the nation's drug users, made up 37 percent of those arrested on drug charges, 55 percent of those convicted, and 74 percent of all drug offenders sentenced to prison. At the same time, the government reported that 80 percent of the country's drug users were white.

Leslie V. Tischauser

Women's World Cup of Soccer Draws Unprecedented Attention

> *The victory of the United States over China in the third women's World Cup tournament was not only a resplendent moment in American sports history, it also represented a sea change in the status of women's sports generally.*

What: Sports; Gender issues; Entertainment
When: July 10, 1999
Where: Pasadena, California
Who:
MIA HAMM (1972-   ), high-scoring forward who became the most famous woman soccer player well before the World Cup
BRANDI CHASTAIN (1968-   ), U.S. player who scored the winning penalty kick in the championship game
GAO HONG (1967-   ), China's goalkeeper

A Defining Moment in Women's Sports

On a blistering hot Saturday afternoon in July, 1999, more than ninety thousand people—the most ever to witness any women's sporting event—filled Southern California's Rose Bowl to capacity to watch the final game of the World Cup of women's soccer. Record-sized crowds and television audiences and the nationwide fervor that had surrounded the entire competition since its inception three weeks earlier all vastly exceeded even the most optimistic expectations of tournament planners.

The significance of the event can be appreciated by comparing it to the first women's World Cup tournament, staged in China only eight years earlier. After the U.S. women won that tournament, they returned to the United States in near-obscurity. By contrast, American interest in the 1999 tournament approached national hysteria. Qualifying games drew record crowds, members of the team, such as prolific goal-scorer Mia Hamm, were becoming media stars, and televised games were drawing even higher ratings than National Hockey League playoff games. Few Americans at all aware of sports could have overlooked this event, and President Bill Clinton himself even attended the game.

The final game itself, which pitted the United States against China, was well played and intense, but also low-scoring. Both teams mounted exciting assaults on their opponents' goals but failed to score in regulation play or extra-time. Consequently, the game went to penalty kicks—exactly as the final of the men's World Cup had been decided in the very same stadium five years earlier.

The climax of the game produced one of the most electrifying moments in sports history: After Brandi Chastain put the winning U.S. shot past the superb goalkeeper Gao Hong, she suddenly tore off her jersey and dropped to her knees with her fists raised in triumph, as her teammates poured onto the field. Pictures of Chastain at that moment made the covers of *Time, Newsweek,* and *Sports Illustrated* simultaneously—a feat never before accomplished by any athlete—male or female—in any sport. The moment left an indelible image of women's power that is unlikely ever to be forgotten.

The Road to Pasadena

The roots of the U.S. women's success were planted in two different fields: youth soccer programs and federal legislation. The proliferation of youth soccer leagues that began with the founding of the American Youth Soccer Organization in 1964 would help to provide future national teams with an unmatched talent pool on which to draw. No other country—including Europe and South America's traditional soccer powers—

U.S. soccer team captain Carla Overbeck raises the trophy after defeating China in overtime during the Women's World Cup Final in Pasadena, California. Brandi Chastain, who scored the winning penalty kick, is at the far right.

has even come close. Of the estimated thirty million women who played soccer in the world in the late 1990's, approximately one-fourth were Americans. By contrast, only ten thousand women played soccer in China. A generation earlier, the number of American women, of any age, playing soccer had been minuscule.

However, the vast pool of youth players could not by itself have given U.S. women world leadership in soccer, were it not for the revolution in college athletics brought about by Title IX of Educational Amendments Act of 1972. That law dictated that

No person in the United States shall, on the basis of sex, be excluded from participation in, be denied the benefits of, or be subjected to dis-

crimination under any educational program or activity receiving Federal financial assistance.

Before Title IX, few women—even Olympic champions—could foresee futures in their sports. Title IX brought about a revolution in college sports programs that gave young women opportunities to raise their play to much higher levels. Virtually all colleges across the nation began providing female athletics with financial support, facilities, and opportunities equivalent to those long provided to male students. The consequent changes can be measured in such bald statistics as scholarships. In the early 1970's, an estimated fifty thousand men attended college on athletic scholarships, compared to fewer than fifty women. By the late 1990's, women were receiving about one-third of all athletic scholarship funds.

Improvements in women's college athletic programs inspired increased participation by girls at lower levels, particularly in girls' high school programs, which saw a ten-fold increase in participation in the three decades following Title IX. Among the programs that benefited most from this change were soccer and basketball.

The most dramatic product of Title IX was the U.S. women's national soccer team. All twenty members of the team had enjoyed athletic scholarships that they could not dreamed of having before 1972. Indeed, so aware of their debt to Title IX were they that they nicknamed themselves the "Title IX Team."

Consequences

In the euphoria following the World Cup, many ranked the event among the greatest triumphs in the history of American sports, and good feelings abounded. Meanwhile, the event raised such questions as what it meant to the future of soccer in America—which had never previously embraced the world's most popular sport so warmly—what it meant to the future of women's sports generally, and what it meant to the future of women's soccer specifically.

The question of whether soccer would finally catch on in the United States still seemed to hinge, in part, on whether the U.S. men's team—whose performance in the 1998 World Cup had been dismal—could ever match the popularity of the women's team by making serious challenge for its own World Cup. At the same time, however, the women's team had proved beyond a doubt that Americans could get seriously behind women's sports—a fact born out by the growing popularity of women's college basketball, as well as professional basketball teams that emerged in the late 1990's.

In 2001, the first women's professional soccer league, the Women's United Soccer Association (WUSA), was launched with eight teams spread across the United States. Drawing upon virtually all members of the U.S. team, as well as top players from around the world—including Brazil's Sisi and China's Sun Wen and Gao Hong—the new league promised to provide the highest-possible level of play. The league wisely set modest goals for attendance, television ratings, and sponsorship during its first season and surpassed all of them. In 2002, the future of both soccer and women's sports in America looked brighter than ever.

R. Kent Rasmussen

China Detains Thousands in Falun Gong

> *The Chinese government launched a nationwide crackdown on the religious group Falun Gong, detaining more than five thousand followers and destroying millons of its publications.*

What: Religion; Human rights
When: July 22, 1999
Where: Beijing
Who:
LI HONGZHI (1952-   ), founder of Falun
   Gong
HE ZUOXIU (1927-   ), physics
   professor

The Falun Gong Movement

On July 22, 1999, the Chinese government outlawed Falun Gong as an "evil cult." It detained more than five thousand Falun Gong followers and began a nationwide campaign against its members. The immediate cause of the crackdown was a mass sit-in staged by Falun Gong on April 25.

Falun Gong (also known as Falun Dafa), literally "dharma wheel method," is a religious group blending Buddhist and Daoist philosophy, Confucian teachings, and qigong exercises (meditative and martial practice). It was founded in 1992 in Changchun, China, by Li Hongzhi, whom followers believe is a Buddha who manifests himself in human form to lead people to enlightenment. The sacred text of the Falun Gong is called Zhuan Falun. Its followers believe that *zhen* (truthfulness), *shen* (benevolence), and *ren* (forbearance) are the law of the universe and to achieve *zhen-shen-ren* is to gain salvation. *Xiulian* (cultivation practice) is believed to be the way to get closer to salvation. *Xiulian* includes reading Zhuan Falun and practicing a set of qigong and meditation exercises.

The religion grew rapidly in China and overseas. By July, 1999, before the crackdown, the number of Falun Gong practitioners had reached millions. Founder Li claimed that the worldwide number of practitioners was in excess of 100 million; some Chinese authorities estimated that the number was between 30 million and 40 million. Falun Gong was one of the hundreds of qigong groups that emerged in the late 1980's and early 1990's in China. The meditative and martial practice known as qigong was based on the *Yijing* (eighth to third century B.C.E.; English translation, 1876; also known as *Book of Changes,* 1986).

Falun Gong began as a legal qigong group under the registration of the Chinese Qigong Scientific Research Institute (CQSRI) but lost its registration in 1996. The institute accused founder Li of delivering too many superstitious and unscientific claims in his lectures to the public. Li, however, claimed that Falun Gong lost its registration because the institute was displeased at his insistence on charging low fees to practitioners. In 1996, Falun Gong registered as a legal association in Hong Kong, and under the one country, two systems principle, it remained a legal group in Hong Kong even after sovereignty reverted to China.

After Li and his family migrated to the United States in 1998, Falun Gong became an international organization. By 1999, it had branches in various cities in Asia, North America, and Europe. However, in China, the group faced increasing criticism in public media from scientists and intellectuals. More than ten times from May, 1998, to April, 1999, Falun Gong protested the media's negative depictions of the group.

The Crackdown

Professor He Zuoxiu, one of the scientists who had been criticizing various qigong groups, published an article in a youth magazine arguing that Falun Gong was unscientific and that young people should not practice it. On April 19, 1999, hundreds of Falun Gong followers assembled in front of the magazine's publisher in protest. On April 25, 1999, more than ten thousand Falun Gong members staged a sit-in outside the residence of the Chinese Communist Party leaders in Beijing. They demanded that party leaders return legal status to Falun Gong and release the forty-five followers arrested during the April 19 protest. Prime Minister Zhu Rongji met with the representatives of the protesters.

No arrests occurred on April 25, 1999; however, party leaders were stunned by the mass sit-in and shocked by the mobilization power of Falun Gong. In June, official media in China published articles criticizing the group as superstitious and pseudo-scientific. In response, Falun Gong organized similar protests in dozens of cities in China; however, these were promptly ended. On July 22,

1999, the Chinese ministry of civil affairs denounced Falun Gong as an illegal organization and charged it with causing more than seven hundred followers to die. The Chinese authorities rounded up more than five thousand followers, ransacked their homes, and destroyed more than two million of the group's publications.

The day after the group was banned, overseas members protested against the crackdown in Washington, D.C., and Ottawa, Canada. In China, thousands of Falun Gong followers from all over the country headed for the capital, Beijing, during the week after the crackdown. Most of them were sent home by the government while on their way to Beijing. On July 29, 1999, the Chinese ministry of public security issued an arrest warrant for Li. U.S. officials refused to return Li, who was living in New York, citing the

A day after the government ordered the arrest of the sect's founder, members of Falun Gong meditate outside the Hong Kong branch of the Chinese news agency Xinhua.

AP/Wide World Photos

3059

lack of an extradition treaty. Some of the detained followers were released when they vowed to leave Falun Gong, and others were sentenced to jail terms ranging from two to eighteen years.

Consequences

After the crackdown, Falun Gong members, individually or in small groups, continued to stage protests from time to time in Beijing. At least ten Falun Dafa International Conferences were held in the United States, Europe, and Hong Kong between August, 1999, and June, 2000. After the order for his arrest, Li did not appear in public or at any of the Falun Gong activities until November, 2000.

The Chinese government continued its campaign against the Falun Gong. Some Chinese officials criticized the Hong Kong government as being too lenient regarding the group's activities. In response, Hong Kong's top security official and chief executive both promised to monitor closely the group's activities in Hong Kong. Although large-scale Falun Gong activities were no longer evident in mainland China, the group was thought to still be operating underground.

Cheris Shun-ching Chan

Morocco's King Hassan II Dies

The sudden and unexpected death of King Hassan of Morocco raised fears within this North African Islamic nation about the country's future political and economic stability and identity.

What: National politics; Government
When: July 23, 1999
Where: Rabat, Morocco
Who:
MOHAMMED V (1909-1961), former sultan, king of Morocco from 1957 to 1961, father of Hassan II
HASSAN II (1929-1999), king of Morocco from 1961 to 1999
MOHAMMED VI (1963-   ), king of Morocco from 1999, eldest son of Hassan II

A Royal Life

The sudden and unexpected death of King Hassan II from a heart attack triggered by acute pneumonia removed a sometimes controversial ruler from the center stage of Arab and world politics. King Hassan ruled Morocco for thirty-eight years, ascending to the throne after the death of his father, Mohammed V, in 1961.

Born on July 9, 1929, in Rabat, Morocco, the eldest son of Sultan Sidi Mohammed, Mawlay Hassan Mohammed ibn Yusuf was a thirty-fifth generation descendant of the Prophet Muḥammad by his daughter Fatima. Information about Crown Prince Hassan's life before he became king is limited to that supplied by royal spokespeople. Hassan was educated at the imperial college at Rabat in Arabic literature, Muslim theology, English, French, and Arabic. He studied law at the University of Bordeaux's extension branch in Rabat and passed the qualifying examinations to practice. Prince Hassan enrolled in a training course in the French navy, serving aboard the battleship *Jeanne d'Arc.*

In 1953, the French removed both the sultan and the crown prince from Morocco for agitating for independence, and the crown prince followed his father, Mohammed V, into exile to Corsica and later Madagascar. The Moroccan people demanded their return, forcing France to relent in 1955. Crown Prince Hassan led Morocco's delegation negotiating independence from France. As crown prince, Hassan was chief deputy, commander in chief of the army, and minister of defense to Mohammed V, his father.

Royal Succession

The sudden death of Mohammed V during minor surgery on February 26, 1961, elevated Hassan to the throne as the seventeenth sovereign of the Alouite Dynasty. His playboy image was quickly altered by his 1961 marriage to a commoner, Lalla Lotifa Amhourok, who remained in strict seclusion in the Muslim tradition, and his refusal to establish a harem. The king had five children: sons Sidi Mohammed (later Mohammed VI) and Mulay al-Rashid and daughters Myriam, Asthma, and Hasna.

King Hassan II spent considerable time and money modernizing Morocco by eradicating illiteracy through the construction of primary schools and the importation of teachers from France. Public works projects reduced the nation's high unemployment rate, but a higher birthrate and a larger number of unemployed college graduates strained the nation's limited economic resources and created serious social tensions.

Throughout the thirty-eight years of King Hassan's rule, he remained the undisputed absolute ruler of Morocco in spite of a series of promulgated constitutions presenting the image of constitutional monarchy in 1962, 1972, 1992, and 1996. Hassan II ruled his nation using the strategy of political dualism, which blended modern forms of democracy with centuries-old practices. Although he repressed dissent, King Hassan increased government accountability, improved the status of women, and maintained the cultural integrity of Morocco's Berber minority population. Political parties were free to organize, and the press became more vocal, although it was not allowed to offend the king.

Hassan II had some enemies within Morocco. He escaped numerous leftist plots to overthrow him and two serious military assassination attempts. The first was at his forty-second birthday party in 1971, at which ninety-eight guests were killed; the second was in 1972, when his official jet was repeatedly strafed upon reentering Moroccan airspace on orders from within his own officers corps in Morocco.

King Hassan distracted his people from growing economic and political crises by launching his 1975 "Green March" into the phosphate-rich former Western Saharan Spanish colony of Rio de Oro on Morocco's southern border. The International Court of Justice at The Hague, Netherlands, validated Moroccan legal claims to the Saharan tribes of Rio de Oro, but the court denied Moroccan claims of territorial sovereignty. The march of 40,000 peaceful volunteers led by Prime Minister Ahmed Osman in November, 1975, gave Morocco control of the land. Eventually 350,000 Moroccans were resettled in the Western Sahara. In response, Western Saharan rebels opposed to Moroccan occupation organized a guerrilla movement with Algerian assistance, calling themselves the Polisario.

Political instability in neighboring Algeria, the Moroccan construction of a 750-mile (1,207-kilometer) sand wall with electronic sensors enclosing more than half of the Western Sahara, and a sizable Moroccan army stationed in the territory strengthened King Hassan's control in the region. In 1997, Morocco and the Polisario agreed to an election in the Western Sahara supervised by the United Nations provided the two sides could agree on who was eligible to vote from among the Saharan tribes of the former Rio de Oro.

Under the leadership of King Hassan, Morocco remained a loyal ally of the West during the Cold War years. The king's success in getting the United States to remove its military bases from Morocco won him great popularity with his people. In the 1991 Persian Gulf War, King Hassan contributed troops at Saudi Arabia's request to end the Iraqi occupation of Kuwait. During his reign, King Hassan was a voice for moderation in the Arab-Israeli dispute, frequently counseling both sides toward accommodation. Morocco officially recognized the state of Israel in 1996.

AP/Wide World Photos

The funeral procession of King Hassan heads toward the mausoleum in Rabat, Morocco.

Consequences

King Hassan II will be remembered as Morocco's most important ruler during the twentieth century. His ability to remain king was enhanced by his claim of descent from the founder of Islam, the Prophet Muḥammad. His international role and pro-Western positions during the Cold War enabled him to delay long-needed social, economic, and political reforms. His sudden death on July 23, 1999, left unresolved the long-standing territorial disputes in the Western Sahara and the future evolution of political parties, an elected parliament, and a free press. At the time of his death, Morocco's social fabric was strained by a severe shortage of jobs for a well-educated population that increasingly chafed under the rule of a king who governed by personal decree. Although King Hassan II guided Morocco through the minefields of the Cold War, his son, King Mohammed VI, had to deftly resolve Morocco's many domestic problems to survive in the twenty-first century.

William A. Paquette

Kansas Outlaws Teaching of Evolution in Schools

The Kansas State Board of Education, containing a majority of Christian conservative members, attracted national and international attention when it adopted new science standards that omitted references to several evolutionary concepts.

What: Biology; Education; Religion
When: August 14, 1999
Where: Topeka, Kansas
Who:
LINDA HOLLOWAY, chair of the Kansas
    State Board of Education
STEVE ABRAMS, member of the Kansas
    State Board of Education
JOHN STAVER, cochair of the committee of
    science educators
BILL GRAVES (1953-   ), governor of
    Kansas from 1998

Evolution and Creationism

On August 14, 1999, the Kansas State Board of Education voted 6 to 4 to adopt a new set of science standards that minimized the teaching of certain evolutionary concepts in the Kansas public schools. Statewide student assessment tests are based on the science standards, which set forth what students should learn.

The teaching of the theory of evolution in public schools has been a major controversy for many years not only in Kansas but also in other states such as California, Texas, Ohio, and Michigan. Evolution, a scientific theory developed by Charles Darwin and others, contends that Earth is billions of years old and that the various life-forms evolved through the process of natural selection. However, some Christian groups oppose the teaching of the theory of evolution in public schools and feel that "creationism," the biblical story of how the earth and its life-forms came into existence, should be taught instead of or in addition to the scientific theory. Numerous courts have forbidden the teaching of creationism in public schools, declaring it unconstitutional. Critics charge that conservative Christians are attempting to circumvent these courts' rulings by attempting to control or influence state and local school boards in order to revise school science standards so that the theory of evolution is no longer among the concepts that students are expected to learn.

New Science Standards Passed

During the mid-1990's, five moderates and five conservatives were on the Kansas State Board of Education, which supervises the state's 304 public school districts. This split along ideological lines caused a number of deadlocked votes on important issues involving teacher certification, testing methods, federal funding programs, and curriculum standards. Failure to secure a majority vote on many major issues weakened the state board's effectiveness. In August, 1999, however, the conservatives on the state board, led by Linda Holloway of Shawnee and Steve Abrams of Arkansas City, gained the support of a newly elected board member, Harold Voth of Haven, regarding the manner in which the state science standards should be written.

A science standards committee, consisting of twenty-seven members, was appointed to recommend the new standards. John Staver, a professor of science education at Kansas State University and cochair of the committee, played a major role in writing the proposed standards, which were based on national science educational standards established by the National Academy of Sciences and which most of the state science committee endorsed. The Christian conserva-

3064

tive members of the state board took issue with the way evolution was presented as a major theme in the national science standards.

Abrams, who also served on the science committee, offered an alternative version that de-emphasized certain evolutionary references. Abrams and his supporters questioned the validity of evolution. They believed that evolution is based on bad science and that macro-evolution, the changing of one species of animal to another as it adapts to its environment, is unobservable, and therefore, unprovable. Additionally, they believed that the creation of Earth and humans was done by divine design.

Several attempts to reach an acceptable compromise between those favoring the adoption of the model national science standards and those wanting evolution minimized or removed were unsuccessful. Governor Bill Graves of Kansas, the university presidents of the six state universities in Kansas, and many Kansas science teachers supported the national science standards model and urged the state board to adopt the standards recommended by the state science committee. Their pleas fail on deaf ears, and on August 14, 1999, the Kansas State Board of Education voted 6 to 4 to adopt a revised set of science standards that minimized the role of evolution, demonstrating the state board's Christian conservative majority view. For example, references to geologic time, common ancestry, and the big bang theory were omitted.

Consequences

News of the state board's vote resulted in a storm of protests and many negative comments, criticizing the board and the state of Kansas as well. Professional science organizations condemned the board for its unprofessional action and viewed the vote as a major step backward in the teaching of science. Many critics believed that the revised science standards would cause Kansas public school students to be ill-prepared for college science classes. Kansas was ridiculed and lambasted in the press and by late night talk-show hosts. Governor Graves was outraged and expressed his disgust and embarrassment over the board's decision.

The members of the Kansas State Board of Education who voted for the revised standards and supporters of the majority vote justified their position by stating that the revised standards allowed local school boards to have more control over what they wanted their students to learn about evolutionary concepts and how life began. In addition, those supporting the new standards held that evolution was indeed properly addressed, citing references to microevolution, or the changes within species, and other evolutionary concepts that were not deleted from the new science standards.

The revised science standards continued to be a hot and controversial issue; however, the election of new board members in 2000 settled the controversy. Three of the four Christian conservatives were replaced on the state board by more moderate members. Although Abrams was reelected, Holloway and another conservative failed in their bids for reelection, and another conservative board member decided not to run again.

On February 14, 2001, the Kansas State Board of Education, now containing a majority of moderate members, voted 7 to 3 to replace those standards adopted in August, 1999, with new ones that restored evolution as a key scientific principle based on the national science standards suggested earlier by the state science standards committee. The controversy is far from being resolved, however, as Christian conservatives have vowed to continue their efforts regarding what they believe should be taught about evolutionary concepts.

Raymond Wilson

Full Crew Leaves Russian Space Station Mir

> *Continuous occupation of the Russian Mir space station ended in August, 1999, leaving the Mir to an uncertain future that included plans for commercial ventures and a possible removal from orbit.*

What: Space and aviation
When: August 27, 1999
Where: Earth orbit
Who:
YURI KOPTEV (1940-   ), Russian space agency chief
SERGEI ZALYOTIN, cosmonaut commander
ALEXANDER KALERI, cosmonaut flight engineer
JEFFREY MANBER, MirCorp president

Continuous Occupation of Mir Ends

Soyuz TM-29 launched from Tyuratam Cosmodrome on February 20, 1999, with commander Viktor Afanasyev, French researcher Jean-Pierre Haignere, and Slovak researcher Ivan Bella aboard. Two days later, they arrived at Mir, where they were greeted by the three cosmonauts already at the station. Bella and two of the cosmonauts on Mir, Gennadi Padalka and Yuri Baturin, returned to Russia in the Soyuz TM-28, landing in northern Kazakstan on February 28. The third Mir cosmonaut, Sergei Avdeyev, stayed on Mir with Afanasyev and Haignere.

On April 27, 1999, British businessman Peter Llewelyn agreed to pay $100 million for a week-long station visit hosted by two cosmonauts, the first commercial venture proposed to extend Mir's life. Russian president Boris Yeltsin supported various commercial proposals despite the preference of the National Aeronautics and Space Administration (NASA) to deorbit Mir (remove it from orbit) shortly. The Russians, however, provided no funds for such ventures. On June 8, the Russian space agency proposed leaving Mir unmanned after August, then later ramming it through Earth's upper atmosphere to destroy it.

On June 20, Sergei Avdeyev exceeded 681 days of total time in space, breaking the record for the most time spent in space set by cosmonaut Valeri Polyakov. Five days later, Russian film director Yuri Kara proposed using Mir as the scene of a science-fiction movie about its abandonment and a renegade cosmonaut who refused to depart. On July 23, two cosmonauts exited Mir to test antennae and retrieve scientific equipment from Mir's hull. The airlock was again opened on July 28, presumably for the last time during Mir's existence.

A ground controller accidentally shut down Mir's main computer on July 30. That computer remained off-line until August 4 so that cosmonauts could finish maintenance work. Ground controllers then decided the crew should exchange the computer with a backup unit that would provide manual docking capability for the next Progress vehicle. Mir's computer was configured for rebooting if another crew arrived.

Soyuz TM-29's hatch was closed on August 27. Undocking came several hours later, ending Mir's continuous occupation after 3,641 days. For the first time since 1990, no one was in orbit. On August 28, 1999, Afanasyev, Avdeyev, and Haignere landed in Kazakhstan.

The Future of Mir

Although Mir was not immediately removed from orbit, on September 15, Russian space agency chief Yuri Koptev reaffirmed there were no plans to extend Mir's existence indefinitely. Mir's fate remained uncertain throughout 1999, but on January 12, 2000, the Russian space agency revealed that cosmonauts Sergei Zalyotin and Alexander Kaleri would perform a 45-day mission to Mir in late March, 2000, without government funding. Rumors spread that actor Vladimir Steklov would accompany Zalyotin and Kaleri. On January 13, Mir's owner, Rocket and Space Corporation (RSC-Energia), acknowledged having received $7 million from the British company

Gold and Appel for the delivery of an experiment to Mir that would be performed by Zalyotin and Kaleri. A Progress vehicle departed Tyuratam Cosmodrome on February 1, carrying consumables and equipment for this next manned expedition. That freighter docked on February 3.

On February 16, MirCorp announced it had paid RSC-Energia between $20 million and $30 million, and another $40 million might be forthcoming to keep Mir operational through December, 2000. Also, the Russian space agency announced that Steklov would be aboard the next Soyuz launch. However, just one month later, the agency voided Steklov's ticket for lack of reimbursement.

Zalyotin and Kaleri departed Tyuratam Cosmodrome on April 4 in Soyuz TM-30. Moscow announced that their mission could be extended fifteen days or more unless problems arose. Zalyotin manually docked Soyuz TM-30 on April 6. Opening Mir's hatch, Zalyotin and Kaleri ended a 223-day unoccupied period. Zalyotin and Kaleri exited Kvant 2's airlock on May 12, attempted

hull-patching repairs, inspected Mir's exterior, and determined the cause of a Kvant 1 module solar panel failure.

When this commercial mission wound down, Russian space agency statements suggested that Mir's useful lifetime was nearly over, but MirCorp president Jeffrey Manber proposed continued commercial sponsorship and announced finalization of a deal to finance another manned expedition in September. Mission Control's deputy director was quoted as committing Mir to inactive status, permitting visitations if funding materialized. Hatches to Mir were closed on June 15, and Kaleri and Zalyotin undocked Soyuz TM-30 several hours later, touching down on June 16.

MirCorp insisted corporate backing existed for ambitious plans such as transport of American multimillionaire Dennis Tito for a two-week Mir visit. Mark Burnett, producer of popular reality program *Survivor*, suggested a program called *Destination Mir*, a contest to select the winner of a ten-day trip to Mir in 2002. Like most commercial proposals, it never materialized.

AP/Wide World Photos

From left to right, flight engineer Sergei Avdeyev, crew commander Viktor Afanasyev, and French astronaut Jean-Pierre Haignere.

3067

Progress M-43 lifted off on October 16 from Tyuratam Cosmodrome on a slow rendezvous to Mir to conserve fuel that could boost Mir's presently low altitude. It also carried food and supplies, should another crew launch in 2001. Russian space agency officials met on October 18 to consider deorbiting Mir early in 2001, as station altitude had dropped considerably. No commercial proposals had generated sufficient funding to maintain Mir, but the Progress freighter docked on October 21.

Consequences

Official Russian statements issued on November 15 suggested that the Russian space agency might dispatch another Mir crew, one including the American Tito. The Russian space agency dismissed that report, although it did not dismiss dispatching cosmonauts to bring about Mir's deorbit in early 2001. The Russian Duma (congress) vigorously condemned the deorbit of Mir orbit. However, the Duma withheld governmental funds to keep Mir aloft, and MirCorp announced on December 12 that it would cease marketing Mir.

On Christmas day, 2000, contact with Mir dropped out unexpectedly. Controllers repeatedly attempted to restore contact, but Mir's solar batteries apparently had lost power. By day's end, all batteries had been recharged. On January 18, Russians launched the Progress M1-5, an unmanned tanker that was to push the station out of orbit. Most of the space station was expected to be destroyed during Mir's entry into the atmosphere; however, remaining debris was expected to fall into a remote area of the South Pacific. On March 22, Mir entered deorbit, and its last remnants fell harmlessly into a 120-by-1,800-mile (193-by-2,896-kilometer) target zone in the South Pacific, as planned, on March 23.

Launched in 1986, Mir was the Soviet Union's last technological hurrah in orbit. The final Soviet flag to be lowered had flown atop a boom on Mir. With Mir's passing, the Soviet space program became history. Russia's manned space program continued, largely tied to the International Space Station instead of a national station.

David G. Fisher

Decline in U.S. AIDS Death Rate Slows

A report from the National Conference on HIV Prevention cautioned Americans that while the numbers of AIDS cases in the United States was continuing to fall, the rate of decline was lower than in previous years.

What: Health; Medicine
When: August 30, 1999
Where: Atlanta, Georgia
Who:
HELENE GAYLE (1955-   ), director of AIDS prevention for the Centers for Disease Control and Prevention
DAVID SATCHER (1941-   ), U.S. surgeon general

Good News and Bad News

On August 30, 1999, health officials at the National Conference on HIV Prevention issued a report about human immunodeficiency virus (HIV)/acquired immunodeficiency syndrome (AIDS) in the United States that contained both encouraging and discouraging news. The good news was that deaths due to AIDS had continued to decline through 1998—the rate of decline was about 20 percent during 1997 and 1998. However, the bad news was that this decline in the death rate was slowing and perhaps beginning to level off.

According to the Centers for Disease Control (CDC) in Atlanta, Georgia, AIDS deaths dropped about 48 percent from 1996 to 1997. Therefore, the August 26 report was seen as a warning about HIV/AIDS. First, the news reminded Americans that HIV/AIDS was still a danger in American society and that no cure had been found, despite the great success of new drugs and treatments during the 1990's. Second, the report suggested that Americans had become complacent about the disease because of substantial progress made during the 1990's. Third, the slowing of the AIDS death rate was interpreted to mean that more attention needed to be given to prevention rather than treatment. Fourth, the report was a reflection that HIV/AIDS was becoming ever more resistant to the new drugs that had become prevalent by 1999.

Although the drop in the rate of decline was disturbing, this report was only the most recent one about the battle against HIV/AIDS, in which much progress had been made during the 1990's. In 1981, when the first case of AIDS was reported by the CDC, AIDS deaths accounted for 5.5 deaths per 100,000 people in the United States. This number rose steadily through the 1990's until it peaked in 1995, when about 50,000 people died of AIDS, at a rate of 15.6 deaths per 100,000 people. However, by 1996, a concentrated prevention campaign and various new drugs dropped the death rate by 23 percent, and the number of people with AIDS fell in that year by 6 percent.

In 1997, the rate of HIV infection fell by 18 percent, and in 1998, it declined by 11 percent. Therefore, both the AIDS death rate and the HIV infection rate continued to decline in the mid-1990's, but at an ever slower rate. In 1997, 21,222 Americans died of AIDS, a rate of 5.9 deaths per 100,000 people, and AIDS was no longer among the top ten causes of death in the United States. In 1998, the number of AIDS death had dropped again to 17,047, at a rate of 4.6 per 100,000 people. In fact, the U.S. Department of Health and Human Services estimated in 1999 that HIV mortality had declined more than 70 percent since 1995. However, an ominous trend was developing during the late 1990's— the rate of HIV infection was leveling off. In 1999, the CDC estimated this rate was holding steady at about 40,000 per year.

According to Helene Gayle, director of AIDS prevention for the CDC, the slowing decline in the AIDS death rates was caused by at least two different factors. First, those already diagnosed with the disease had been reached by 1999, but

3069

U.S. surgeon general David Satcher addresses the National Conference on HIV Prevention.

many other Americans were not yet aware of being infected with HIV and had not yet been tested for it. Second, HIV had developed resistance to various drugs, and drug combinations were not as effective as they had previously been. Gayle urged that the report "wake us up and shake us out of some of the complacency that I think has developed around HIV."

More Bad News

Another disturbing trend noted in the 1999 report from Atlanta was that among certain groups of Americans, the decline in HIV infection and AIDS deaths was even slower than that experienced by the rest of the population. According to U.S. surgeon general David Satcher, in the mid-1980's, as a percentage of total AIDS cases reported, African Americans accounted for about 25 percent, women made up 8 percent, and Hispanics represented about 14 percent. By 1997-1998, these percentages had risen to 45 percent for African Americans, 23 percent for women, and 22 percent for Hispanics.

Gayle noted that although African Americans in 1999 made up 13 percent of the total popula-

tion, 50 percent of new HIV infections were among African Americans. This trend could be traced back to the mid-1990's. For example, even though the rate of AIDS deaths declined by 13 percent in the African American population in 1996, the decline among white Americans during the same time was 32 percent. From July, 1998, to June, 1999, African American men accounted for 58 percent of all the AIDS cases reported among American men. By 1999, AIDS was the leading cause of death among African Americans aged twenty-five to forty-four and African Americans accounted for 63 percent of HIV infections among people aged thirteen to twenty-four. Moreover, the 1999 report showed that women were being infected and dying of AIDS at much higher numbers than when the epidemic was first reported in the early 1980's. According to Gayle, in 1999, 30 percent of new HIV infections were among women, compared to 8 percent at the beginning of the AIDS epidemic. The CDC tracked a decline of 10 percent in AIDS deaths in 1996, but this was a much lower rate than the 25 percent decline among men. From July, 1998, through June 1999, women accounted for 10,841 of the 47,083 AIDS cases reported, or 23 percent. In addition, of these 10,841 AIDS cases among women, 80 percent were among African American and Hispanic women.

Consequences

One lesson from the 1999 Atlanta report that researchers emphasized was the need for increased focus on efforts to encourage preventive behavior even as researchers continued to pursue more effective drug treatments and possible cures. Gayle, appearing on the Public Broadcasting System's *The NewsHour with Jim Lehrer*, reminded the nation, "We've got to remember that there is no magic bullet, and it is always going to be better to prevent somebody from getting HIV than to go through the heartache and the complexity of having a very serious disease."

Robert Harrison

East Timor Votes for Independence

Residents of the eastern half of the island of Timor voted overwhelmingly for independence from Indonesia, rejecting the latter's offer of local autonomy.

What: Civil war; Political independence
When: August 31, 1999
Where: East Timor
Who:

José Alexandre "Xanana" Gusmão (1946-   ), leader of the National Council for Timorese Resistance (CNRT)

Suharto (1921-   ), president of Indonesia from 1967 to 1998

Bacharuddin Jusuf Habibie (1936-   ), president of Indonesia from 1998 to 1999

José Ramos-Horta (1949-   ), East Timorese political spokesperson

Carlos Filipe Ximenes Belo (1948-   ), Roman Catholic bishop

A Separate History

On August 31, 1999, residents of the eastern half of the East Indian island of Timor voted for a complete break with the large island nation of Indonesia. Indonesia had proposed that East Timor remain a part of Indonesia but receive a degree of local autonomy (political independence) based on the territory's history.

Portuguese explorers had opened trade with Timor in the early sixteenth century. Beginning in the early seventeenth century, however, their activities were limited by traders and troops from the Netherlands, who eventually colonized most of what is now Indonesia. Between 1859 and 1914, the Netherlands and Portugal negotiated the boundaries separating their respective territories. Portugal received the eastern half of the island, the enclave of Ocusse on the northwest coast, and two smaller islands. Over the years, most East Timorese converted to Roman Catholicism, in contrast to their Muslim neighbors.

Indonesia gained its independence from the Netherlands in 1949, but East Timor remained Portuguese. The Portuguese revolution of April, 1974, encouraged revolutionaries in most of Portugal's colonies, including East Timor, to fight for the independence of their own lands. However, before the East Timorese could gain complete control, troops under the direction of Indonesian president Suharto invaded the territory in late 1975, annexing it the following year.

An estimated 100,000 East Timorese died at the hands of Indonesians during the first year of occupation. Another 150,000 would die directly in clashes or indirectly from torture, disease, and starvation over the next twenty-five years, bringing the death toll to nearly one third of the territory's population. Another third would languish in "resettlement camps."

José Alexandre "Xanana" Gusmão, a journalist turned guerrilla leader, was captured in 1992 and sentenced to life in prison in the Indonesian capital of Jakarta. During his subsequent years of imprisonment, Gusmão emerged as a popular symbol of the East Timorese struggle.

No End to Violence

Events in the late 1990's raised hopes that the period of repression might be coming to an end. Two figures prominent in the East Timorese resistance movement—Roman Catholic bishop Carlos Filipe Ximenes Belo and exiled spokesperson José Ramos-Horta—received the Nobel Peace Prize for 1996. Indonesian president Suharto resigned under pressure in 1998. On January 27, 1999, Suharto's successor, Bacharuddin Jusuf Habibie, announced that if the East Timorese rejected continued political union, they might receive complete independence. Gusmão was transferred from prison to house arrest the following month and freed soon after.

The United Nations Mission in East Timor

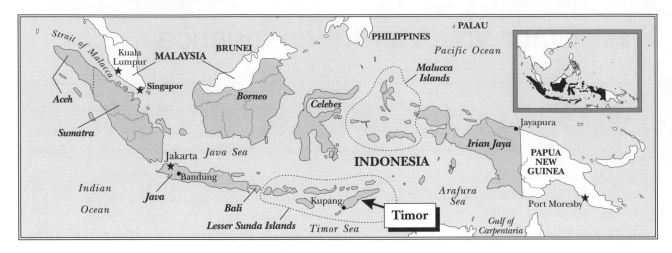

(UNAMET) was set up to supervise a referendum, but escalating violence between people favoring Indonesian rule and East Timorese nationalists led to two postponements. When the vote was finally held on August 30, 1999, more than 98 percent of eligible voters took part, and more than 78 percent of them rejected union with Indonesia.

The vote did not bring peace, however. Enraged with the results, pro-Indonesian militias massacred hundreds of East Timorese and forced hundreds of thousands more from their homes. Several thousand refugees, among them Bishop Belo, fled to Australia, where Ramos-Horta had lived in exile since 1975. Gusmão briefly took refuge in the British embassy in Jakarta.

A United Nations peacekeeping force landed in Timor in late September, 1999. The last pro-Indonesian troops left the eastern part of the island in October, making it possible for Gusmão to make a triumphal return to Timor. That same month, however, pro-Indonesian forces attacked Ocusse, forcing tens of thousands of its inhabitants into concentration camps in the western part of the island.

Consequences

East Timor achieved independence in late 1999, but the devastation caused by pro-Indonesian troops and sympathizers was so great that the territory was unable to function without outside aid. Although a convention of rival East Timorese political parties held in Portugal in 1998 had created a framework for independent government, the United Nations Transitional Administration in East Timor (UNTAET), was created to help the new country operate. Gusmão assumed de facto leadership, although he rejected running for president in elections scheduled for mid-2001.

Experts believed that it would be decades before the country returned to normal, although there were hopes that the nation's coffee harvest might quickly return to prewar levels. In addition, East Timorese representatives asked Australia to renegotiate a treaty governing drilling rights to oil and natural gas reserves in the Timor Sea lying between the two countries.

Critics inside and outside East Timor deplored the acquiescence of many other nations—including Canada and the United States—in the atrocities committed by economically powerful Indonesia. They pointed out that the Indonesian invasion was more brutal than the Iraqi invasion of Kuwait in 1990 and the conflicts that destroyed Yugoslavia in the 1990's—situations that generated far greater concern among the world community.

As the East Timorese struggled to establish a viable country, revolutionaries in other parts of ethnically diverse Indonesia were asserting their causes. The Aceh People's movement claimed much of the large island of Sumatra. Another independence movement was active in Irian Jaya, the western half of the large island of New Guinea. Violent clashes between native Dayaks and immigrant Madurans were common on the large island of Borneo. Observers believed that East Timor might be only the first of many Indonesian territories to win its independence.

Grove Koger

Viacom Buys Columbia Broadcasting System

Viacom, the cable network company and owner of Paramount film and television studios, Nickelodeon, and MTV, officially bought the CBS Corporation in the spring of 2000, creating a huge conglomerate with annual revenues at $20 billion.

What: Business; Communications
When: September 7, 1999
Where: New York City
Who:

MEL KARMAZIN (1944-   ), president of
   CBS Corporation
SUMNER REDSTONE (1923-   ),
   chairman of Viacom

Viacom and CBS

On September 7, 1999, Viacom, a huge media company, announced plans to buy the CBS Corporation. Although its name was not widely known among consumers, Viacom owned numerous cable channels, including MTV (Music Television), Showtime, VH1 (Video Hits 1), The Movie Channel, and Nickelodeon. It also owned Paramount Pictures (film production), the United Paramount Network (UPN), Blockbuster entertainment, Simon and Schuster publishing company, and King World Productions syndicated television (producers of *The Oprah Winfrey Show*).

UPN had been considered a struggling cable channel for a while but still maintained loyal viewers with such shows as *WWF Smackdown!* and *Star Trek: Voyager.* Viacom also owned five amusement parks and an Internet group that included Nick.com, mtv.com, cbs.sportlines.com and vh1.com. In August, 1999, Viacom had become the largest U.S. media company to establish an Internet-only company with the formation of MTV Interactive. Viacom's purchase of the CBS Corporation was expected to create a company that would play a leading role in mainstreaming Internet radio format and the selling of music online.

The CBS Corporation owned Infinity Broadcasting, with a radio group of more than 160 stations, sixteen television stations, cable stations including The Nashville Network (TNN) and Country Music Television (CMT), and the Columbia Broadcasting System (CBS) network, one of the original three networks, along with the American Broadcasting Company (ABC) and National Broadcasting Company (NBC). The network still maintained high ratings with many of its shows, including the popular news program *60 Minutes* and the late-night talk show *The Late Show with David Letterman.*

The Buyout

Companies that own television stations operate under the guidelines of the Federal Communications Commission (FCC), an independent U.S. government agency. The FCC regulates interstate and international communications by radio, television, wire, satellite, and cable. In the past, the FCC had limited duopolies—ownership of more than one television station in the same market—and limited the ownership of both television and radio stations in certain larger markets. In August, 1999, the FCC relaxed ownership rules to allow TV duopolies in certain markets. This appeared to open up the way for media company mergers.

However, mergers involving media companies attract critics who say that when fewer companies own more of the media, competition is weakened and consumers' choices are limited. One of the more vocal protests in this instance involved the League of Women Voters of the United

CBS president Mel Karmazin (right) grasps the hand of Viacom chairman Sumner Redstone.

States, the Alliance for Better Campaigns, and six other national public interest groups. In February, 2000, they asked the FCC to block the merger on the ground that CBS had not fulfilled its public interest obligation (under FCC guidelines) to give political candidates recommended air time in the days before an election.

The American Association of Advertising Agencies also protested the merger because of concern over the elimination of competition in the six markets in which UPN and CBS networks would coexist and concerns about the overcrowding of television commercials.

Viacom closed the $36 billion acquisition of the CBS Corporation in May, 2000, with the approval of the FCC. The deal resulted in one of the largest media companies in the world, worth about $91 billion. In terms of annual revenues, only Time Warner and Disney were larger. When first combined, the Viacom-CBS companies, later known simply as Viacom, maintained 41 percent coverage of U.S. television households. In other words, 41 percent of all households with a television had access to at least one of the company's many stations.

Consequences

The FCC gave the new company a year to sell enough television stations to comply with rules that bar a single company from owning stations that have a combined reach of 35 percent of television-viewing households. The FCC also required Viacom to divest itself of UPN. Under the federal guidelines at the time, a company could not own two broadcast networks; Viacom owned both CBS and UPN. Viacom chairman Sumner Redstone had said that keeping UPN in the company was essential for the cable channel's survival.

CBS president Mel Karmazin and other broadcasters were pushing for the FCC to raise the television station ownership cap to at least 49 percent of the nation's households. The Tribune Company and others also wanted the commission to lift the ban that prohibited new ownership of a newspaper and television station in the same market.

Soon after the deal was made, Viacom began cross-programming and cross-promotion. Popular Nickelodeon weekday shows such as *Blue's Clues* and *Kipper* showed up on CBS on Saturday mornings. UPN began airing reruns of MTV's *Celebrity Deathmatch* on weekday evenings. In January, 2001, MTV produced CBS Sports' Super Bowl halftime show. The company planned to use CBS music-oriented programming on MTV and VH1. Cross-promotion meant that cable shows could be advertised on CBS, and CBS shows on Viacom's cable channels, including MTV. The company hoped to use these advertisements to attract the attention of MTV's primary audience (teenagers) to CBS television shows. By any definition, Viacom had become a media powerhouse.

Sherri Ward Massey

School Busing Ends in North Carolina

> *The school system that thirty years earlier had pioneered school busing for desegregation was ordered to halt its busing program by a federal judge who ruled that forced integration was no longer necessary because intentional discrimination had disappeared.*

What: Civil rights and liberties; Education; Social reform

When: September 10, 1999

Where: Charlotte, North Carolina

Who:

JAMES SWANN (1959-　　　), subject of original busing case

JAMES McMILLAN (1916-1995), federal district judge in original Swann case

BILL CAPACCHIONE, parent whose suit led to new court decision

ROBERT D. POTTER, federal district judge who reviewed the case in 1998

Swann v. Charlotte-Mecklenburg Board of Education

In 1965, more than a decade after the Supreme Court's landmark school desegregation decision, only 2 percent of black schoolchildren in Charlotte, North Carolina, attended integrated schools. To maintain this racial separation, 60 percent of all students were actually bused to schools far from their homes. Due to the injustice of this practice, the National Association for the Advancement of Colored People and other civil rights groups called for parents to challenge the system. Vera and Darius Swann wanted their six-year-old son, James, to attend Seversville Elementary School, one of the few integrated schools in the city. That school happened to be closest elementary school to their home, but the city school board refused to admit James to it.

The Swanns filed a lawsuit but lost. They continued to appeal until their case reached the desk of federal district court judge James McMillan in 1969. When he examined the family's case he concluded that Charlotte's school board was plainly in violation of the law and that the board would probably not desegregate unless forced by the courts. The precedent had been set with the 1968 Supreme Court decision *Green v. County School Board of New Kent County*, in which the Court had ruled school boards had an "affirmative duty" to ensure "racial discrimination would be eliminated root and branch."

In the Swann case Judge McMillan ruled that Charlotte-Mecklenburg had been operating two separate school systems, or "dual systems"—one for black children and one for white children. He ordered the system to desegregate and required it to draw up a plan to take steps—not necessarily busing—to do so. He appointed an outside referee to devise a desegregation strategy, which entailed the busing of black children to previously white schools, and vice versa.

Mandatory busing began in the Charlotte-Mecklenburg school system began on September 9, 1970. The plan sent black children to white neighborhood schools through half their elementary school years and white children to black neighborhood schools for the other half. This plan desegregation would last for two decades.

Enraged, many white families protested the busing plan, sometimes violently. Judge McMillan received death threats and was hanged in effigy. The office of the Swanns' attorney was firebombed. Black students were beaten at school. Some white parents sent their children to private schools, while others formed citizens' groups to agitate against busing. Nevertheless, the U.S. Supreme Court upheld McMillan's order in its 1971 decision in the case of *Swann v. Charlotte-Mecklenburg Board of Education*. This case cleared the way for busing for desegregation nationwide.

Aftermath

A thirty-three-year-old housewife named Maggie Ray began convening local activists, black and white, pro- and anti-busing, to work out an alternative solution—and also endeared herself to the judge at a backyard barbecue. Her informal committee, called the Citizens Advisory Group, persuaded Judge McMillan to put it in charge of the busing plan. In 1974, the school board forged a compromise with the committee on an arrangement that won McMillan's approval. In 1975, McMillan ruled that the school board was properly implementing the desegregation court order, and removed the case from active court supervision.

In 1976 and again in 1981, Charlotte-Mecklenburg parents claimed that assigning students to certain schools because of their race violated the U.S. Constitution. The courts disagreed, saying the school system had not yet fully desegre-gated. In 1986 the Supreme Court ruled that busing in a city could end when school officials decided that segregation no longer characterized the community's schools. In 1991, for example, the Court ruled that Oklahoma City could end its school busing plan, even though integration had not been achieved. The Court stated that everything had been done that could be done to eliminate the effects of decades of discrimination.

Meanwhile, the Charlotte-Mecklenburg schools maintained a mix of about 60 percent white and 40 percent black students over the next two decades. The presence of whites, especially affluent ones, in once all-black schools lifted those schools' quality, since white parents' complaints about outdated textbooks, sub par teachers, and dilapidated facilities were heeded by school officials.

Local NAACP president Dwayne Collins speaks at the AME Zion Church in Charlotte, North Carolina, after a judge's decision ending busing.

Consequences

In September 1997, Charlotte-Mecklenburg parent Bill Capacchione sued the school system, charging that his daughter (who was half white and half Hispanic) was twice denied entrance to a magnet school because she was not black. In the following October, the Swann family's original attorneys announced that they would join the case to fight Capacchione's lawsuit, saying that the school system had still not fully desegregated and thus should not be released from court-ordered desegregation.

U.S. District Judge Robert Potter reactivated the *Swann* case in March of 1998 and consolidated it with Capacchione's lawsuit. Judge Potter was a former aide to North Carolina senator Jesse Helms and a Ronald Reagan appointee who in 1969 swore he would fight school busing.

Six parents joined the case with Capacchione in April of 1998, claiming that race-based policies influence everything from the schools to which students were assigned to where schools were built. The parents argued that the Charlotte-Mecklenburg schools were fully desegregated; therefore the continued use of race-based policies was unconstitutional. They pointed out that the system was not working because black students in Charlotte-Mecklenburg schools still scored substantially lower than white students on standardized achievement tests. In 1998, blacks averaged 850 points, compared to 1,043 for white students on the SAT. The 193-point disparity is somewhat worse than the rest of North Carolina, where there is a 189-point gap.

The Swann attorneys asked Judge Potter in August of 1998 to allow three African American families to join their side. Potter agreed. The case was a class-action suit filed on behalf of Charlotte-Mecklenburg's black children.

Judge Potter ruled against the Charlotte-Mecklenburg schools in favor of Capacchione and the other parents on September 10, 1999. He ordered the school district to pay all the expenses of the parents who had brought the lawsuit. This decision effectively ended school busing in the very location where it began.

Calvin Henry Easterling

Scientists Produce Most Proton-Rich Nucleus

Physicists produce nickel-48, the most proton-rich nucleus, a feat that represents an international breakthrough in nuclear physics.

What: Physics
When: September 12, 1999
Where: Grande Accelerateur National d'Ions Lourds, Caen, France
Who:
Bertram Blank, nuclear physicist, Center for Nuclear Studies, Bordeaux-Gradignan, France
Marek Lewitowicz, nuclear physicist, Grande Accelerateur National d'Ions Lourds
C. Lucian Borcea, atomic physicist, Institute of Atomic Physics, Bucharest, Romania
Robert Grzywacz, nuclear physicist, University of Tennessee

A Major Breakthrough

On September 12, 1999, researchers at the Grande Accelerateur National d'Ions Lourds (GANIL) in Caen, France, succeeded in producing four nuclei of nickel-48, the most proton-rich nucleus ever made. Because the mode of radioactive decay of this nucleus is apparently a form of radioactivity never before observed, it should provide important insights into the fundamental nature and understanding of nuclei.

Similar to the electrons that fill different energy shells outside the nucleus of an atom, neutrons and protons are arranged in energy shells inside the nucleus. The nucleus can be pictured as an onion consisting of layers of protons and neutrons, which are collectively referred to as nucleons. Each time a nuclear shell has the maximum possible number of nucleons it can hold, the nucleus is very stable. With the completion of more shells, the nucleus acquires even greater cohesion. These maximum numbers are termed magic numbers. Based on the shell model of the nucleus, and in general agreement with experimental data, magic numbers occur when either the number of neutrons or the number of protons in the nucleus equals 2, 8, 20, 28, 50, 82, 114, 126, or 184.

When both the number of protons and the number of neutrons are magic numbers, the nuclei are doubly magic, possessing full shells of protons and full shells of neutrons. For example, the helium-4 nucleus with two protons and two neutrons is doubly magic and is very stable. However, being doubly magic alone does not guarantee stability of a nucleus. For example, nickel-56 has 28 protons and 28 neutrons, making it doubly magic, but it is not stable. Until late 1999, only nine doubly magic nuclei had been discovered by nuclear physicists. Scientists still sought the most interesting double-magic holdout, nickel-48, containing 20 neutrons and 28 protons.

Magic Nickel-48 Produced

During the late 1990's, researchers at GANIL were successful in producing two of nickel's lightest isotopes that are doubly magic, but it was not until mid-September, 1999, that the very difficult task of producing nickel-48 was finally achieved. Results were confirmed and announced in February, 2000.

Directed by Bertram Blank, Marek Lewitowicz, C. Lucian Borcea, Robert Grzywacz, and some of their colleagues, the experiment at GANIL consisted of an ordinary nickel target that was bombarded with a beam of highly charged nickel-58 ions for a period of ten days. On September 12, 1999, at least two, and possibly four, nickel-48 nuclei were observed as products of the nuclear collisions. The new nuclei were identified by their time of flight after emerging from the nickel target and traveling through a series of selectors and detectors. Besides nickel-48, three other proton-rich isotopes, iron-45, chromium-42, and nickel-49, were also generated in the nuclear collisions at GANIL.

In a plot of proton number versus neutron number for all known nuclei, nickel-48 falls on the extreme outer limit of nuclear stability, at the edge where nuclei can no longer be held together by the strong nuclear force. The newly produced nickel-48 nucleus undergoes radioactive decay in a fraction of a second (a lower limit estimate of 0.5 microseconds), and if it had not been for the stability generated by its doubly magic numbers, it probably would have never been observed.

Because only a few nickel-48 nuclei were observed, there were not enough events to make a detailed comparison with existing nuclear models. To do this will require experiments with higher statistics so that the exact half-life of nickel-48 can be determined. The fact that nickel-48 was observed contradicts previous nuclear models that predicted that nickel-48 was unstable and should last for less than one microsecond, the typical time it takes for collision products to travel from the stationary nickel target to the detectors in the GANIL experiment.

Consequences

Although the role of magic numbers was established for stable nuclei in 1949, the production of nickel-48 extends the applicability of the magic number scheme to the extreme case of the most proton-rich nucleus ever observed. Furthermore, the production of nickel-48 allows nuclear physicists the opportunity to study a form of radioactive decay that had long been sought but never observed. Because the nickel-48 experiment at GANIL was optimized for observation of the doubly magic nucleus, the mode of decay was not detected. Because single-proton decay is energetically impossible for nickel-48, it is expected to decay by emitting two protons, a form of radioactivity never before observed. Observing this decay mode may answer the question of whether protons and neutrons are correlated in pairs within the nucleus.

Blank and his colleagues planned future experiments to study the energy distributions and angular correlations of the two emitted protons from nickel-48, allowing researchers to determine whether they are correlated as they leave the nucleus. Furthermore, as improvements continue in producing higher beam intensities and accelerations of the bombarding ions, corresponding higher production rates of nickel-48 will follow and allow accurate determination of the half-life and other important properties of the most proton-rich nucleus ever observed.

Particle accelerator facilities that existed around the world in 1999 were not capable of producing any other doubly magic nuclear configurations. Among the lighter nuclei, the remaining candidates would have too many protons compared with neutrons to produce stability. For the super-heavy elements, magic numbers had not been accurately predicted. If and when they are, future experiments, similar to those at GANIL, can be designed to observe these nuclei and deduce further insights into the fundamental nature and understanding of nuclei.

Alvin K. Benson

Mars-Orbiting Craft Is Destroyed

The Mars Climate Orbiter, a $125 million robotic spacecraft designed to investigate weather on Mars, disappeared as it was to go into an orbit around Mars, probably because a human or software error took it too close to the planet, causing it to disintegrate in the Martian atmosphere.

What: Space and aviation; Astronomy
When: September 23, 1999
Where: Mars
Who:

RICHARD COOK, Mars Climate Orbiter project manager

CARL PILCHER (1947-   ), science director for Solar System Exploration, National Aeronautics and Space Administration (NASA) headquarters, Washington, D.C.

DANIEL MCCLEESE, chief scientist for the Mars Exploration Directorate at Jet Propulsion Laboratory (JPL)

RICHARD W. ZUREK, Mars Climate Orbiter project scientist

The Loss and Its Causes

When the Mars Climate Orbiter fired its main engine to go into orbit around Mars early on September 23, 1999, at about 2 A.M. Pacific daylight time on Earth, the Orbiter's mission seemed to be heading for a brilliant success. All information from the spacecraft up to that point had seemed normal. The engine burn began as planned five minutes before the spacecraft passed behind the planet as seen from Earth. Flight controllers never heard from it again. Only after intensive investigation and review did it become clear that the Orbiter was doomed even before the craft's launch in December, 1998.

Lockheed Martin Astronautics in Denver, which built the craft, had expressed data about the thrust of the engine in pounds. However, scientists at the Jet Propulsion Laboratory (JPL) in Pasadena, California, had assumed that thrust data were expressed in metric units, as newtons, a metric unit of force required to accelerate one ki-

logram one meter per second per second rather than as pounds of force. The mission team members failed to discover that the transformation of pounds of force to newtons had not occurred. Consequently, when the main engine fired, the calculations were incorrectly made, and the Mars Climate Orbiter approached the planet at an altitude of about 37 miles (60 kilometers) rather than the expected 93 miles (150 kilometers). Therefore, mission scientists believe, the orbiter either was torn apart by the atmosphere, denser at 37 miles than at 93 miles, or it burned up.

An internal review conducted by the Mars Climate Orbiter Mission Failure Investigation Board identified eight contributing factors that led directly or indirectly to the loss of the spacecraft. However, the heart of the problem was the failure of systems engineering at National Aeronautics and Space Administration (NASA) to detect the basic incompatibility of the two systems of measurement, English units versus metric units.

Aims of the Mission

When a team of scientists announced in 1996 that a meteorite believed to have come from Mars contained what could be the residue of ancient microbes, the public became intrigued by the possibility of past or present life there. Many scientists believe that the key to understanding whether life could have evolved on Mars is determining the history of water on the planet. The Mars Orbiter Mission was part of a large overall scientific effort undertaken to learn about the geology and the climate of Mars and especially what happened to the water on Mars. Ultimately the goal was to assess the possibility of life on Mars.

By designing and sending a number of specialized probes to Mars, people were learning much

3081

Illustration of the Mars Climate Orbiter, before it was to go into orbit around Mars.

about the planet in the solar system that is most like Earth. Water ice has been detected in the permanent cap at Mars' north pole, and scientists think it may also exist in the cap at the south pole. Much water may be trapped under the surface—either as ice or, if near a heat source, possibly in liquid form well below the surface. The Mars Surveyor 1998 missions were designed to investigate theories of the planet's climate-history and its water resources and to investigate the question of life on Mars by gathering evidence of climate change and its evolving effects on the distribution of water.

Consequences

The destruction of the Mars Orbiter spacecraft, while a disappointment to everyone in the NASA and JPL scientific communities, taught the scientists and technicians a great deal about

the challenges of integrating production and mission teams on a project of such complexity. Although not without disappointments and setbacks, the history of human interest in and exploration of Mars is a record of continuing advancement. On July 4, 1997, for example, Mars Pathfinder touched down on a windswept, rock-laden ancient flood plain. Two months later, Mars Global Surveyor went into orbit, sending back pictures of towering volcanoes and gaping chasms that were sharper and clearer than any seen before. The Mars Orbiter was launched in December, 1998, and lost on September 23, 1999. However, in January, 1999, another orbiter and lander were launched to Mars. Every 26 months over the next decade, when the alignment of Earth and Mars is suitable for launches, still more robotic spacecraft are scheduled to join them at the red planet.

These spacecraft carry varied payloads, ranging from cameras and other sensors to rovers and robotic arms. Some of them have their roots in different NASA programs of science or technology development. However, they all have the goal of understanding Mars better, primarily by delving into its geology, climate, and history. Each mission brings new knowledge. For example, the Mars Global Surveyor spacecraft, which mapped the red planet, transmitted gravity and topography data that permitted scientists to derive a picture of the crustal structure of Mars along longitude zero degrees east and to infer that the northern lowlands was probably a zone of high heat flow early in Martian history, reflecting vigorous convection of the Martian interior. According to Mars Global Surveyor teams, this rapid heat loss could have released gases trapped within the planet to the atmosphere and brought underground ice or water to the surface, helping to produce a warmer, wetter climate than now exists on Mars.

Theodore C. Humphrey

Japan Has Its Worst Nuclear Accident

A radiation leak at a uranium processing plant killed two workers, posed a severe health risk to the surrounding population, and raised international questions about nuclear safety.

What: Energy; Environment; Disasters
When: September 30, 1999
Where: A uranium processing plant in Tokaimura, Japan, about 70 miles (112 kilometers) northeast of Tokyo
Who:
HISASHI OUCHI (1964-1999) and
MASATO SHINOHARA (1970-2000), nuclear fuel production plant workers
YUTAKA YOKOKAWA (1945-   ), nuclear fuel production plant supervisor

Disaster Strikes

On September 30, 2000, three workers at the JCO fuel production plant in Tokaimura—Hisashi Ouchi, Masato Shinohara, and Yutaka Yokokawa—were preparing uranium oxide for use as fuel in an experimental nuclear reactor when disaster struck. Under Yokokawa's supervision, Ouchi and Shinohara made the mistake of adding too much medium-enriched uranium to a nitric acid solution. The result was a nuclear chain reaction and the emission of dangerous levels of neutrons and gamma rays. At about 10:35 A.M.—almost as soon as the chain reaction began—Ouchi and Shinohara became seriously ill, both displaying signs of severe radiation poisoning. Within seven months, the two men would die of multiple organ failure, making them Japan's first nuclear accident fatalities. Yokokawa would survive only because he was in another room—and hence was not as severely affected by radiation—when the reaction began.

The danger, however, did not remain within the confines of the plant. Indeed, the reaction continued until 6:30 A.M. on October 1, spilling radiation into the surrounding area for about twenty hours. At least 439 people—including nearby residents, the paramedics who first arrived on the accident scene, those who were working close to the plant, and workers who engaged in the cleanup—were exposed to radiation. The ultimate effects and true scale of this exposure are not yet known. The following November, a study of urine samples by St. Marianna School of Medicine suggested that at least eight people residing near the plant may have undergone deoxyribonucleic acid (DNA) damage because of the radiation.

In April, 2000, Murdoch Baxter, editor of the *Journal of Environmental Radioactivity*, concluded that at least fifty workers within 100 meters (328 feet) of the facility were exposed to 100 or more millisieverts of radiation—one hundred times more exposure than is recommended by the International Commission on Radiological Protection. Dudley Goodhead, director of the Radiation and Genome Stability Unit at Harwell in Oxfordshire, Great Britain, noted that "with good fortune," there will be no local rise in cancer as a result of the accident but added that the effects of high-energy neutrons on the human body are not well understood.

Questions of Safety and Negligence

The Tokaimura accident raised not only scientific and medical questions but also questions of safety and of illegal actions on the part of JCO, the company that owned the plant. Indeed, JCO's situation in the market had become so bleak—sales had dropped 47 percent during the 1990's as a result of its failure to successfully compete with foreign suppliers of nuclear fuel—that the company started emphasizing production over safety. JCO went so far as to issue its employees a secret manual that outlined such truncated procedures as mixing uranium oxide in stainless-steel buckets instead of using the precision dissolving container designed for that purpose. This was said to cut the time needed to dissolve the materials in half.

The plant's director, Kenzo Koshijima, placed the fault for the accident on Ouchi, Shinohara, and Yokokawa, claiming that they had ignored safety procedures and gotten too "creative." His estimate of the situation, however, was almost immediately called into question by the discovery of the secret manual. It seemed that the three workers had simply been following orders when they mixed dangerous radioactive materials in buckets. This is not to say that the workers did not contribute to the accident. Among other gaps and lapses, it was later revealed that Yokokawa, the immediate supervisor of the failed mixing procedure, did not even know the meaning of "criticality," the point at which a nuclear chain reaction can occur.

The actions of the Science and Technology Agency were also called into question. Though JCO had been using unsafe and illegal procedures since at least 1996, agency inspectors had apparently never discovered (at least, had not reported) any serious violations by the company. There had been inspections, but these were always announced well in advance and never occurred when the plant was operating. The Science and Technology Agency was criticized by Japan's Nuclear Safety Commission for its failure to uncover the negligent policies at the JCO plant that led to the Tokaimura accident, including the existence of a secret manual. The commission, Japan's primary policy-making body on nuclear safety, issued this criticism although it was the organization ultimately responsible for overseeing the Science and Technology Agency.

Consequences

By mid-December, 1999, in response to the accident and the questions left in its wake, the Japanese government had passed two new laws: one requiring periodic inspections at fuel production facilities, the other delegating decisions in emergency-response situations to the general government. However, even these laws might not prevent future disasters. For example, it was the mayor of Tokaimura who evacuated the accident area; government officials thought evacuation unnecessary. Under the emergency-response bill, the evacuation probably would not have occurred and many more people might have been exposed to dangerous levels of radiation.

Though processing plants like the one in Tokaimura are not common in Asia outside of Japan, there are many such plants in the West, including seven similar facilities in the United States. The fact that such an accident happened at a well-designed Japanese plant has been seen

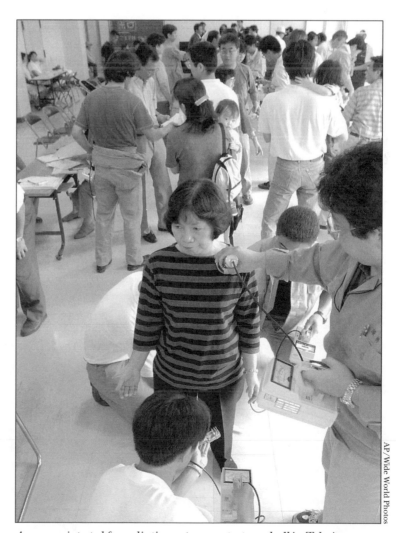

A woman is tested for radiation exposure at a town hall in Tokaimura.

AP/Wide World Photos

by many as an indication that no country, no matter how state of the art its nuclear facilities are, is immune from the dangers inherent in producing and using nuclear energy. Indeed, the international response to the Japanese disaster, seemingly in recognition of this fact, focused on safety at domestic nuclear facilities. In the United States, for example, President Bill Clinton ordered the Pentagon and the Energy Department to evaluate the vulnerability of domestic nuclear fuel production. However, not all experts per-

ceived such actions as resolving the looming question of nuclear safety. David Lochbaum of the Union of Concerned Scientists said, "The technology in Japan is very similar to ours. The controls are also similar, and those controls can break down, and there can be accidents. I think our technology can be equally vulnerable." In the wake of the Tokaimura accident, he and many others were left with some residual doubts as to the safety of nuclear energy.

Jeremiah R. Taylor

Clarkson Becomes Governor-General of Canada

Noted journalist and reporter Adrienne Clarkson, the daughter of Chinese refugees, was the first member of an ethnic minority appointed the queen's representative in Canada.

What: Government
When: October 7, 1999
Where: Ottawa, Ontario
Who:
ADRIENNE CLARKSON (1939-   ),
   broadcast journalist
JOHN RALSTON SAUL (1947-   ), author
   and essayist, Clarkson's husband
JEAN JOSEPH-JACQUES CHRÉTIEN
   (1934-   ), prime minister of
   Canada from 1993

A Controversial Appointment

On September 8, 1999, after Romeo LeBlanc, governor-general of Canada, had announced his resignation because of ill health, Prime Minister Jean Joseph-Jacques Chrétien announced in the foyer of the House of Commons that the queen had accepted the appointment of Adrienne Clarkson as the country's twenty-sixth governor-general. She was to be the second woman, first Asian-born person, first refugee, and first of a visible minority to become the British monarch's representative in Canada. She was also the first vice royal who was neither French nor English, though she was fluent in both official languages.

The prime minister pointed out that Clarkson was admired and respected across Canada for her strong intellect, personal integrity, warmth, and humanity. "Her appointment," he said, "is a reflection of the diversity and inclusiveness of our society, an indication of how our country has matured over the years." Chrétien also revealed that he had been heavily lobbied by female Liberals to appoint another woman to the office once held by Jeanne Sauve.

The appointment was controversial to those Canadians who saw Clarkson as a liberal leftist who had been a spokesperson for cultural and economic nationalism, even after she had ended her illustrious career as a broadcast journalist, host, director, and producer of numerous award-winning Canadian Broadcasting Corporation television shows. What compounded the controversy was the fact that her husband, the award-winning and best-selling writer John Ralston Saul, was also regarded as an even greater radical. The couple were noted for their opposition to the Free Trade Agreement between Canada and the United States, and Clarkson had earlier been part of a delegation investigating human rights infringements in Chile. Saul was a noted critic of corporatism, conformism, and self-serving power elites.

In addition, Clarkson is of Chinese extraction; some racial purists saw her as being a member of an "inferior race" and, therefore, unworthy of appointment to the royal office. In contrast to this view, some Canadians believed her and her husband to be "pampered snobs," members of the upper class who own a medieval stone house in France in addition to an expensive Toronto home and, therefore, unworthy of the honor of the office of governor-general.

The Installation

For Clarkson's installation as governor-general, the viceroy inspected a guard of honor in front of Peace Tower at flag-festooned Parliament Hill in downtown Ottawa. The surrounding streets were lined by two hundred officer cadets of the Royal Military College, Kingston, and the Canadian Forces Band provided the music. Attended by Heritages minister Sheila Copps, Clarkson and

3087

AP/Wide World Photos

Governor-General Adrienne Clarkson.

her husband proceeded to the swearing-in ceremony in the red carpet senate chamber. The eight hundred invited dignitaries included officials of Ottawa (the lieutenant governors, supreme court justices, the diplomatic corps, cabinet ministers, privy councillors, senators, members of parliament) and specially invited guests (Clarkson's ninety-two-year-old father, William Poy; her elder brother Neville, a retired plastic surgeon; and her two once-estranged daughters by her first husband, Stephen Clarkson).

After a fanfare, the secretary to the governor-general read the queen's commission, the legal document appointing Clarkson. Justice L'Heureux Dube then administered the oath of allegiance, the oath of office as governor-general and commander in chief, and the oath as keeper of the great seal of Canada. After assenting orally to each oath, Clarkson signed her name to each of the solemn promises. The vice regal salute was played inside the chamber and a twenty-one-gun salute was fired by thirty members of the field

regiment of the Royal Canadian artillery. Taking her seat on the red-cushioned, carved-wood throne, and flanked by her chief of defense staff, her commissioner of the Royal Canadian Mounted Police, and her five aides-de-camp in waiting, the viceroy listened to a prayer offered by a close prelate-friend and then to music by Inuit, French, and English performers of her own choosing. She then received the collars of the Order of Canada and the Order of Military Merit.

In her inaugural address, which revealed an acute consciousness of history, Clarkson stressed that she would encourage English and French Canadians to know one another better. She said, "As John Ralston Saul has written, the central quality of the Canadian state is its complexity. It is a strength and not a weakness that we are a permanently incomplete experiment built on a triangular foundation—Aboriginal, Francophone, and Anglophone." She promised to crisscross the country by airplane, train, kayak, car, and canoe, and she announced regular public levees, beginning in Calgary on October 16 and in Vancouver on November 21. Clarkson then signed the government of Canada Golden Book and a proclamation before being taken with her husband by horse-drawn carriage to Rideau Hall, their new residence, where a gala reception was to follow in the evening.

Consequences

Having played radical roles in the country as cultural critics, Clarkson and Saul received divided responses from the Canadian public. Despite her many distinctions in journalism, the arts, and public service (including the position of first agent-general for Ontario in Paris to promote the province's business and cultural interests), Clarkson's candor and political convictions generated some hostility. However, the governor general quickly asserted that the role of the

Crown can well involve speaking out on matters of national concern. "Standing apart from the everyday political fray does not mean not having ideas," she said. She stated her intent to provoke debate on controversial subjects much as the prince of Wales had stirred discussion on modern architectural trends in Great Britain.

Acknowledging that the couple's dynamism and philosophy would invigorate Rideau Hall, the media cautioned that Clarkson would do well to win the hearts of the public before attempting to influence their minds. Her critics wonder whether she and her husband would listen rather than lecture to the diverse public. However, as she expressed in her installation speech, Clarkson and her spouse were deeply committed to the harmonious relationship of francophones and anglophones, the sheltering of human rights, and the promotion of the arts. She saw her role as a journey that must be undertaken with an intelligence that includes knowledge of one's responsibility toward society. In this way, she and her husband were able to reconcile their socialistic ideology with Canadian democracy.

Keith Garebian

World's Six Billionth Inhabitant Is Born

> *According to United Nations data, the world's six billionth person was born on October 12, 1999. Some people expressed alarm over this landmark event, believing that Earth was becoming dangerously overpopulated; others saw strength in growing numbers.*

What: Environment
When: October 12, 1999
Where: Sarajevo, Bosnia-Herzegovina
Who:
ADNAN MEVIC (1999-   ), son of
    Jasminko and Fatima Mevic

Booming Population Growth

On October 12, 1999, Earth's population reached six billion according to United Nations figures. The individual named as the six billionth person, in a largely symbolic gesture by the United Nations Population Fund, was an unnamed baby boy born to Jasminko and Fatima Mevic, Bosnian Muslims, in Sarajevo, Bosnia-Herzegovina, at 12:01 A.M., on October 12. By U.N. figures, the world population had doubled in only forty years, and the world population jumped from five billion to six billion people in only twelve years. Experts estimated that if the rate of population growth slowed slightly, the seven billionth person would be born sometime in 2012. They also calculated that if the current rate of growth continued, the population would reach twelve billion by 2050. Many people believe that rapid population growth is one of the major problems facing humankind. The population explosion of the twentieth century was made possible by better hygiene and medical care and huge increases in food production.

At the end of the twentieth century, the population was growing fastest in the world's poorest countries. Nearly nine of every ten babies were born in Asia, Africa, and Latin America. More than half of all births occurred in six countries—India, China, Pakistan, Nigeria, Bangladesh, and Indonesia—with India and China alone accounting for 40 percent of this gain. Not all countries were growing. For example, in 2000, sixteen European countries were experiencing population declines. Europe was the first continent in the modern era to achieve zero population growth. Russia, which spans portions of both Europe and Asia, was experiencing the fastest nonwartime population decline in modern history. Japan also was approaching zero population growth.

Too Many, or Too Few?

At the time of the U.N. announcement, some scientists believed that the "population explosion" was nearly over. They pointed to a recent sharp decline in the rate of population growth. World population was growing at a record high rate of 2 percent a year during most of the 1970's. However, by the end of the twentieth century, the annual growth rate was about 1.3 percent. Although the growth rate was declining, it was applied to a steadily increasing base population. Therefore, this 1.3 percent increase translated to an annual population gain of about 85 million, about 10 million people more than were being added four decades earlier.

Some experts believed that there was not enough food, water, space, energy, or other resources to provide adequately for six billion people. In contrast, others thought that humans are the most valuable resource. They viewed human ingenuity as the most important key to achieving well-being for a growing human population. In supporting their belief, they pointed to such things as increased agricultural production, longer life expectancy, and a higher standard of living for most of the world's people, all improvements that have accompanied the twentieth century's population explosion.

Both points of view have some merit. However, both contain important flaws. For example, many of those worried about overpopulation use the idea of carrying capacity to illustrate their concern. They believe that each of Earth's natural environments—desert, rain forest, humid plains, forested mountains, and fertile plains—has limits to its ability to carry population growth. Therefore, the more people, the fewer resources will be available to each individual.

In a global and increasingly highly technical society, numbers of people alone mean very little. If each of the world's six billion people occupied an area of several square feet, humans would occupy a space no greater than an average Midwestern county. People alone are not the problem. Rather, a people's culture, or way of life, is what determines their ability to provide for themselves. Do they have the technology and capital to make their land and resources productive? Are they able to purchase needed resources and raw materials from global markets? Are their land and other resources used wisely? Are the country's human resources well developed and constructively used? Is each individual able to achieve his or her maximum potential? Do they have good government that supports both its citizens and a strong and stable economy?

These are just some of the factors that determine whether a country is overpopulated. Some countries have very dense populations (people per square mile). Yet in nations such as Germany, the Netherlands, France, England, Japan, and Singapore—despite being some of the most crowded places on Earth—citizens enjoy a high standard of living. On the other hand, some countries with very low populations and population densities suffer extreme poverty. By any measure, they have too many people for the land, resources, and national economy to support.

Consequences

By noting that the world population had reached six billion, the United Nations alerted people to some significant facts. Experts believed that if the trends evident in 1999 did not change,

the human population would continue to grow, although at a slower pace than during past decades. Some areas would continue grow at rates higher than the world average. It also highlighted the fact that when a country's population grows faster than its economy, that country almost certainly will remain poor. It will be unable to gain the financial resources to purchase much-needed food, energy, and other materials on world markets, and its people will have a meager standard of living.

In some developed countries, experts viewed declines in population growth as cause for alarm. A drop in births has caused these countries to have rapidly aging populations, with fewer young people to fill less-skilled and lower-paying jobs. Many of these countries have had to encourage immigration from other lands to fill the void in the labor force. The resulting cultural, ethnic, and social changes sometimes created uncertainty and conflict within these societies.

Throughout history, humans have shown a remarkable ability to adjust and adapt when confronted with challenges. The human population may always have represented a condition of too many people for the economy and technology of the time. Evidence for this belief is found in the very slow rate of population increase throughout most of history. By end of the twentieth century, enough space, food, water, energy, and other resources and raw materials existed to provide all Earth's six billion people with an adequate standard of living. The problem was not the number of people but distribution of resources. Future human well-being may be determined by the ability of people and nations to obtain those things needed to make their lives comfortable and secure. This requires that governments, economic systems, and people work together in order to function well in the global economy, regardless of whether their populations are increasing or declining.

Charles F. Gritzner

Pakistani Government Is Ousted by Coup

The democratically elected Pakistani government of Prime Minister Nawaz Sharif was ousted by a military-led coup, which brought to power Pervez Musharraf, who declared a state of emergency and suspended the constitution and the national assembly.

What: Government; Coups; Military; Coups

When: October 12, 1999

Where: Islamabad, Pakistan

Who:

PERVEZ MUSHARRAF (1943-   ), general and chief of the army staff

NAWAZ SHARIF (1948-   ), prime minister of Pakistan from 1997 to 1999

KHWAJA ZIAUDDIN, lieutenant general and head of Pakistan's Inter-Services Intelligence

MAHMOOD AHMED, lieutenant general and commander of Pakistani army's Tenth Corps

Fragile Democracy

Since Pakistan's establishment in 1947 through the partition of India into two countries, democracy has never gained a solid foothold in the country. By 1999, Pakistan had spent twenty-five years of its fifty-one years of existence under military rule. Democracy has faltered in Pakistan, partly because of severe social and economic divisions in Pakistani society, with people from Punjab dominating the other parts of the country, and partly because of the Pakistani people's dissatisfaction with democratically elected leaders.

Another reason for the reoccurrence of military rule is Pakistan's troubled relationship with its neighbor, India. Since its establishment, Pakistan has fought three fully declared wars in 1947, 1965, and 1971 and one undeclared war in 1999 with India. India-Pakistan relations were further strained during the Cold War, when Pakistan sided with the United States and India with the Soviet Union.

When Nawaz Sharif became prime minister in 1997 after Prime Minister Benazir Bhutto was dismissed in 1996 on charges of corruption, the Pakistani people were hopeful that Sharif would improve their standard of living. However, Sharif's administration soon ran into difficulties. Sharif was faced with a corruption inquiry after trying to enact legislation curbing the president's power to appoint military leaders to replace those elected by the people of Pakistan.

The inquiry ended when President Farooq Leghari, a Bhutto ally, resigned in December, 1977, and Mohammad Rafiq Tarar became president in 1998. When India conducted nuclear weapons tests in May, 1998, Sharif directed tests of nuclear weapons in Pakistan within the same month. The nuclear weapons tests incurred criticism from many nations, which placed economic sanctions on both nations. These punitive measures made life more difficult for ordinary people in Pakistan and set in motion the events that eventually resulted in the coup against Sharif's government. Sharif's fate was probably sealed by his opposition to the Pakistani army-sponsored invasion of the Indian-controlled portion of Kargil, in disputed Kashmir, in 1999. Under pressure from the international community, Sharif forced the army to call off its support for the invaders. Many powerful officers in the army felt that Sharif had let them down.

An Airborne Coup

When Pervez Musharraf, chief of the army staff, went to Sri Lanka on an official trip, Sharif decided to take action against him in his absence in an effort to lessen the resentment that was building against Sharif in the army. While Musharraf was on board a Pakistan International Airline flight from Sri Lanka to Karachi, Sharif fired him and appointed Lieutenant General

Khwaja Ziauddin, the head of Pakistan's Inter-Service Intelligence (ISI), as the new chief of the army staff.

Sharif had anticipated that the army would accept the reshuffle. However, army units loyal to Musharraf, who directed them through telephone from the airplane, soon sprung into action. Two supporters of Musharraf, Lieutenant General Mohammed Aziz and Lieutenant General Mahmood Ahmed, who commanded the army's Tenth Corps close to Islamabad, were crucial to the general's success. At their order, the army took over television stations, airports, and other vital institutions. The army also arrested Sharif, his brother Shahbaz, and Ziauddin.

When he took power, Musharraf declared a state of emergency; suspended the constitution, the national assembly, and the Pakistani congress; and appointed himself chief executive. By a special decree, he ensured that his actions could not be challenged by any court of law. By not dismissing President Tarar or the assembly, Musharraf retained the option of ushering in civilian rule at a time of his own choosing.

The people of Pakistan seemed genuinely relieved that the government of Sharif had fallen and that the military had taken control—an indication of how unpopular the Sharif government had become. In the eastern city of Lahore, Sharif's hometown and center of support, people cheered the new military ruler and burnt Sharif's pictures. Sharif was eventually sentenced to life imprisonment on charges of hijacking and terrorism. However, after the Saudis put pressure on Musharraf, Sharif was sent into exile in Saudi Arabia.

Consequences

When the army seized control in Pakistan, most experts worried that the region would be-

Pakistani troops guard Prime Minister Nawaz Sharif, who is inside the Pakistan Television headquarters in Islamabad.

3093

come unstable because they feared a resurgence of strife between Pakistan and India. Musharraf was the head of the Pakistani army when Pakistan invaded India in 1999, and his position on Kashmir was known to be in direct opposition to that of India. The anxiety of these experts was heightened by the knowledge that both Pakistan and India possess nuclear weapons.

The fear that the situation could deteriorate in a short time period created a desire in most of the international community to maintain a dialogue with Pakistan. To this end, U.S. president Bill Clinton visited Pakistan in his trip to South Asia in 2000. With the suspension of Pakistan from the Commonwealth, the group of fifty-four countries connected to Britain by their colonial history, and the discontinuation of the South Asian Association for Regional Cooperation, few

venues remained in which India and Pakistan could discuss bilateral issues.

As expected, under the leadership of Musharraf, Pakistan's hard-line stance against India became more rigid. Musharraf appointed Abdus Sattar, an anti-India hawk, as the new foreign minister. On the economic front, however, there was some good news. Corruption was significantly down, and ordinary Pakistanis were much more satisfied with their economic opportunities. Many countries feared that a poor economic situation in Pakistan might lead it to sell its nuclear expertise in exchange for foreign currency. To avoid such a possibility, many members of the international community planned to maintain relations with Pakistan.

Amandeep Sandhu

Philip Morris Company Admits Smoking Is Harmful

> *For the first time, a major U.S. cigarette manufacturer publicly acknowledged the negative medical effects of its product.*

What: Business; Health
When: October 13, 1999
Where: Miami, Florida
Who:
ROBERT KAYE, Miami-Dade Circuit Court judge
MICHAEL SZYMANCZYK (1949-   ), president of Philip Morris

Lighting Up

October 13, 1999, marked a major break in the history of the tobacco industry. After years of insisting that no study had ever conclusively proved any link between smoking and health problems, tobacco executives made a clear and unambiguous statement that smoking tobacco does indeed negatively affect people's health.

For the first half of the twentieth century, smoking was a common and important adult social activity, even a badge of adulthood. Many adults, including athletes, smoked, and smoking was considered to have important benefits, including calming the nerves and aiding concentration. During World War II, cigarette companies gave free packages of cigarettes to American servicepeople overseas, and their donations were regarded as noble and patriotic.

In the 1950's, attitudes toward smoking began to shift with the emergence of the first research to link smoking with lung cancer. In 1964, the U.S. Surgeon General's Report formally linked smoking with lung cancer and other health problems. Not long afterward, the federal government began tightening down on tobacco companies' ability to advertise their products. In 1968, the Fairness Doctrine led to numerous television public service announcements on the dangers of smoking, a counterweight against the cigarette advertisements that urged people to smoke. Three years later, in 1971, the Federal Communications Commission (FCC) banned cigarette advertising from television and radio.

In the 1980's, public opinion grew increasingly negative toward smoking, particularly as the dangers of environmental tobacco smoke became known. No longer was smoking considered to be a purely private matter. Smoking in public came to be seen as affecting everyone else in the area and was treated accordingly. A steadily growing range of businesses, government offices, and other public places began to restrict or ban smoking within their confines. Smokers had to go outside to indulge their habit.

Stubbing It Out

As public opinion shifted, tobacco companies remained steadfast, insisting no link had been proven between their product and such health concerns as lung cancer, high blood pressure, and heart attacks. They asserted that smoking was a private decision, not a public health issue, and they could not be held accountable for people's illnesses when there was no way to be certain that those problems were not caused by other factors. Therefore, there was no basis on which to bring a product liability lawsuit against them.

This changed when states began to press lawsuits to recover the cost of Medicare and Medicaid claims of smokers from the cigarette manufacturers. One of the most important cases occurred in 1999 in Miami, Florida. In the Miami-Dade Circuit Court, Judge Robert Kaye heard a lawsuit on the behalf of more than 500,000 Floridians whose illnesses, the state claimed, had been caused by smoking. Many of the plaintiffs attended, accompanied by tanks of oxygen to help

3095

lungs ravaged by cancer and emphysema or equipped with electronic devices to replace cancer-destroyed larynxes. Others brought photographs of family members killed by illnesses attributed to smoking.

At first the cigarette companies held to their usual defenses. No proven link existed between smoking and these illnesses, only inferences based on circumstantial evidence. Smoking was an individual decision, and no one was ever forced to smoke a cigarette.

However, on October 13, Michael Szymanczyk, president of Philip Morris, the world's largest tobacco manufacturer, came to the witness stand. Testifying under oath, he made a statement that departed from decades of tobacco industry defenses: Yes, smoking was indeed bad for one's health, and it caused cancer and other life-threatening illnesses.

Consequences

With this admission, the tobacco companies could no longer claim that their products did not cause illnesses when used as intended. When it was time for the jury to decide, Judge Kaye told them that they need not consider themselves limited by the net worth of the cigarette companies involved in determining a sum for punitive damages. The jury in the Miami case awarded $145 billion in damages to the ill Florida smokers, one of the largest judgments in history.

With the precedent established by this case, other courts found it easier to decide in favor of people who were suing tobacco companies to re-cover the costs of illnesses caused by smoking. Another major suit involved nonsmoking flight attendants who were attempting to recover damages caused by having to inhale other people's smoke in the cabins of the airplanes on which they served. Even the federal government launched a major investigation of possible corrupt business practices among the tobacco companies.

At the same time, it became known that a number of senior Philip Morris executives had sold off large number of shares of stock in their company in August of 1999, shortly before Szymanczyk's testimony. This raised the question of whether they knew in advance that the admission was going to be made and whether based on this knowledge, they decided to act in order to protect their own wealth, both from any judgment and from the probable effects of the admission on the value of their company.

Not everyone praised the lawsuit that brought forth the admission of harm from Philip Morris. Some political commentators, including those at Overlawyered.com have warned that the long-term negative effects of these rulings could actually outweigh the health benefits. According to these experts, the rulings set a dangerous precedent of "legislation by litigation," using lawsuits to create results that would be better brought about by statutes. If objectionable organizations such as tobacco companies could be bankrupted and destroyed through lawsuits, they argued, no company or individual would be safe. Nuisance suits could ruin anyone financially.

Leigh Husband Kimmel

Enzyme Is Linked to Alzheimer's Disease

Researchers announced that they had identified an enzyme that had long been suspected of being a key factor in the development of Alzheimer's disease. The enzyme is associated with the production of protein plaques in the brain, which may cause the disease.

What: Health; Medicine; Genetics
When: October 21, 1999
Where: Amgen, Thousand Oaks, California
Who:
MARTIN CITRON and
ROBERT VASSAR, leaders of the research team at Amgen

The Long Search

In the October 21, 1999, issue of *Science*, a group of researchers from Amgen, a biotechnology company in Thousand Oaks, California, reported that they had successfully identified an enzyme called beta secretase. Many scientists have believed for years that this enzyme facilitates the onset of Alzheimer's disease. Its identification may lead to the development of drugs that will slow the progression of the disease or possibly even reverse its effects.

Alzheimer's disease is a progressive dementia that afflicts about four million Americans. It causes memory loss that worsens until the effects are extreme, and eventually, it results in death. One of the structural changes that occur in the brains of Alzheimer's sufferers is the accumulation of amyloid plaques—protein deposits on brain cells—that some scientists believe cause the cells to die, resulting in memory loss. The brains of people who do not have Alzheimer's contain almost none of this plaque. The protein that makes up these plaques is created when a larger protein, called the amyloid precursor protein (APP), is cut into smaller sections by two enzymes named beta secretase and gamma secretase. For five years until October, 1999, numerous research teams had unsuccessfully attempted to discover the sources of beta secretase and gamma secretase. There were claims that beta secretase had been identified, but they were not supported conclusively by further study.

The Amgen Research

The Amgen study found that the specific sequence of the cleaving process that segments the APP involves beta secretase making the first cut and gamma secretase the second. One of the resulting APP segments is the protein that becomes amyloid, the substance that forms the plaques. The Amgen researchers thought that because beta secretase makes the first cut on the APP, stopping production of this enzyme might inhibit amyloid production in the brain. If amyloid plaques do, indeed, cause Alzheimer's disease, then halting production of this enzyme might prevent the disease or slow its progression. Therefore, these researchers mounted a two-year search for the gene that instructs cells to produce beta secretase.

The Amgen team conducted this research by drawing from the catalog of human genes and adding one hundred genes at a time to cells until they observed increased amyloid production. Then they repeated the process multiple times, using subsets of the genes that they had added. Through a long process of elimination, they created a smaller and smaller gene pool until they finally identified the single gene that produces the beta secretase enzyme. They renamed the enzyme BACE, which stands for beta-site APP-cleaving enzyme. These scientists were convinced that BACE is beta secretase, because, as its name implies, when synthesized BACE was introduced to cells, it cut the APP at the same positions as beta secretase would, possessed all the same known characteristics, and increased amyloid production. Furthermore, reduced BACE levels caused decreased amyloid production. Amgen concluded that the next logical step was the development of BACE inhibitors, which might prove to be a major breakthrough in Alzheimer's research.

There is a strong consensus within the scientific community that BACE is the beta secretase enzyme, not only because the evidence from the Amgen study is convincing, but also because at least two other research companies have independently reported the same findings. Elan Pharmaceuticals in South San Francisco, California, and SmithKline Beecham Pharmaceuticals in King of Prussia, Pennsylvania, announced the identification of beta secretase at a professional neuroscience meeting shortly after Amgen published the results of its study.

Consequences

Amgen's identification of the beta secretase enzyme is an important breakthrough in Alzheimer's disease research. Pharmaceutical companies were already working on the development of beta secretase inhibitors before the enzyme was identified, but the ability to use BACE in these efforts vastly improved the chance of finding an effective inhibitor. Amgen spokespeople said it would be at least seven years before such a drug would be commercially available. If these inhibitors are developed, scientists can determine whether their application will stop the formation of amyloid plaques in the brain and if inhibiting plaque production slows or stops the progression of Alzheimer's disease or prevents its development in patients without the condition.

In the event that BACE is not found to be the cause of amyloid plaque formation, its identification still will have had enormous impact on Alzheimer's disease research efforts. Among scientists who believe that amyloid plaques cause the disease, there are differing professional opinions regarding what engenders plaque production. Amgen's discovery has stimulated the research efforts of scientists who favor alternative theories of plaque formation and inhibition. For example, not long after BACE was identified, several research teams announced that they had identified gamma secretase, the enzyme that makes the second cut on the APP, which results in amyloid formation. These researchers were convinced that the best way to prevent plaque formation is to develop gamma secretase inhibitors. Other plaque-related research focused on the development of a vaccine to eliminate plaque deposits.

If the cause of and cure for Alzheimer's disease are eventually found, Amgen's identification of BACE will be viewed as an important event, whether it directly led to those discoveries or it eliminated one probable cause, thereby allowing research efforts to focus on alternatives.

Jack Carter

Federal Court Rules Microsoft a Monopoly

In the first court battle between the federal government and Microsoft, a district court judge ruled that Microsoft was a monopoly—a decision beginning a new era of government intervention in the high-tech computer industry.

What: Business; Computer science
When: November 5, 1999
Where: Federal District Court for the District of Columbia, Washington, D.C.
Who:
DAVID BOIES (1941-), lead attorney for the Justice Department
BILL GATES (1955-), chief executive officer of Microsoft
THOMAS PENFIELD JACKSON (1937-), federal district court judge for the District of Columbia

From Obscurity to Dominance

In 1980, International Business Machines Corporation (IBM) entered into an agreement with Microsoft, a small software company, to produce an operating system for the new IBM personal computer. Over the next twelve years, this small company moved from obscurity to being the largest software company in the world. In 1992, Microsoft dissolved its partnership with IBM by introducing Windows 3.1. The success of Windows 3.1, Microsoft Word, and other software releases such as Windows 95 gave Microsoft a dominant position in some software markets. This caused some people to wonder if the company had become too large and if its sheer size might stifle competition.

Many recognized that a new type of operating system would ultimately replace the desktop operating systems of the 1990's. This new operating system would include integrated access to the World Wide Web as well as support for the convergence of video, audio, and data on desktop computers. Microsoft recognized this trend and initiated a long-range project to develop such an operating system. The new .NET run-time achieved this by adding features to Microsoft's operating systems, allowing access to data, information, and entertainment resources anywhere in the world.

Although it is unlikely that any single company will dominate the various ways of accessing the Internet, it is certainly true that Microsoft appeared to be in a good position to do this in 1995. When Microsoft sought to incorporate Internet Explorer in Windows 95.b, some became concerned that Microsoft might be able to use its dominant position in desktop operating systems to become equally dominant as an Internet interface. Microsoft won the right to include Internet Explorer in Windows 98 after the District of Columbia Court of Appeals overturned an earlier decision of Judge Thomas Penfield Jackson.

The Lawsuit and the Appeals Process

When Microsoft first introduced Windows 95, its Internet Explorer browser was packaged on a separate compact disc (CD) from the base operating system. By the release of Windows 98, Microsoft not only included the Internet Explorer browser on the same CD as the operating system but also had moved most of the key components of Internet Explorer into the operating system itself. Although it produced a better operating system, this created a real problem for browser companies. Their browsers would either need to use the same basic components as Internet Explorer or would have to load many extra software components.

Netscape developed the first major browser for the World Wide Web and was the dominant browser for UNIX, Macintosh, and Windows 3.1 computers before Internet Explorer. Netscape complained to the Justice Department that Microsoft's decision to incorporate most of Internet Explorer into its operating system was

3099

AP/Wide World Photos

Joel Klein (center), assistant attorney general for the Antitrust Division, comments on the Microsoft ruling, accompanied by Attorney General Janet Reno (left).

done just as much to put Netscape out of business as to improve the quality of the Windows operating system. The Justice Department had been concerned about Microsoft's dominance in the desktop operating system area for some time and decided to file an antitrust suit against Microsoft based on Netscape's complaint. Nineteen state attorneys general joined this suit as well.

The Justice Department filed its case in the U.S. District Court for the District of Columbia on May 18, 1998. Judge Jackson heard the case without a jury. David Boies, who participated in an earlier IBM antitrust suit, joined the Justice Department and prosecuted the case. The case reached a head on November 5, 1999, when district court judge Thomas Penfeld Jackson issued his preliminary judgment in the case, ruling that Microsoft was a monopoly that stifled innovation and hurt consumers. Over the next seven

months the company and the government engaged in unproductive negotiations and continued their battle in court.

After a long and contentious trial, Judge Jackson rendered his decision on June 7, 2000. The final decision found that Microsoft did have a monopoly in desktop operating systems. The decision further stated that Microsoft had used its desktop operating system monopoly to damage Netscape by making it difficult for Netscape to be included as a browser for Windows. The remedy that Judge Jackson selected was to break up Microsoft into two companies. One company would be devoted to operating systems and the other to applications. Judge Jackson specifically stated that the Internet Explorer browser was to be part of the applications company.

Microsoft won the first round of appeals when the U.S. Supreme Court refused to fast track the case (hear the case without submitting it to the

3100

District Court of Appeals). On June 28, 2001, the District Court of Appeals upheld the verdict that Microsoft had a monopoly in desktop operating systems. However, the court overturned Judge Jackson's remedy of splitting Microsoft into two companies and required that a new judge from the District of Columbia be appointed to determine the appropriate remedy. The case was then reassigned to Judge Colleen Kollar-Kotelly.

On September 6, 2001, the Justice Department announced that it would not pursue a remedy of breaking Microsoft into two companies but would vigorously pursue a course of seeking to create more competition in desktop operating systems. Although Microsoft benefited from this decision because it could remain a single company, it appeared that the Justice Department planned to center on the key issue of Windows being a monopoly rather than whether Microsoft used anticompetitive practices in bundling Internet Explorer with the Windows 95 operating system. Microsoft had indicated that it planned to file an appeal to the Supreme Court of the United States to overturn the monopoly decision. On November 2, 2001, the Department of Justice and Microsoft announced that they had worked out a settlement, which they presented to the district court for approval. The proposed consent decree outlined a variety of changes that Microsoft promised to make in how it would develop and distribute its software, allowing smaller software companies greater opportunities to develop competing products. At the same time, the agreement allowed Microsoft itself to continue developing new products, including new versions of Windows.

This case had a major effect on the relationship between the high-tech computer companies and the federal government. The new high-tech computer industry was revealed to be subject to antitrust laws. Federal legislation of the computer industry (or portions of it) became a real possibility. As a result of this, the high-tech computer companies became involved in political processes just like most other industries.

The antitrust suit hurt Microsoft. The company was reported to be having trouble recruiting and retaining top-quality computer scientists. In addition, it was forced to direct valuable resources away from product development to fight the antitrust suit, and some of its customers began to look at competitors products, just in case. Thus, regardless of the outcome of the lawsuit, it appeared to have fostered competition in desktop operating systems.

George Martin Whitson III

Australia Votes Against Becoming a Republic

Australians voted to keep Queen Elizabeth II as sovereign mainly because of flaws in the proposed new constitutional model.

What: Politics; Government
When: November 6, 1999
Where: Australia
Who:

JOHN HOWARD (1939-), prime minister of Australia from 1996

KERRY JONES (1956-), leader of Australians for a Constitutional Monarchy

MALCOLM TURNBULL (1954-), leader of the Australian Republican movement

PETER COSTELLO (1957-), pro-republic federal treasurer

Falling Short of the Threshold for Change

Australians usually hold federal elections every three years. However, in November, 1999, a special vote was held to decide who would be Australia's head of state. Officially federated in 1901, Australia became an independent country during the twentieth century. However, like many former British colonies, Australia retained its constitutional connections with the British crown. In 1999, Queen Elizabeth II was, in legal terms, queen of Australia as much as she was of Great Britain. Because the queen did not live in Australia, even her ceremonial role was limited; she was represented by an Australian governor-general and, in any event, had no executive power.

Australia had long been content with this situation, if not enthusiastic about it. However, in 1991, lawyer Malcolm Turnbull helped found the Australian Republican movement. In 1993, Prime Minister Paul Keating announced his intention to make Australia a republic by the end of the decade. Advocates of a republican Australia said that it was time for Australians to have a democratically elected head of state, born in Australia. Australia, they argued, was now a multicultural nation with close links to Asia. Abolishing the monarchy, it was said, would also help the goal of reconciliation between the often maltreated Aborigines and white Australians. Finally, becoming a republic would rid the country of the humiliations of its convict past, when Britain would exile its prisoners to the huge southern continent. Australia, the host of the 2000 Olympics in Sydney, would show the world it was in synchrony with the times.

Advocates of a republic made clear that they were not motivated by personal antagonism to the queen, who was widely respected in Australia. It was thought, though, that the martial problems of the heir to the throne, Prince Charles, did lessen the popularity of the throne as they did in Britain itself. Republicanism received an apparent setback in 1996 when Keating was defeated by the monarchist John Howard, but Howard permitted a constitutional convention in 1998.

At the convention, a sharp divide emerged between Australians who wanted a directly elected president with considerable power and those who wanted the new presidency to be only a modified version of the governor-general position—appointed and with only ceremonial power. In February, 1998, a compromise, offered by retired judge Richard MacGarvie, called for the president to be selected by Parliament from a short list chosen by prominent Australians. Many politicians accepted the compromise because they did not want a new president to rival the role of the prime minister. However, many who were hoping for a popularly elected president were disappointed.

Howard announced on May 12, 1999, that a referendum on the republic would be held on November 6. Voters could choose either to approve or disapprove the proposed republic;

there was no third option. Howard also proposed a preamble to the constitution addressing the Aboriginal issue, which many did not feel went sufficiently far. Though the Labor Party was nearly united in support of the republic, Howard's Liberal Party was divided, with Peter Costello, the treasurer and Howard's likely successor, strongly in favor. However, the campaigns for and against the referendum were led by nonpoliticians: Turnbull and Kerry Jones, head of Australians for a Constitutional Monarchy, respectively.

For several years, a republic had seemed inevitable, but later voter surveys had shown skepticism. As results trickled in from east to west, it became clear that voters had rejected the republic. In traditionally conservative states such as Southern Australia, Western Australia, and Queensland, the rejection was by a wide margin. In more urbanized states such as New South Wales, the race was much closer. The referendum barely lost in Victoria, home to many Irish Australians with a hereditary resentment of any British connection.

Howard Versus "the Ascendancy"

Howard, flush from his perceived success in the East Timor peacekeeping mission, saw the result as a personal triumph. However, his proposed preamble to the constitution failed by an even wider margin than the republic proposal. Many members of his own Liberal Party, including his potential successor, Costello, had been in favor of the republic. For the opposition Labor Party, almost entirely committed to the republican cause, the results were more bitter. Many had supposed the republic proposal would command broad support across the Australian voting public. How-

ever, the proposal was often identified with what the Australian poet Les Murray, himself an early supporter of republican dreams, termed "the Ascendancy"—the group of wealthy, highly educated professionals, often living on the urban east coast of Australia, who, though hugely powerful in cultural terms, did not make up a majority of the voters. Rural voters and some working-class voters were far less in favor of Australia's becoming a republic. In traditional Labor constituencies such as the western suburbs of Sydney, voter sentiment was unexpectedly antirepublican.

Why had Australia kept the queen in the face of the expected inevitability of the republic? Many Australians resented the fact that the president would not be popularly elected but chosen by Parliament. The image of a "politicians' republic" robbed the movement of its populist appeal. Though the predicted divisions held up—younger voters and non-British migrants prone to vote in favor of the referendum, older Australians and those of British descendent likelier to vote against it—there were sufficient exceptions to boost the opposition tally.

Consequences

Though the referendum failed, it was still generally thought that most Australians wanted a republic—just not the particular model that was on the ballot. The issue would not be on the front burner in the immediate future. The republicans would have to change their agenda and bring a wider cross-section of the public into the movement. However, few supposed that the issue of the Australian republic was permanently settled on November 6.

Nicholas Birns

Hereditary Peers Are Removed from British House of Lords

Fulfilling its promise in its 1997 election manifesto, the Labour Party secured the expulsion of 655 heredity peers from the House of Lords and created the Wakeham Commission to propose a second set of reforms to make the Lords more representative.

What: Political reform; Government
When: November 12, 1999
Where: Great Britain
Who:
TONY BLAIR (1953-   ), leader of the British Labour Party
BARONESS JAY OF PADDINGTON (1939-   ), leader of the House of Lords

The Tradition of the House of Lords

For 734 years, possessors of high clerical offices and great landed estates with hereditary titles were guaranteed a major role in government as members of the House of Lords, the upper house of Parliament, Great Britain's legislature. Throughout most of its history, the House of Lords provided the majority of cabinet ministers and high government officials. In the nineteenth century, electoral reform moved the nation from a constitutional monarchy with a republican structure to a modern democracy with a ceremonial sovereign. The powers of the House of Lords, with its hereditary peerage, came under increasing attack as the electorate expanded, the House of Commons became dominant, and lifetime peers were added to secure the passage of major democratic reforms. In 1911, the House of Lords lost its power to veto legislation and was left with only the power to delay the enactment of legislation for two years. In 1949, this was further limited to a one-year delay as statutes gaining approval in the Commons became law after two readings in the House of Lords.

The composition of the Lords was affected by the Appellate Jurisdiction Acts of 1876 and 1887, which provided for an expansion of life peerage.

In 1907, the first act to end the automatic membership of hereditary peers was introduced. The Life Peerage Act of 1958 allowed 400 to 500 additional male and female life peers, and five years later, female heredity peers were allowed to attend. A major effort to reform the Lords in 1967 and 1968 was frustrated. Of the nearly 1,300 peers in the House of Lords in 1999 when this reform was enacted, 750 were hereditary peers (including 16 women), 26 were clerics, and 505 were life peers (including 80 women).

The issue of further reform of the House of Lords surfaced as Great Britain faced new concerns such as the devolution of powers to constituent "countries" of Britain, its membership in Europe's Common Market, the emergence of a third major political party, and the consequences of post-World War II demographic changes that left the nation without its empire but with large groups of the "peoples of the British Isles" who were neither Europeans nor members of the Church of England. During the last three decades of the twentieth century, Conservative governments dominated Britain and filled the Lords with their supporters. The Labour Party, which until 1992 had advocated the abolition of the House of Lords, softened its stand under party leader Tony Blair and proposed modest reforms such as the failed 1994 bill, which would have let the eldest child of either sex inherit the title.

In its 1997 election manifesto, which preceded its overwhelming victory at the polls, Blair's Labour Party promised to end, by statue, the right of hereditary peers to sit and vote in the House of Lords as a "self-contained reform." It also promised a second set of reforms to ensure that party appointees to life peerages would "more accurately reflect the proportion of votes cast at the previous general election." This two-step process promised to restructure the Lords into a body of life peers that was more "independent" and more representative of the nation, with no single party allowed to attain a dominating majority.

Enacting Reform

In Parliament, Conservative members of the House of Lords obstructed the Labour Party's reform of the welfare bill and blocked the passage of thirty-nine of the party's social programs. In retaliation, a discussion of a bill for the removal of hereditary peers was initiated in the Lords on October 18, 1998, by Baroness Jay of Paddington, the lord privy seal. While some Liberal, Labour, and Conservative members supported reform, the majority of the nearly two hundred peers who addressed the issue opposed it or proposed alternatives. The queen's November speech referred to the removal of the hereditary peerage as "the first stage in a process of reform," and on December 2, 1998, the Labour government and the Conservative Party leadership implemented a deal to exempt ninety-two peers from the reform until the second stage of reform.

The reform bill was introduced in the Commons on January 19, 1999, and its introduction was followed by the publication of a White Paper, "Modernizing Parliament: Reforming the House," and the creation of a royal commission chaired by Lord Wakeham to propose further reforms. The full commission was appointed on February 8, 1999, and the reform bill received its first reading in the Lords on March 17, 1999. After much maneuvering, an amendment to define those to be excluded (forty-two Conservative, twenty-eight Independents, three Liberal Democrats, and two Labour plus two royal appointees and fifteen deputy speakers or committee chairpersons) was approved during the committee stage on May 11, 1999. On June 17,

the government announced the creation of forty new life peers to give Labour a slight majority after the reform. The election for the fifteen officials occurred on October 27-29, and the remaining exempted seventy-five hereditary peers were selected on November 5, 1999.

The bill received the monarch's assent on November 11, 1999, and ended the participation of 655 hereditary peers on the last day of the session.

Consequences

In December, in acknowledgment of their former service, the hereditary peers were granted

Queen Elizabeth II opens a new session of Parliament, minus hundreds of hereditary peers in the House of Lords.

AP/Wide World Photos

access to one gallery of Parliament and given limited use of the Lord's Library and Refreshments Department. In addition, their right to sit on the steps of the throne, when invited, was recognized.

The Wakeham Commission issued its report, *A House for the Future,* on January 20, 2000. It called for a unnamed, new upper house with 550 members, a minority of whom would be elected, with the rest selected by a process to be determined by a new, independent appointments commission, which would ensure that 20 percent of the members are independent and that the rest reflect the results of the previous general election. It provided three radically different models for conducting elections of either 65, 87, or 195 members using proportional representation. Winners would serve for fifteen years and be eligible for a second fifteen-year term. Women would make up 30 percent of the membership, with representation provided for all ethnic minorities. Representation would be provided for all Christian and non-Christian faiths, with the Church of England getting sixteen of the twenty-one places for Christians. Law Lords would be retained. The report was strongly attacked by all opposition parties and criticized by many in the government. As a result of these criticisms, as the Labour government entered the final year of the Parliament that brought the removal of the hereditary peerage, the government had not submitted the Wakeham report to a joint committee of both Houses, as proposed in the report.

Sheldon Hanft

Ukraine Votes Down Return to Communism

> *Incumbent president Leonid D. Kuchma of the Ukraine won a tough victory over his communist opponent in the country's second election after its independence from Soviet rule.*

What: National politics
When: November 14, 1999
Where: Ukraine
Who:

LEONID D. KUCHMA (1938-   ), president of the Ukraine from 1994

PETRO SYMONENKO (1952-   ), leader of the Ukrainian Communist Party

OLEKSANDER MOROZ (1944-   ), leader of the Ukrainian Socialist Party

GEORGIY GONGADZE (1969-2000), editor of *Ukrainska Pravda*

Searching for Democracy

After four hundred years of Russian and Soviet domination, the Ukraine finally became an independent country in 1991 with the breakup of the Soviet Union. Seeking to establish a democracy on Western models, elections were held for parliament and president in 1994. Leonid D. Kuchma, a former member of the Communist Party Central Committee running as an anticommunist democrat and Ukrainian nationalist, became the country's first president elected under the new Constitution for a five-year term.

However, Kuchma found it difficult to deal with the country's economic problems, many of which were the products of the changeover from a socialist to a free-market system and of the political problems of disentangling from Russia. Critics complained of Kuchma's high-handed tactics of silencing opposition and the corruption that wracked his government. Journalists in particular protested Kuchma's terrorist methods of silencing the opposition press. Kuchma used the tax laws to shut down newspapers or to force them to rely on foreign and private sources. Several papers did in fact close. The journalists believed that Kuchma gave tacit approval to bombings of newspaper offices and assaults on reporters and editors. Many editors stayed away from controversy lest they fall victim to Kuchma's wrath.

When Kuchma's term ended and the Ukraine prepared for the presidential elections of 1999, many opposition candidates appeared. More than one hundred parties sprung up in the Ukraine, and thirty-two competed in the presidential election. In the parliamentary elections of the previous year, no party had won a majority, but the communists won the most seats, 124 of the 450. Other left-of-center parties moved the parliament in a direction opposition to Kuchma's policies. The president's party, the People's Democratic Party, won only 28 seats.

A Close Victory

On October 31, 1999, Ukrainians went to the polls to elect a president. Early predictions put the race at a tie between the two leading candidates, Kuchma and the Communist Party leader and candidate Petro Symonenko. Only two of the other candidates, Socialist leader Oleksandr Moroz and former prime minister Yevhen Marchuk, presented a challenge to the frontrunners. These candidates had made agreements with other candidates who promised to throw them their support to block Kuchma and Symonenko. The president's major support was among the so-called new Ukrainians, businessmen who benefited from his strides toward a free-enterprise economy, the middle class, and miners with whom Kuchma curried favor. The elderly, called "the vanguard of the left," supported Symonenko. State professional employees such as teachers, hospital workers, and librarians disliked Kuchma's policies and supported other candidates, as did the farmers.

As election day approached, Kuchma pulled ahead of his opponents, but he did not have enough support for the necessary 50 percent

3107

AP/Wide World Photos

Petro Symonenko, leader of the Ukrainian Communist Party.

majority. Before election day, 25 percent of the electorate was undecided. With a turnout in excess of 75 percent, their votes would prove decisive. Kuchma won 36.49 percent of the total, Symonenko received 22.24 percent, and the remaining candidates together received 41.27 percent. The Ukrainian constitution required a runoff between the leading candidates to be held two weeks later, November 14. In the runoff, Kuchma won by a substantial majority, with 56.31 percent to Symonenko's 37.76 percent. The president did very well in the western districts of the country and was able to hold his own elsewhere.

Consequences

The victory, despite Kuchma's unpopularity, showed that the people of the Ukraine and Eastern Europe continued to mistrust the Communist Party. However, under Kuchma, the Ukraine

continued to suffer economic and political problems. Critics complained of widespread corruption. One of the most vocal was opposition journalist Georgiy Gongadze, who edited the internet newspaper *Ukrainska Pravda* and believed that Kuchma himself was the source of Ukraine's crime and corruption. Gongadze also believed that Kuchma was changing the constitution to concentrate more and more power into his hands, creating a dictatorship. Gongadze disappeared in September, 2000, and his decapitated body was later found in a shallow grave in Kiev.

President Kuchma was implicated in Gongadze's murder. His opponents produced a tape in which Kuchma appeared to be ordering the journalist's death. Kuchma insisted that he was innocent, and the tape was a fraud. An opposition group, the Salvation Forum, called on Kuchma and other government leaders to resign as did the international financier, George Soros, well-known for his contributions in helping the new countries of Eastern Europe change to democracy and market economies. Soros also called on Western governments to condemn the president. Kuchma called his Ukrainian critics professional revolutionaries and fascists. However, Ukrainian and international opposition to his presidency mounted.

The president's opponents nicknamed the situation "Kuchma-gate" in reference to the scandal that forced the resignation of U.S. president Richard Nixon in 1974. The government began to harass participants in peaceful demonstrations against Kuchma's rule. The Socialist leader

Moroz came to the United States seeking the government's help in ousting the Ukrainian president, but Moroz, whose socialist rhetoric often attacked U.S. businesses investing in the Ukraine, was not popular in the United States. The U.S. government at first refused to commit itself to either party, simply urging the Ukrainian authorities to complete their investigation of the murder and the tapes. However as President Kuchma increased harassment against protesters, U.S. president George W. Bush sternly warned him to follow constitutional and legal procedures or lose U.S. aid.

Frederick B. Chary

United States and China Reach Trade Accord

Chinese and U.S. representatives signed a broad-ranging bilateral trade agreement believed to be an important stepping stone in China's path to joining the World Trade Organization.

What: Business; International relations
When: November 15, 1999
Where: Beijing, China
Who:
CHARLENE BARSHEFSKY (1951-   ), U.S. trade representative
SHI GUANGSHENG, Chinese foreign trade minister

Thirteen Years of Struggle

On November 15, 1999, after six days of round-the-clock negotiations, Chinese foreign trade minister Shi Guangsheng and U.S. trade representative Charlene Barshefsky signed an agreement between their two governments on China's application to join the World Trade Organization. The agreement was heralded by both sides as a major step forward in China's thirteen-year struggle to join the organization.

The World Trade Organization is the successor to the General Agreement on Tariffs and Trade (GATT), which was established by the United Nations in 1947. Although the Chinese government signed the agreements establishing the GATT, civil war in China interfered with its involvement in global organizations. The Communist forces defeated the Nationalists and drove them off the mainland, but the United Nations decided to recognize the Nationalist faction on Taiwan as the legitimate government of China. This decision forced the communist government in Beijing to withdraw from most global organizations, including the GATT. When the Beijing government was allowed to reenter the United Nations in 1972, doors began to open for China to rejoin other global organizations.

In September of 1982, the Chinese government began the long process of reentering the GATT by applying for observer status. Then, on July 11, 1986, Beijing formally applied for the restoration of its founding member status in the GATT, beginning thirteen years of negotiations aimed at achieving this goal. In 1995, the World Trade Organization formally replaced the GATT, and China renewed its efforts to join this global trade club.

Agreeing to Trade Freely

The World Trade Organization is a partnership of 134 nations. The heart of the World Trade Organization is free trade—trade without barriers such as tariffs and quotas. Trade practices in China, however, are ripe with restrictions, tariffs, quotas, and other trade barriers. Although numerous meetings had been held in Geneva, Switzerland, to discuss the possibility of Chinese membership in the World Trade Organization, it became increasingly clear over time that China needed the support of individual global trade leaders to gain membership. The Chinese government opened negotiations with the United States, Canada, the European Union, and others in the hope of gaining support from these economic superpowers.

The administration of U.S. president Bill Clinton sent its top trade negotiator, trade representative Barshefsky, to meet with Chinese leaders. Her mission was to get the Chinese government to agree to sweeping changes in its foreign trade policy. In effect, the United States was asking China to turn its economy upside-down. To the amazement of many observers, the Chinese government agreed to nearly every U.S. demand. To the Chinese leadership, a promise to open Chinese markets to the United States was simply the price it needed to pay for American support in their attempt to join the World Trade Organization.

The new trade agreement opened the door for the United States to support China's World Trade Organization membership by calling for reductions in tariffs and quotas and, in some cases, complete removal of existing trade barriers. According to the agreement, Chinese industrial tariffs would fall from 24.6 percent to 9.4 percent and would be eliminated on high-tech imports. Agricultural tariffs would drop by more than half and export tariffs would be removed. Chinese nationals would have access to U.S. banks and financial institutions, and many other markets would open up to free exchange.

Consequences

Trade experts expect the agreement to have a profound influence on the economies of both nations after China enters the World Trade Organization. The United States is expected to benefit from increased agricultural sales, particularly in corn, cotton, rice, soybeans, and wheat.

Foreign trade minister Shi Guangsheng signs the trade agreement.

The Chinese automobile market is to be opened to U.S. distributors, and the massive telecommunications market is to be opened to partial foreign ownership. U.S. banks are to be allowed to enter the Chinese financial markets and to deal directly with Chinese consumers. U.S. businesses wishing to enter Chinese markets will no longer be required to work with a Chinese partner that controls the majority of the resulting joint venture's stock. Even Hollywood is set to benefit, as the limits on imports of foreign films are to be lifted to allow twenty foreign films to be released in China each year.

The economic benefits for the United States seemed clear. However, critics in the U.S. government focused on political issues. Both conservative Republicans and liberal Democrats criticized the agreement as not going far enough in demands for improved human rights conditions, more democracy in China, or a promise that China will not use force to regain control over Taiwan. The Clinton administration, on the other hand, insisted that by opening up trade with the Chinese people, the agreement would foster improvements in all these fields. The administration said that if the agreement is successfully implemented, it would expose the Chinese people to the type of democratic values the critics hope to see.

China was likely to benefit from the removal of textile tariffs, opening the door for additional sales to the lucrative U.S. market. After quotas are lifted on Chinese imports in 2005, China's share of the U.S. textile market is expected to rise from 12 percent to 30 percent. In addition, experts believe that opening the door to foreign ownership is also opening the door to foreign investment, something the Chinese economy desperately needs if it is to modernize. As a member of the World Trade Organization, China would enjoy broad access to foreign markets in exchange for following a long list of fair trade rules. The agreement looked good for both sides, on paper. However, whether this agreement will lead to Chinese membership in the World Trade Organization rests solely on the Chinese government's ability to turn promises into realities.

Richard R. Pearce

3111

AP/Wide World Photos

Coast Guard Rescues Cuban Refugee Elián González

The rescue of a Cuban boy off the coast of Florida caused a seven-month legal battle between his relatives in Miami and his father in Cuba, possibly affecting U.S. immigration policy and the image of Cuban Americans and their influence.

What: International relations; Human rights; Law
When: November 25, 1999
Where: Miami, Florida
Who:
ELIÁN GONZÁLEZ (1993-   ), Cuban boy rescued off the Florida coast
LÁZARO GONZÁLEZ, Elián's great-uncle
JUAN MIGUEL GONZÁLEZ, Elián's father
JANET RENO (1938-   ), U.S. attorney general

Rescue

On November 22, 1999, Elián González, a five-year-old Cuban boy, and thirteen other Cubans left Cardenas, Cuba, bound for the United States on a sixteen-foot (five-meter) motorboat. At 10 P.M., the boat capsized and Elián's mother, Elizabeth Brotons, and at least ten of the other passengers drowned.

On November 25, Elián was found floating on an inner tube three miles (nearly five kilometers) from Ft. Lauderdale, Florida, and was taken to Hollywood Regional Medical Center. He was later released to his great-uncle, Lázaro González, by the Immigration and Naturalization Service (INS) until his immigration status could be determined.

Lázaro González and Elián's cousin Marisleysis González wanted to keep the boy in the United States, but Elián's father, Juan Miguel González, demanded that the boy be returned to him in Cuba. After González and his wife divorced, Elián had been living with his mother, although the father, who had remarried, maintained a relationship with the boy. González be-

lieved that he had the right to claim his son, who was allegedly taken out of Cuba without his father's knowledge. Elián's stateside relatives argued that his mother had died bringing her son to the United States in search of freedom and that she would have wished for her child to live in the United States.

On December 8, Cuban president Fidel Castro demanded that Elián be returned to Cuba. After Castro's statement, Lázaro González submitted Elián's political asylum application to the INS. A policy created in 1994 grants any Cuban who makes it to land the right to apply for asylum and to stay in the United States. However, anyone caught before reaching land is sent back to Cuba.

Political Battle

On January 5, the INS announced that Elián belonged with his father and must be returned to Cuba. However, Elián's Miami relatives' attorneys pleaded for the attorney general to reconsider the case. In Miami, in response to the INS statement, hundreds of Cuban American protesters blocked intersections and cut off access to the port. Many were arrested.

Following the protests, Lázaro González filed petitions for temporary custody of Elián in a Florida state court. Attorney General Janet Reno denied González's request to overturn the decision of the INS commissioner. On January 22, Elián's Cuban grandmothers came to the United States and met with Reno to appeal for Elián's return to Cuba. They met with Elián but returned to Cuba without the boy.

Subsequently, Elián's Miami relatives argued for an asylum hearing. U.S. district judge K. Michael Moore was assigned to hear the case, and

U.S. government lawyers asked Moore to dismiss the asylum lawsuit. Demonstrators tied up traffic outside the court, and others gathered outside the González home in support of the family. Judge Moore dismissed the lawsuit, and the INS informed the Miami relatives that it would revoke Elián's legal status in the United States and strip them of their right to care for him if they did not hand over the boy once the appeals were exhausted.

In March, Castro announced that he would send Juan Miguel González to the United States to pick up his son. The U.S. State Department approved six visas for González, his wife, and their infant son as well as Elián's cousin, his teacher, and pediatrician to travel to the United States. On April 6, they arrived in Washington, D.C. Following more demonstrations, Miami Dade mayor Alex Penelas called on the Cuban American community for peace.

Reno met with Elián's Miami relatives and ordered them to surrender the boy. However, the relatives defied the order and obtained a court order to keep Elián in the United States. The federal appeals court extended the court order until a May hearing. Nevertheless, on April 22, after long negotiations with Elián's Miami relatives and their lawyers, Reno gave orders for federal agents to seize Elián from their home. In response to the surprise 5:15 A.M. raid in which Elián was removed from the home, people in Miami rioted in the streets for two days. More than 268 people were arrested.

On May 11, the Eleventh U.S. Circuit Court of Appeals in Atlanta heard oral arguments from lawyers for Elián's father and Miami relatives. However, on June 1, the Atlanta federal court ruled Elián was not entitled to a political asylum hearing and upheld a Miami federal judge's ruling that Elián's father had the right to speak on his behalf. After several requests from the Miami relatives' attorneys, the court advised them that the injunction preventing Elián from leaving the United States would expire at 4 P.M. on June 28, and on that day, Elián, his father, and the rest of his family and friends returned to Cuba.

Consequences

For Castro and his people, Elián represented a kidnapped child who should be returned to his father. The streets of Cardenas, Elián's place of birth, were filled with posters demanding Elián's return to Cuba. Castro's ultimatums and his handling of the situation may have enhanced his stature among his followers. However, some experts charge that Castro used the situation to distract human rights advocates from 1999 political and human rights problems in Cuba and his own people from severe economic problems during 2000. For Cuban Americans in Miami, Elián was a symbol of Cuban suffering and a poster child for the anti-Castro movement. They felt that he would have access to a more prosperous life in the United States and that once he returned to Cuba, he would live a life of communist brainwashing.

The legal battle over Elián was expected to affect the development of future U.S. immigration policy and the way in which Cuban Americans and their influence are viewed. The government may strengthen its power to deny asylum; alternatively, the INS may develop new regulations on children applying for asylum.

Although many Cuban Americans thought that trying to keep the boy in the United States was the right thing to do, influential voices questioned whether they should have made a five-year-old boy into a symbol of the battle against Castro. Others thought that the Elián saga weakened the influence of Cuban Americans in the U.S. government. Finally, the case raised an important question of who should decide a child's future—the parents or government officials.

José A. Carmona

Protesters Disrupt Seattle Trade Conference

More than 50,000 protesters disrupted a summit meeting of the World Trade Organization in Seattle, Washington, demanding that the organization be more open and responsive to labor and environmental concerns.

What: Economics; International relations
When: November 29, 1999
Where: Seattle, Washington
Who:

PAUL SCHELL (1927-   ), mayor of Seattle

NORM STAMPER, Seattle police chief

RALPH NADER (1934-   ), political activist

Why Seattle?

Seattle, Washington, is a U.S. showcase for the global trade economy. Such global giants as Microsoft Corporation and Boeing are located in the greater Seattle area, and thousands of jobs in the area depend on exports. Being one of the most trade-dependent regions of the United States helped Seattle win the right to host the conference of the World Trade Organization (WTO) even though it faced competition from such cities as Dallas, Texas; Denver, Colorado; and Detroit, Michigan.

More than 130 countries belong to the WTO. It is the ultimate decision maker in trade disputes that arise out of various trade agreements and treaties, including the General Agreement of Tariffs and Trade (GATT) and the North American Free Trade Agreement (NAFTA). Much antagonism toward the WTO arises from its decisions modifying or setting aside national laws relating to labor standards or the environment on the basis that such national laws are discriminatory or unfairly burden world trade. For example, a U.S. law prohibited the importing of shrimp if caught in nets that did not allow for the escape of sea turtles. This law was found to be discriminatory and was set aside in a decision by the WTO. Decisions such as this and the perception by some that the WTO is a closed and undemocratic organization controlled by wealthy countries and multinational corporations triggered the demonstrations at the conference.

Days of Rage

By November 29, 1999, the eve of the start of the conference, more than fifty thousand demonstrators had gathered in Seattle. Some were labor union members who held a rally of more than twenty thousand at the Space Needle. Thousands of supporters of social activist Ralph Nader, together with their environmental allies, were also present in Seattle as were supporters of conservative leader Pat Buchanan. Besides groups from North America, there were contingents from Vietnam, Taiwan, Mexico, and Tibet. Simultaneous protests were also held in other cities throughout the world.

Along with these dedicated protesters who sought to make known their grievances were the less committed, who came to Seattle looking for action and excitement. Some deliberately sought to create chaos and looting. A small group of anarchists—those opposed to established government—came equipped with spray paint, hammers, and M80 firecrackers. Their targets were retail stores associated with large global corporations or their products. Dozens of store windows were smashed. A series or explosions racked the downtown area and fires were set in trash receptacles. This scene of lawlessness was an invitation to the local criminal element to join in and begin looting.

The intent of the organized demonstrators was to prevent delegates from entering the Washington State Convention Center where the WTO meeting was being held. They succeeded in doing this for at least six hours, delaying the start of the conference. Fearful of the chaos, the delegates remained trapped in their hotels. These included such visiting dignitaries as United Nations secretary general, Kofi Annan and U.S. secretary of state Madeleine Albright. Later, U.S. president Bill Clinton made several appearances in the Seattle area. He denounced the violence but at the same time sided with the causes of the protesters while citing the benefits of expanded trade and the need for successful trade negotiations.

In an attempt to regain control of the streets, the police, backed by armored vehicles, used rubber bullets and exploding canisters of pepper spray and tear gas. A state of emergency was declared and a 7 P.M.-to-dawn curfew imposed. The National Guard was called out. Protest-free zones were set up around the convention center and hotels. All demonstrations were banned in these

zones, and violations of this ban resulted in many arrests.

Consequences

The city of Seattle certainly was the loser from these occurrences. Several evenings of violent images on television news tarnished its image as a convention city. More than $3 million in property damage was sustained a result of the rioting and demonstrations. An estimated $10 million loss of retail sales was incurred. There was also the substantial expense of the heavy police presence. Seattle mayor Paul Schell and police chief Norm Stamper were heavily criticized. The demonstrators and their supporters claimed police brutality and a gross disregard of their civil rights. Those not sympathetic to the demonstrators claimed that the mayor and police chief should have been better prepared and not have allowed the demonstrations to get out of hand. It was claimed that certain problems should have

Protesters chant and wave signs on the streets of downtown Seattle.

3115

been anticipated, and among other things, barricades should have been in place establishing safe corridors between the hotels and the convention center. As a consequence, the mayor suffered politically, and the police chief resigned.

The WTO conference accomplished nothing other than the usual expressions of goodwill. There was no real narrowing of differences, and the delegates returned home, having undergone a hostagelike experience. Unlike in previous years, cities did not compete to host the next WTO meeting scheduled for November, 2001. Only one country, Qatar, made a formal offer to host the conference.

The demonstrators may have won a point by raising the consciousness of the American people as to the WTO and the trade issues. Before the event in Seattle, most Americans were not aware of the WTO and its activities. However, the news coverage of the events was certainly not favorable to the protesters or their cause.

Gilbert T. Cave

Northern Ireland Gains Home Rule

After nearly thirty years of direct rule by Great Britain, Northern Ireland achieved a measure of self-government upon implementation of the Good Friday Agreement.

What: Political independence; International relations
When: December 2, 1999
Where: Northern Ireland
Who:
GERRY ADAMS (1948-   ), leader of Sinn Féin
DAVID TRIMBLE (1944-   ), leader of the Ulster Unionist Party
JOHN HUME (1937-   ), leader of the Social Democratic and Labor Party
GEORGE MITCHELL (1933-   ), former U.S. senator

The Irish Problem

After centuries of political control by Great Britain, most of the island of Ireland moved toward autonomy in 1922,when twenty-six counties were established as the Irish Free State, becoming the independent Republic of Ireland in 1949. However, six counties in the northeastern part of the island—Antrim, Armagh, Down, Fermanagh, Londonderry, and Tyrone—remained part of the United Kingdom, as the country of Northern Ireland.

A slight majority of Northern Ireland's population is Protestant; many of them favor remaining united with Great Britain and, therefore, are known as Unionists. They have had continuing conflict with Northern Ireland's Roman Catholic Republicans, who still hope to be reunited with the Irish Republic. Neighborhoods, schools, and other social institutions typically are segregated along religious lines. Firebombings, shootings, beatings, and other violence have cost lives and created a climate of fear and tension among Northern Ireland's residents. The civil unrest also discouraged tourism and investment in the country, unlike in the republic, where foreign investments fueled unprecedented growth, and tourism is a major component of the economy.

In the 1990's, many people in Scotland and Wales were seeking greater autonomy for their regions from Great Britain. In Westminster, such calls for more self-government throughout the United Kingdom, along with the rising costs of policing the conflict in Northern Ireland, set the stage for possible changes in Northern Ireland. Most political leaders in Northern Ireland wanted to see peace in their country, but there were numerous conflicting demands among the various factions. In 1995, U.S. president Bill Clinton, of Irish descent, hoping to broker peace, sent former U.S. senator George Mitchell to work with both sides to seek a solution to the conflict.

The Good Friday Agreement was signed on April 10, 1998. Neither side got all that it wanted: Republicans had to give up hopes of reunification in the foreseeable future because the agreement emphasized that only the consent of a majority in Northern Ireland could lead to a united Ireland; Unionists had to accept a lessening of British influence. On May 22, 1998—in the first popular vote by all the people of Ireland in eighty years—the agreement was strongly endorsed on both sides of the border. In Northern Ireland, with a turnout of 81.1 percent, the vote was 71.1 percent for and 28.8 percent against; in the Republic, 56.3 percent voted, with 94.3 percent for and 5.6 percent against the agreement.

Terms of the Agreement

The agreement called for an assembly to be seated in Northern Ireland as well as the establishment of the British-Irish Intergovernmental Conference, the North-South Ministerial Council, the British-Irish Council, and several implementation bodies that would cooperate on a cross-border, all-island level. Power would be devolved (transferred) formally by the British gov-

AP/Wide World Photos

Northern Ireland Secretary Peter Mandelson (left) and Irish foreign minister David Andrews toast the new agreement.

ernment to the Northern Ireland Assembly and Executive, ending nearly thirty years of direct rule by England. The Government of Ireland Act, claiming British jurisdiction over all Ireland, would be repealed. However, all the people in Northern Ireland could hold both British and Irish citizenship.

The 108-member assembly would be elected by proportional representation, to ensure the sharing of power among the various factions. Committee chairs, ministerial posts, and committee assignments would be allocated in proportion to party strength. Decisions requiring only a simple majority would require a majority of both Republicans and Unionists, or approval by 60 percent of the assembly with at least 40 percent of one party. The executive would consist of a first minister, a deputy first minister, and up to

ten department ministers, all allocated on a proportional basis.

The North-South Ministerial Council would consist of ministers representing the Republic and Northern Ireland and six implementation bodies to deal with trade and business development, language, aquaculture and marine matters, programs regarding the European Union, food safety, and inland waterways. Cross-border cooperation would be promoted in transportation, agriculture, education, health, environment, and tourism through existing agencies.

The British-Irish Council would include representatives from the British and Irish governments, the newly devolved institutions of Northern Ireland, Scotland, and Wales, and the Isle of Man and the Channel Islands. Council members were charged to cooperate on matters of mutual

3118

interest, such as environmental, cultural, health, and educational concerns. The British-Irish Intergovernmental Conference was set up to discuss and resolve matters not under the control of the devolved government.

Other matters covered by the Good Friday Agreement were human rights; promotion of the Irish language, which had been forbidden earlier under British rule; and economic development in Northern Ireland. Weapons decommissioning was required of all paramilitary organizations within two years, and Britain's police force in Northern Ireland, the Royal Ulster Constabulary, was to be reviewed. Finally, a number of political prisoners from each side would be released.

Consequences

Elections were held for the new assembly on June 25, 1998. The two leading moderate parties received the greatest number of seats: The Ulster Unionist Party (UUP) led by David Trimble had twenty-eight seats, and the Republican-leaning Social Democratic and Labor Party (SDLP) led by John Hume won twenty-four. The more radi-

cal Democratic Unionist Party (DUP) led by Reverend Ian Paisley had twenty seats, and its nemesis Sinn Féin, led by Gerry Lázaro, won eighteen. Parties with between one and six seats were the Alliance Party, Northern Ireland Union Party, United Unionist Assembly party, Northern Ireland Women's Coalition, Progressive Unionist Party, UK Unionist Party, and Independent Unionist. Because the UUP had the largest number of votes, the first minister was chosen from that party; the deputy first minister was from the second-ranked party, the SDLP.

Conflict over forming the executive before the completion of the arms decommissioning delayed implementation of the agreement. In September, 1999, Mitchell returned to mediate among the factions. On December 2, 1999, power was officially devolved from Great Britain to Northern Ireland, and the executive's first formal meeting took place in Belfast, Northern Ireland, at Stormont, which had been the site of the Northern Ireland's government under England from 1920 until Britain took direct control in 1972.

Irene Struthers Rush

Ireland Amends Its Constitution

After a referendum and the implementation of the Good Friday peace agreement in Northern Ireland, the Irish constitution was changed to reflect new political realities.

What: International relations; Political reform
When: December 2, 1999
Where: Ireland
Who:
BERTIE AHERN (1951-   ), taoiseach (prime minister) of Ireland from 1997
MARTIN MANSERGH (1946-   ), special adviser to the taoiseach on Northern Ireland affairs

A Country Divided

In March, 1920, Ireland was partitioned by an act of the British parliament in London. During the years of British rule over Ireland, two groups had emerged; Nationalists who sought an independent Ireland and who constituted over 80 percent of the population, and Unionists who considered themselves politically British and did not wish to sever Ireland's ties with London. The Government of Ireland Act (1920) created two separate states: the Irish Free State (later the Republic of Ireland), which had a 90 percent Nationalist majority, and Northern Ireland, where Unionists were in a majority. Within Northern Ireland there existed a sizeable minority (about one-third of the population) who were Nationalist in politics and sought to be reunited with their fellow countrymen in the south.

Opposing Views on the Irish Constitution:

While devising Ireland's new constitution in 1937, Prime Minister Eamon de Valera (whose title was taoiseach), sought to affirm Nationalist Ireland's objections to the partition of the country. Article 2 of the constitution stated that the "national territory consists of the whole island of Ireland, its islands and territorial seas." Article 3 then sought to reconcile the aspiration toward

Irish unity with the reality of partition. It did this by stating that the Irish government had a right to legislate for the whole of Ireland; however, unless it explicitly said that it was doing so, the its laws passed would apply only to southern Ireland. While these provisions were important in securing support for the new constitution in the south, and particularly within de Valera's own party, Fianna Fáil, they went largely unnoticed in Northern Ireland and in Great Britain.

Unionists first began to take serious note of the articles during the 1980's. Since 1969, a brutal war had developed between the Nationalist Irish Republican Army (IRA), the British army and a variety of Unionist paramilitary bodies. Unionists began to view the articles as a threat and argued that they gave a political justification to the IRA to use violence to force the Unionist population of Northern Ireland to accept Irish unity. In short, Unionists saw the territorial claim as aggressive, and demanded its removal.

The issue posed some problems for the ruling party in the Irish Republic, Fianna Fáil which had traditionally viewed itself as the guardian of Nationalist rights. The party's leader, Bertie Ahern, realized that most Nationalists, north and south were opposed to deleting or fundamentally altering the substance of the constitution's articles 2 and 3. They agreed that the constitution should have a clear definition of the national territory and that removing the existing definition would send the wrong signals to Nationalists in Northern Ireland, many of whom felt that they had already been abandoned by the Dublin Government.

The Good Friday Agreement

On April 10, 1998, the Good Friday Agreement was negotiated by the most of the main political parties in Northern Ireland with the assistance of the British and Irish governments. New political structures were devised for Northern Ireland, but the implementation of these depended on changes to articles 2 and 3 of the Irish constitution.

One of the most controversial amendments to articles 2 and 3 involved the insertion of the "consent principle" into the constitution. For many years, Nationalists had argued that as Northern Ireland had not been established by reaching a consensus among the people of Ireland, Unionist consent was not necessary to achieve the reunification of the country. Some Nationalists preferred to depict the "consent principle" as the "Unionist veto," as it gave Unionists the right to prevent progress towards Irish unity. They feared that the removal of articles 2 and 3 might make northerners less Irish in a constitutional sense and even deprive them of Irish citizenship.

The proposed changes took account of Unionist objections while satisfying Nationalists that their rights would not be eroded. The citizenship rights of people in Northern Ireland were enshrined in the constitution as the changes shifted the constitutional emphasis from claiming the territory of Northern Ireland, which was considered dated and open to misinterpretation, to claiming its people. Therefore the new article 2 stated that it was the entitlement and birthright of every person born on the island of Ireland to be part of the Irish nation.

The revised article 3 stated that it was the firm will of the Irish nation, in harmony and friendship, to unite all the people, irrespective of identity or tradition, who shared the territory of Ire-land. Moreover, it stated that a united Ireland could only come about by peaceful means with the consent of the majority of the people in both parts of the island. Until then, the application of laws enacted by the Dublin parliament would be limited to the twenty-six counties and there was no mechanism provided for extending those laws to include Northern Ireland. The new article 3 also facilitated the establishment of cross-border bodies with executive powers which could legislate for the whole of Ireland.

Consequences

Parliamentary parties in the Irish Republic were united behind the proposed changes and the date of the referendum was set for 22 May 1998, the same day that voters in Northern Ireland were due to vote on the Good Friday Agreement. More than 1.5 million people turned out to vote (56.3 percent of the electorate). Of this number, 94.6 percent favored the proposed changes and only 5.6 percent opposed it. The emphatic nature of the result did not surprise anyone. Had it failed to pass the future of the Irish peace process would have been jeopardized and few in the south wished for such an eventuality. As the changes to the Irish constitution were dependent on the implementation of the Good Friday Agreement, the old articles 2 and 3 remained intact until December 2, 1999, when a new power-sharing government was established in Northern Ireland.

Donnacha Ó Beacháin

Evidence of Martian Ocean Is Reported

A team of scientists at Brown University presented topographical measurements that indicated that an ocean had once existed on Mars.

What: Astronomy; Space and aviation; Technology
When: December 9, 1999
Where: Providence, Rhode Island
Who:
JAMES W. HEAD III (1941-   ), planetary geologist at Brown University
MICHAEL H. CARR (1935-   ), scientist at U.S. Geological Survey, Menlo Park, California

Oceans Away

Through a study of the topography of Mars, planetary geologist James W. Head III and a team of scientists at Brown University discovered evidence consistent with there having once been rivers, lakes, and perhaps even vast seas on the red planet. Head and his colleagues analyzed topographical measurements taken by the Mars orbiter laser altimeter, an instrument on the Mars Global Surveyor. They tested the hypotheses that oceans or other bodies of water had existed on Mars, set forth in 1989 and 1991 by two teams of scientists, based on images taken by the Mariner and Viking spacecrafts in the mid-1970's and other evidence.

According to the Brown scientists, the measurements reveal a variety of valleys or channels and a possible shoreline that might have been shaped by flowing water. In addition, they found evidence that the northern lowlands of Mars may have harbored oceans 2 billion years ago. A shallow sea, hundreds of meters deep, could once have covered parts of the planet.

On the Shores of a Distant World

The laser altimeter aboard the Mars Global Surveyor, which had been orbiting the planet since 1997, detected a border region around the northern lowlands of Mars that has a nearly constant elevation and could be a shoreline. Two billion years ago, Mars was wetter and warmer than it is today. According to a team of scientists at the Los Alamos National Laboratory, as the liquid seas evaporated, they left telltale traces of salts that could have originated only in oceans that once covered the dry basin in the north.

In addition, the area adjacent to the shore is very smooth, as if it had been covered over by layers of sediment, such as that which might filter down to the bottom of a body of water. The existence of terraces suggests the gradual rise and fall of water, which would leave irregular deposits of materials that would form sinuous lines across the parched surface once the water had permanently receded. Because of the effects over time of wind and sand erosion, the topographic data gathered are somewhat unclear. Michael H. Carr of the U.S. Geological Survey suggested that the oceans may have dried up as long as 3.8 billion years ago.

Scientists continue to find sandy gravel ridges called eskers, streamlined teardrop-shaped hills, apparent seepage marks in the walls of canyons, dried craters, and other topographical features similar to those associated with flowing and still bodies of water on Earth. The mounting evidence supports the theory that Mars was once a wet world. Carleton Moore, a researcher at Arizona State University, found charged particles of water-soluble ions in the pristine interior of a Martian meteorite, ALH48001, recovered in the desert near Nakhla, Egypt, in 1911. He believes brine evaporated within the meteorite and left high concentrations of sodium, chloride, magnesium, sulfates, and fluoride, similar to what is found in terrestrial oceans.

Most of the water on Mars is believed to have just evaporated over the eons. However, some of it may have drained into deep underground aquifers, become trapped in thick permafrost just below the red dust, or hovered in the thin, cold, dry atmosphere. Liquid water cannot form or remain in that state because of the frigid temperatures on Mars, but huge volumes of liquid seem to have left their mark across the planet.

Conditions on the planet now include perpetual dust storms, carbon dioxide snow, and almost no air pressure. However, some hundreds of millions of years ago, Mars had a thick carbon dioxide atmosphere, and it could have conceivably been two to three times as warm as it is today because of the greenhouse effect, which would have lasted for an indeterminate period. The spin axis of Mars also tends to tilt randomly, causing extreme temperature variations unsuitable for the retention of liquid water.

Consequences

Theories regarding climate change and axial irregularities will have to be further tested as scientists continue to search for answers. Future findings could hold the key to proving or disproving the existence of the vanished rivers, lakes, and seas. Scientists believe that various forms of water might occasionally be liberated from the solidly frozen permafrost by meteors striking the surface, tectonic movements,

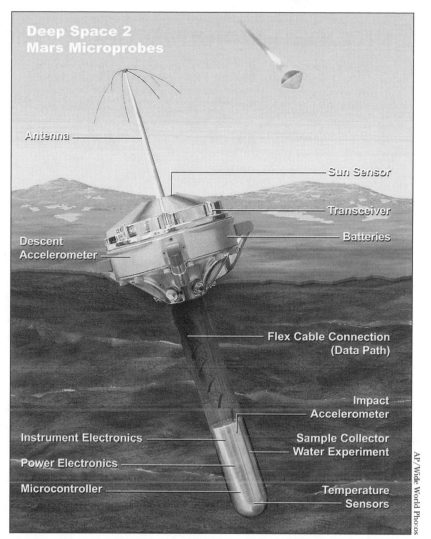

Artist's rendering of a probe penetrating the soil of Mars. Two such probes were crashed into the surface of Mars—much like lawn darts—in December, 1999.

sporadic volcanism, or other events. Wherever the water went, scientists hope to find evidence of it and of water on Mars.

Michael W. Simpson

European Union Accepts Turkey as Candidate for Membership

The European Union decided to allow Turkey as well as Cyprus, Malta, and a number of Eastern and Central European countries to become candidates for membership in the union, raising the number of members from fifteen to twenty-eight.

What: International relations; Economics; Political reform
When: December 11, 1999
Where: Helsinki, Finland
Who:
PAAVO LIPPONEN (1941-   ), prime minister of Finland from 1995
BÜLENT ECEVIT (1925-   ), prime minister of Turkey from 1999
ABDULLAH OCALAN (c. 1948-   ), leader of the Workers Party of Kurdistan

The European Union

On May 9, 1950, Robert Schuman, the foreign minister of France, approached West Germany with a plan to unite certain elements of German and French production under a single multinational organization to control vital resources. In this case, the resources were coal, steel, and iron, without which, it was believed, it was impossible to go to war. Italy, Belgium, the Netherlands, and Luxembourg were also invited to join the organization, which was to be known as the European Coal and Steel Community (ECSC). The reorganization of Europe into a multinational economic unit has moved forward ever since by extending the community to all products, services, and peoples and by adding new members (in five waves) on the basis of demonstrated commitment to democracy and peace. Because of its expanded economic interests, in 1958, the ECSC was renamed the European Economic Community (EEC). In 1995, it became the European Union, reflecting the increasing integration of its growing number of members. In 1973, the United Kingdom, Ireland, and Denmark joined.

In 1981, Spain and Portugal were added; in 1986, Greece; in 1995, Austria, Finland, and Sweden. On December 11, 1999, the European Union accepted Bulgaria, Cyprus, Malta, Lithuania, Latvia, Estonia, Slovakia, Poland, Hungary, Romania, Slovenia, the Czech Republic, and Turkey as candidates for membership.

Turkey's Candidacy

Turkey, which straddles the point where Europe and Asia meet, is slightly smaller than Texas and Louisiana combined. About 97 percent of its land area is located in Asia and 3 percent in Europe, around the city of Istanbul. Despite its overwhelming proportion of land in Asia, in the twentieth century, Turkey has been oriented toward the West. A member of the North Atlantic Treaty Organization (NATO), it joined the Organization for European Economic Cooperation (OEEC) in 1963.

In 1973, Turkey signed a treaty with the EEC (later the European Union) to begin a customs union. Under this treaty, goods shipped between Turkey and the EEC would not be taxed at the border. This customs union was eventually completed in 1995. Opinion polls in Turkey consistently demonstrated that two-thirds of Turks were in favor of joining the European Union permanently. Public opinion pressured the Turkish government to apply for full membership in 1987. Its application was turned down; among the reasons that triggered the EEC's refusal were Turkey's human rights track record and its large population of Muslims, whose culturally based attitudes might not coincide with those of the overwhelming Christian populations of other members.

On December 11, 1999, Paavo Lipponen, prime minister of Finland and president of the summit meeting of the European Union, announced that Turkey (along with twelve other countries) had been accepted as a candidate for membership. Although candidacy does not ensure gaining membership, Turkey's prime minister, Bülent Ecevit declared his satisfaction at being accepted. He was confident that Turkey had resolved, or soon would, the problems that had blocked its earlier acceptance.

Consequences

There are four main problems that need to be resolved before Turkey may join the European Union: a peaceful settlement with the Kurdish minority, the abolition of the death penalty, reduction of the rate of inflation, and a peaceful settlement of the territorial dispute between Turkey and Greece over the island of Cyprus.

Of the nearly 25 million Kurds in the world, 15 million live in Turkey, where they constitute about 25 percent of the population. Since World War I, there have been dozens of Kurdish uprisings. All were crushed by the military, often without any respect for human rights. In 1984, a Marxist-led group called the Workers Party of Kurdistan (PKK) began an armed struggle against the government. Led by Abdullah Ocalan, the PKK has committed many acts of terrorism, and in fifteen years of fighting in Turkey, nearly 40,000 lives have been lost, most of them because of retaliation by the Turkish government. A just resolution to this continuing internal struggle is essential if Turkey is to become a peaceful eastern extension of the European Union.

The second problem concerns the standards of human rights in Turkey, which must be brought up to approximate Western European levels. The European Union requested that Turkey abolish the death penalty, improve conditions in prisons, and ensure fair and impartial justice. Ocalan, the PKK president, was captured and condemned to be executed after a speedy trial; European Union members viewed his case as an example of the unfair social conditions in Turkey. However, after Turkey became a candidate for European Union membership, the Turkish government made no move to execute this terrorist, suggesting to some Europeans that Turkey may be moving toward higher standards of human rights.

The third problem is inflation, which was about 75 percent in 1998, 65 percent in 1999, and about 36 percent in 2000. These rates make long-term economic development almost impossible. The United States and most of the countries of the European Union have a yearly inflation rate less than the monthly inflation rate in Turkey. The Turkish government has attempted to control this problem and has made some improvements, but the rate of inflation remains dangerously high.

The fourth problem concerns civil unrest in Cyprus stemming from a dispute between Greece and Turkey over control of the island. In 1974, the Turkish army invaded Cyprus, gaining control over 38 percent of the island. Turkey has refused to recognize the legal title of Greece to Cyprus and other small islands, even though that title was established by the Treaty of Lausanne of 1923 and the Treaty of Paris of 1947. For full membership in the European Union, this dispute and the other three problems were required to be settled peacefully before 2004.

Denyse Lemaire

Prime Minister Bülent Ecevit.

AP/Wide World Photos

3125

Astronauts Renovate Hubble Telescope

Seven astronauts aboard the space shuttle Discovery, *launched December 19, 1999, successfully restored the "vision" of the Hubble Space Telescope, shut down since November 13, at its orbiting site 350 miles (583 kilometers) above Earth.*

What: Astronomy; Space and aviation
When: December 24, 1999
Where: Orbit around Earth
Who:

JOHN H. CAMPBELL, Hubble project director of the National Aeronautical and Space Administration (NASA) Goddard Space Flight Center

CURTIS BROWN, JR. (1956-   ), former U.S. Air Force pilot and commander of the space shuttle *Discovery*

STEVEN SMITH (1958-   ), former electrical engineer and veteran of three earlier Hubble servicing missions

JOHN GRUNSFELD (1958-   ), an astrophysicist specializing in black holes and X-ray pulsars

The Rescue Mission

A crew of seven astronauts aboard a special launch of the space shuttle *Discovery* was lifted into space on the evening of December 19, 1999, on a mission to restore the "vision" of the Hubble Space Telescope. The Hubble had been "blind" for two months after three of its gyroscopes, the instruments that allow the flying telescope to point precisely at targets in deep space, failed. On November 13, a fourth gyroscope had failed, prompting the decision to launch the service mission as soon as possible.

During their mission, guided by John H. Campbell, Hubble project director, the crew replaced all six gyroscopes, installed a 486-microprocessor computer and new battery components, upgraded the guidance system, and performed other maintenance. As a result, the Hubble was

pronounced "better than new" because the added electronics allowed it to operate at a higher level than it did the original equipment. Had the repair mission not been successful, the Hubble would have been reduced to a piece of space junk that wasted tens of millions of taxpayer dollars.

The failure of its gyroscopes was not the Hubble's first near catastrophe. In December, 1993, a team of four spacewalking astronauts installed corrective optics to fix a flawed primary mirror. In February, 1997, astronauts installed two new instruments: a space telescope imaging spectrograph (STIS) and a near infrared camera and multi-object spectrograph (NICOMOS), which failed in 1999. Such equipment failure occurs because the environment in which the Hubble operates is very harsh, including temperature variations of several hundred degrees and the possibility of impacts from micrometeors.

The Replacement Ballet

In the December, 1999, mission, four members of the international seven-man crew, in teams of two, took turns conducting the four extra-vehicular activities (EVAs), each lasting six hours. Astronaut Steven Smith worked with John Grunsfeld, and Michael Foale, who had spent 134 days on Mir (the Russian space station) in 1997, was paired with Claude Nicollier of Switzerland, who had operated the robotic arm during the 1993 mission. Grunsfeld, Foale, and Nicollier all have advanced degrees in astrophysics.

Commander Curtis Brown, Jr., had the delicate task of maneuvering the 230,000-pound (105,000-kilogram) shuttle close enough to the Hubble for the astronauts to perform the needed work and of doing so without damaging the solar arrays of the Hubble, solar panels that gather solar energy to power the telescope. Then Smith and Grunsfeld took turns guiding each other into the bowels of the telescope itself, where there was barely enough room to fit an astronaut in a bulky spacesuit. Once inside, they had to disconnect three boxes containing the old gyroscopes and replace them with three new boxes, each worth $5 million to $10 million. They also installed a kit to prevent the batteries from overcharging and, later, a new transmitter to replace a failed one, but replacing the six gyros was their most important task.

On the next day, Foale and Nicollier replaced Hubble's original computer with one equipped with the sturdy 486 microprocessor, early 1990's technology but a move that assured reliability and reduced operating costs. They also replaced one of Hubble's three onboard fine guidance sensors that guide the Hubble's telescope precisely to its viewing target and added a solid-state recorder. Each team member conducting the actual EVA was guided by his partner, who was inside the ship with the checklist keeping an eye on everything. As Smith said, "The guy inside . . . conducts the ballet."

Although the mission was judged a complete success, concerns for the possibility of year 2000 problems caused the astronauts to head for home before performing additional tasks such as placing wallpaperlike sheets on patches of Hubble's skin that had been damaged by the sun.

The Hubble Space Telescope, as seen from the space shuttle Discovery's *cargo bay.*

AP/Wide World Photos

Consequences

By May, 2000, it was clear that the repair mission was a significant success, and the platform was stable. One mission of the improved Hubble was to examine the Eskimo Nebula, so called because it looks like a face surrounded by a fur parka. The Hubble's image suggests that the "fur" may be giant comets flowing away from a dying central star, an observation that astronomers will find important in theorizing the origins of the star and its characteristics.

Hubble's next portrait was the massive cluster of galaxies known as Abell 2218 located around 2 billion light-years away in the constellation Draco. The cluster acts like a magnifying glass, making visible light from even more distant objects. Such gravitational lensing, created by Abell 2218's gravity, magnifies the light of even more distant galaxies; furthermore, the portrait was in color, unlike the black-and-white images taken by Hubble in 1994, and therefore provides additional insights into the nature of the universe.

Theodore C. Humphrey

Panama Takes Full Control of Panama Canal

At midnight on December 31, 1999, the United States transferred control of the Panama Canal to the Republic of Panama, as obligated under the 1977 Panama Canal Treaties.

What: International relations; Political independence; Transportation
When: December 31, 1999
Where: Panama City, Panama
Who:

JIMMY CARTER (1924-   ), president of the United States from 1977 to 1981

BILL CLINTON (1946-   ), president of the United States from 1993 to 2001

MIREYA MOSCOSO (1946-   ), president of Panama from 1999

MADELEINE ALBRIGHT (1937-   ), U.S. secretary of state from 1997 to 2001

History of the Canal

In 1903, the United States and the newly created Republic of Panama signed the Hay/Bunau-Varilla Treaty, under which the United States acquired the right to build and operate the Panama Canal for all time. The canal, once built, provided an important strategic advantage to the U.S. Navy over its potential adversaries. By using the canal rather than sailing around the tip of South America, the United States could move fleets between the Atlantic and the Pacific Oceans quicker than any possible adversary.

Operation of the canal meant that the United States also owned the Canal Zone, the narrow strip of land down either side of the canal. This fifty-mile-long (eighty-kilometer-long) zone became sovereign U.S. territory, a physical extension of the United States. The Panama Canal and the Canal Zone essentially split the independent nation of Panama into two separate halves.

The United States built military bases inside the Canal Zone to protect the canal from attack. The presence of U.S. troops in the middle of their country was irritating to the Panamanian people. The government of Panama tried unsuccessfully for decades to restore national sovereignty by reestablishing Panamanian control over this territory. U.S. presidents repeatedly decided that only U.S. control could guarantee reliable operation of the canal in peacetime and wartime.

Treaties and Transition Period

In 1977, U.S. president Jimmy Carter determined that U.S. national interests would not be at risk if the canal and the Canal Zone were eventually returned to Panama. The Carter administration entered into negotiations with the Panamanian government. An agreement between the negotiators was finally reached. On September 7, 1977, President Carter and General Omar Torrijos, president of Panama, met in Washington, D.C., in the headquarters of the Organization of American States (OAS) to sign two separate treaties, the Panama Canal Treaty and the Neutrality Treaty.

The Panama Canal Treaty was the basic agreement, detailing the arrangements for the day-to-day operation and military protection of the Panama Canal. The treaty also specified the transition period of twenty years between U.S. and Panamanian control of the Canal Zone. The effective dates for this treaty were October 1, 1979, to December 31, 1999, the final transfer date.

The Neutrality Treaty addressed the permanent neutrality of the Panama Canal. The purpose of this second treaty was to ensure that no unfriendly nation could gain control of the Panama Canal or deny the use of the canal to all the nations of the world. This treaty obligated Panama to protect the canal. The treaty also allowed the United States to act militarily, if necessary, to keep the canal open for unrestricted use in the future.

Spanish king Juan Carlos (left) shakes hands with former U.S. president Jimmy Carter as Panamanian president Mireya Moscoso (seated) watches at a ceremony transferring control of the canal to Panama.

The people of Panama accepted these two treaties on October 23, 1977. The U.S. Senate approved the Neutrality Treaty on March 16, 1978. The Senate later approved the Panama Canal Treaty on April 18, 1978. Both treaties went into effect on October 1, 1979.

During the twenty-year transition to Panamanian control, the United States controlled the Canal Zone and oversaw daily operations of the canal. The Panama Canal Commission, an official unit of the U.S. federal government, managed the operation. During these same years, the Panama Canal Commission planned several major engineering projects to modernize the canal. The Panamanian government was to supervise completion of improvements not finished at the time of transfer.

A ceremony commemorating the transfer of the Panama Canal was held on December 14, 1999. Because President Bill Clinton was unable to attend the historic event, former president Jimmy Carter, the U.S. signer of the treaty, and Secretary of State Madeleine Albright headed the delegation. Panama president Mireya Moscoso represented the Panamanian people.

Consequences

The transfer of control of the Panama Canal demonstrated a commitment by the United States to foster a more balanced approach to the region and individual nations such as Panama. The transfer was favorably received by most countries in the region, some of which were glad to see the removal of what some viewed as a remnant of U.S. imperialism. Removal of the U.S. military presence resulted in lost jobs for many

Panamanians, and some feared that their absence would have a negative effect on efforts to fight the drug trade in the region.

To the Panamanian people, the return of the Panama Canal was a recognition of their nation's sovereignty. However, in addition to its symbolic importance, the canal has strategic value because of the key role it plays in international trade. In the world market, goods will continue to be moved by sea in the twenty-first century. By fostering that movement, making it easier and less expensive, the Panama Canal provides tremendous benefit to industrial nations and their trading partners.

The security of the Panama Canal continues to be a major concern. Upon transfer, Panama signed concessions with a private corporation from Hong Kong to oversee seaport operations at both ends of the canal. This caused U.S. leaders to worry that the People's Republic of China might close the canal in the event of hostilities with the United States. Regardless of who controls the canal, the national security and economic interests of the United States were expected to be linked to the Panama Canal for many years to come.

Michael S. Casey

Russian President Yeltsin Resigns

> *Boris Yeltsin, who led Russia's independence from the Soviet Union and oversaw its partial democratization, resigned from the presidency after eight years in power. His move facilitated Russia's first peaceful and democratic transfer of executive power.*

What: Government; Political reform
When: December 31, 1999
Where: Moscow, Russia
Who:
BORIS YELTSIN (1931-   ), president of Russia from 1991 to 1999
MIKHAIL GORBACHEV (1931-   ), last leader of Soviet Union before Yeltsin
VLADIMIR PUTIN (1952-   ), former Russian prime minister and successor to Yeltsin as president

Russian Independence and Democratization

A committed member of the Soviet Union's Communist Party since 1961, Boris Yeltsin gained steadily in power and influence during the Cold War. It was a difficult period, with the Soviet Union and the United States locked in an international contest for military and political superiority. The Soviet Union, whose economy was under the control of the government and not subject to the discipline of market forces, found it increasingly difficult to keep up with the technological advances in weapons systems that the Cold War demanded. In addition, because its leadership consisted of unelected party bosses, the government lacked the legitimacy that could be conferred by democratic elections. This required that the government maintain strict censorship and other controls over the population to maintain stability. Moreover, because the Soviet Union was the world's largest country, spreading across eleven time zones and drawing in more than one hundred linguistic, ethnic, and national groups, it was constantly faced with the threat of ethnic and national unrest.

In 1985, a new Soviet leader, Mikhail Gorbachev, began to call for major reforms in the Soviet Union's economic system, foreign policy, and even its political system. Essentially, Gorbachev reckoned that the Soviet Union was in danger of being ruined financially by the arms race. He determined that only by such measures as introducing market reforms, negotiating new arms control treaties, and injecting a measure of democracy could the Soviet Union survive the challenges posed by the United States under the leadership of President Ronald Reagan.

Gorbachev's reforms encountered considerable opposition from the entrenched Soviet bureaucracy. In response, Gorbachev brought new allies into the leadership in 1986. Among these was Yeltsin, then head of the Communist Party's regional leadership in Sverdlovsk. Soon, however, it became evident that Yeltsin wanted the country's reforms to move faster and further than envisioned by Gorbachev. In particular, Yeltsin desired that the people be given more meaningful opportunities for political participation and that capitalism be more fully adopted. In addition, Yeltsin increasingly became associated with the cause of Russian (as opposed to Soviet) nationalism. Russia was one of fifteen republics that made up the Soviet Union. Although it was by far the largest republic, many Russians resented the control of their people and the oppression of their thousand-year-old culture by the Communist Party of the Soviet Union.

Throughout the late 1980's, the Soviet Union was caught in a cycle of implementing new reforms to neutralize potential unrest, which instead served only to fuel demands for further reforms. By 1990, Yeltsin separated himself from the Soviet leadership and established his political base in Russia. By the end of 1991, the nationalistic forces in the various Soviet republics succeeded in dismantling the Soviet Union, which was replaced by fifteen newly independent countries. Yeltsin emerged as the elected president of the first independent Russian state in hundreds of years.

The End of the Yeltsin Presidency

Yeltsin was immediately faced with the need to stabilize his government, secure his country's borders, and develop foreign relations with Russia's new neighbors. These were difficult times, given the instability and uncertainty of the post-Soviet, post-Cold War world. However, Yeltsin managed with skill and luck. After some conflicts with the Russian legislature, Yeltsin promoted a new Russian constitution, which was adopted by a vote of the people in 1993.

The new constitution emphasized property rights, civil rights, and democracy. It also established more clearly the powers of the president.

Among other provisions, the constitution limited any one person to no more than two four-year terms as president. In 1996, Yeltsin ran for a second term. Although the country had managed a measure of political stability and a rebirth of a free society, increasing economic problems and a loss of Cold War-era glory led a large number of voters to oppose Yeltsin. He nonetheless managed to win a second term after a runoff ballot against a Communist Party leader.

As the 1990's drew to a close, speculation arose that Yeltsin would attempt to circumvent the constitution and run for a third term in 2000. At the same time, however, Yeltsin's heart condition and other problems threatened to cut short even his second term. A constantly shifting inner circle of heirs-apparent in Yeltsin's Kremlin fueled speculation about Russia's impending post-Yeltsin era.

Finally, at the end of 1999, six months before the scheduled presidential election, Yeltsin re-

Russian president Boris Yeltsin announces his intention to resign on a televised broadcast.

signed the presidency. Executive power passed to Prime Minister Vladimir Putin, as required by the constitution. Yeltsin virtually dropped from public sight, as had Gorbachev some eight years earlier.

Consequences

Yeltsin's early resignation from the presidency only heightened the interest in the presidential elections, which had to be held within ninety days of his resignation. Because Putin had been handpicked as Yeltsin's prime minister, his elevation to the presidency lacked popular approval. The election therefore would be a major test of Russia's resilience as a post-Soviet democracy: Would Russia's first constitutional transfer of executive power be validated by the voters? If Putin did not win the election, would he abide by the people's decision? As it turned out, Putin was elected in his own right in March, 2000. He won 53 percent of the vote, handily beating his closest competitor by almost 25 percentage points. Because Putin received a clear majority, no runoff election was necessary.

Clearly Russia under Yeltsin passed some of the toughest tests of democracy and independence. Although the country still faced serious challenges, including economic decline, widespread crime and corruption, and ethnonational conflict, it managed to establish new, democratic institutions and unleash long-forgotten freedoms for the Russian people. Yeltsin's stewardship was responsible for much of this. His decision to resign in 1999 paved the way for his handpicked successor to go before the voters at an opportune time. However, more important, Yeltsin's willingness not to push for a third term, which some of his advisers had urged him to do, avoided a constitutional crisis and offered the Russian government continued opportunities to distance itself from the Soviet legacy.

Steve D. Boilard

Hybrid Gas-Electric Cars Enter U.S. Market

Two major automobile manufacturers began marketing low-polluting, energy efficient hybrid gas-electric cars in the United States.

What: Energy; Engineering;
Environment; Technology;
Transportation
When: January, 2000
Where: United States
Who:
TOM ELLIOTT (1940-   ), executive vice
president, American Honda Motor
Company
HIROYUKI YOSHINO (1939 -   ), president
and chief executive officer, Honda
Motor Company
FUJIO CHO, president, Toyota Motor
Corporation

Announcing the Hybrids

Not only did Honda Motor Company show off its new Insight model at the 2000 North American International Auto Show in Detroit, it also announced expanded use of its new technology. Hiroyuki Yoshino, president and chief executive officer of Honda, said its integrated motor assist (IMA) system would be expanded to other mass-market models. The system basically fits a small electric motor directly on a 1-liter, 3-cylinder internal combustion engine. The two share the workload of powering the car, but the gasoline engine does not start up until needed. The electric motor is powered by a nickel-metal hydride (Ni-MH) battery pack, with the IMA system automatically recharging the energy pack during braking. Tom Elliott, Honda executive vice president, said it was a continuation of the company's philosophy of making the latest environmental technology accessible to consumers. The $18,000 Insight is a sporty two-seat car that uses many innovations to reduce weight and improve performance.

Fujio Cho, president of Toyota Motor Corporation, was also speaking at the Detroit show, where his company was showing off its new $20,000 hybrid Prius. The Toyota Prius relies more on the electric motor and has more energy storage than does the Insight, and it is a larger four-door, five-seat model. The Toyota hybrid system divides the power from a 1.5-liter gasoline engine and directs it to drive the wheels and a generator. The generator either powers the motor or recharges the batteries. The electric motor is coupled with the gasoline engine to power the wheels under normal driving. The gasoline engine supplies average power needs, with the electric motor helping the peaks; at low speeds, the power is all electric. A variable transmission seamlessly switches back and forth between gasoline engine and electric motor or applies both of them.

Variations on an Idea

Automobiles generally have a gasoline (or diesel) engine for driving, an electric motor for starting up, and a way of recharging the batteries that power the electric motor and other devices using the motion of the engine and vehicle. In a completely electric car, the gasoline engine is eliminated, and the batteries recharged from stationary sources. In a hybrid car, the mix between the gasoline engine and electric motor is changed so that the electric motor handles some or all of the driving. This is at the expense of an increased number of batteries or other energy storage devices. Possible in many combinations, hybrids couple the low-end torque and regenerative braking potential of electric motors with the range and efficient packaging of gasoline, natural gas, or even hydrogen fuel power plants. The return is greater energy efficiency and reduced pollution.

With sufficient energy storage and a large enough motor, an electric motor can actually

3135

propel the car from a standing start into motion. When the driver accelerates past a certain speed, the gasoline engine, which is more efficient at higher speeds, kicks in. However, these gasoline engines are smaller, lighter, and more efficient than those in typical cars, having been designed for average rather than peak power needs, reducing pollution and considerably improving fuel economy. The batteries in hybrid vehicles are recharged partly by the engines and partly by regenerative braking; a third of the energy from slowing the car is turned into electricity. What finally made hybrids feasible at reasonable cost were the developments in computer technology, which allow sophisticated controls to coordinate electrical and mechanical power.

One way to describe hybrids is to split them into two types: parallel, in which either of the two power plants can propel the vehicle, and series, in which the auxiliary power plant is used to charge the battery, rather than propel the vehicle. The Insight is a simplified parallel hybrid using a small, efficient gasoline engine. The electric motor assists the engine, providing extra power for acceleration or hill climbing, helps provide regenerative braking, and starts the engine but cannot run the car by itself. The Prius is a parallel hybrid with a power train that allows some series features. Its gasoline engine runs only at an efficient speed and load and is combined with a unique power splitting device. This device allows the car to operate like a parallel hybrid—motor alone, engine alone, or both. It can act as a series hybrid with the engine charging the batteries rather than powering the vehicle. It also provides a continually variable transmission using a planetary gear set that allows interaction between the engine, the motor, and differential that drives the wheels.

Consequences

In January, 2000, Honda and Toyota began stateside marketing of gas-electric hybrids that offer more than 60 miles per gallon (96 kilometers per gallon) fuel economy and meet California's standards for super ultra-low-emission vehicles. These cars provide an environmentally friendly form of private transportation without the inconvenience of fully electric cars, which go only about 100 miles (160 kilometers) on a single battery charge and require special adaptations such as kerosene-powered heaters. Other manufacturers have begun following suit. Ford promised a hybrid sports utility vehicle by 2003; other automakers, including General Motors and DaimlerChrysler also announced development of alternative-fuel and low-emission vehicles. For example, the ESX3 concept car used a 1.5-liter, direct injection diesel combined with a electric motor and a lithium-ion battery.

American automakers planned to offer some full hybrids, cars capable of running on battery power alone at low speeds, but they reportedly were focusing more enthusiastically on electrically assisted gasoline engines called mild hybrids. Full hybrids typically increase gas mileage by up to 60 percent; mild hybrids by only 10 percent or 20 percent. The mild hybrid approach uses regenerative braking with electrical systems of a much lower voltage and storage capacity than those found in full hybrids, a much cheaper approach. However, there is still enough energy available to allow the gasoline engine to turn off automatically when a vehicle stops and turn on instantly when the accelerator is touched. Because the mild hybrid approach only adds $1,000 to $1,500 to a vehicle's price, it is likely to be used in many models. Full hybrids cost much more but achieve more benefits.

Stephen B. Dobrow

U.S. Mint Introduces Sacagawea Dollar Coin

Although the Sacagawea coin proved popular with the American public, it raised controversies regarding the spelling of its honoree's name and assessing her historical legacy.

What: Economics; Social reform
When: January, 2000
Where: United States
Who:
SACAGAWEA (1788-1812), Shoshone guide to the Meriwether Lewis and William Clark expedition
GLENNA GOODACRE (1939-   ), sculptor who designed the coin
PHILIP N. DIEHL (1952-   ), director of the U.S. Mint

Conception

The Sacagawea dollar coin was the U.S. Mint's second late twentieth century attempt to put one dollar coins in circulation. Minted from 1979 to 1981, the Susan B. Anthony dollar coin was never popular with the U.S. public. Not only were its size and reeded edge overly similar to those of the twenty-five-cent piece, but it was also dubbed "the Carter quarter" in 1980 by presidential candidate Ronald Reagan in reaction to the period's high inflation. Moreover, the selection of Anthony as the first historical woman to appear on a U.S. coin failed to create a consensus in an era still divided on the issue of women's rights.

To address these problems, the United States Dollar Coin Act of 1997 authorized the minting of a new coin to replace the Susan B. Anthony dollar. The law stipulated that the coin be gold in color and have a smooth edge to distinguish it from the quarter. Treasury Secretary Robert Rubin established a Dollar Coin Advisory Committee to propose a fitting replacement for Anthony.

The advisory committee, chaired by U.S. Mint director Philip N. Diehl, narrowed the list to American Red Cross founder Clara Barton; Girl Scout founder Juliette Gordon Low; civil rights activist Rosa Parks; Native Americans Pocahontas and Sacagawea; flag designer Betsy Ross; Congress members Shirley Chisholm, Jeanette Rankin, and Margaret Chase Smith; first woman governor Nellie Tayloe Ross; Underground railroad founder Harriet Tubman; abolitionist Sojourner Truth; poet Emma Lazarus; first ladies Eleanor Roosevelt and Martha Washington; and the mythical Lady Liberty.

On July 9, 1998, the advisory committee selected Sacagawea to appear on the new coin. Her selection reflected renewed interest in the Lewis and Clark expedition, subject of popular historian Stephen Ambrose's *Undaunted Courage* (1996) as well as Ken Burns's television documentary *Lewis and Clark: The Journey of the Corps of Discovery* (1997). Because no historical representation of Sacagawea existed, the advisory committee chose to fuse allegorical and historical figures for the coin's design. The committee's official recommendation proposed a depiction of "Liberty, represented by a Native American woman, inspired by Sacagawea."

Success

Twenty-three artists were invited to submit designs for the new coin. With the help of Native American leaders, professional numismatists, and the public (via an interactive Web site), the U.S. Mint selected seven designs, which were forwarded to the U.S. Commission on Fine Arts for the final selection. For the obverse (front) of the coin, the commission chose a design submitted by sculptor Glenna Goodacre, best known for designing the Vietnam Women's Memorial in Washington, D.C. Goodacre's coin design featured a portrait of Sacagawea, bearing an infant on her back, in a three-quarters pose facing the viewer.

For a model, Goodacre used Randy'L He-dow Teton, a Shoshone college student. The selection for the reverse design, submitted by Thomas

The model for the Sacagawea coin, Randy'L He-dow Teton, a Shoshone college student, stands in front of a copy of the coin.

Rogers, depicted an eagle in flight surrounded by seventeen stars (representing the number of states in the Union at the time of the Lewis and Clark expedition). Launched with a special ceremony at the Philadelphia mint, the Sacagawea dollar coin went into production on November 18, 1999, making it ready for circulation in January, 2000. Sparked by a $40 million advertising campaign—as well as the need for larger denomination coins in vending machines—the new dollar coin was an immediate success. Within its first year alone, more than 700 million were circulated (compared with the 100 million Susan B. Anthony coins in circulation over the two decades of its existence).

Consequences

This success only fanned the flames of controversy surrounding Sacagawea's historical legacy.

Primarily known to the world through the dozen or so references made to her in Lewis and Clark's journals, she remains a shadowy figure whose origin is unclear and whose role in the expedition is contested. Even the spelling of her name stirs debate. Since the publication of Lewis and Clark's journals in 1811, her name had been spelled "Sacajawea"; however, in the 1990's, many scholars opted for "Sacagawea" as the preferred spelling. Hidatsa-Mandan Native American groups, on the other hand, prefer "Sakakawea." Although her name has been translated as "Birdwoman" from the Shoshone, some scholars have proposed "Boat-launcher" as the actual translation, a less poetic name that may reveal a more mundane role for her in the Lewis and Clark expedition.

Moreover, the standard history of Sacagawea—as promoted by the U.S. Mint—portrays her as a

Shoshone girl captured by a rival Hidatsa band and subsequently sold to French-Canadian trader Toussaint Charbonneau (whose child she bore at age fourteen—about which time, he and she served as guides to Lewis and Clark). The official history depicts Sacagawea as an indispensable member of the expedition who not only helped navigate rivers and locate food but also served as interpreter and intermediary with Native American groups (indeed, the presence of a woman with a child within the expedition probably did more than anything to highlight its peaceful nature). However, despite her near-mythic status as a virtual "savior" to the expedition, some revisionist historians have tended to downplay her importance to the expedition. Also, Hidatsa-Mandan groups have disputed her Shoshone origin, claiming her as one of their own—and thereby denying the claim that their ancestors were any party to slave-trading in Sacagawea's case.

Some critics of U.S. treatment of Native Americans have pointed to Sacagawea more as a symbol of victimization than of heroism. Little more than a slave, the teenage mother was compelled to take part in a punishing cross-continent expedition with her infant child. Moreover, her subsequent life shows her ultimate rejection by white society. According to one tradition, in the years following the expedition, she had no means to support herself and had to give up custody of her son to William Clark before she died in Dakota Territory in 1812.

Luke A. Powers

Computers Meet Y2K with Few Glitches

Despite fears of catastrophic breakdowns in power, water, food, bank, and credit systems, few problems were reported as the year 2000 (Y2K) began. Public and corporate officials and computer experts took credit for preventing major disasters, but some critics wondered if potential problems had been overestimated.

What: Technology
When: January 1, 2000
Where: Worldwide
Who:
GRACE MURRAY HOPPER (1906-1992), developer of an early word-recognition programming system
JOHN A. KOSKINEN (1939-   ), chair of the President's Council on Year 2000 Conversion

Origin of the Problem

In the early days of computer programming in the 1950's, when Rear Admiral Grace Murray Hopper and her colleagues were searching for a language to simplify computer use, resources such as memory and storage space were severely limited and costly. Addition of these items added to the already large size of a computer system. For example, the Electronic Numerical Integrator and Computer (ENIAC), completed in 1946, weighed thirty tons. At first, storage was on punch cards. Each card had only eighty spaces for data. To save space and costs, programmers regularly left out the first two digits of a year. The year 1946, for example, was recorded as "46." Despite technological advances, this habit was not changed. Therefore, as late as 1997, experts estimated that 65 percent of the hardware sold could not register the year 2000 as the start of a new century. Computers would register only the digits 00, which would be interpreted as the year 1900. Experts predicted that systems dependent on accurate computer dating would break down. Some predicted a domino effect in which breakdown of one computer system would lead to breakdowns of connected systems, possibly worldwide.

Reactions

By the late 1990's, reactions to the potential problem ranged from near hysteria to serious, organized planning efforts. Mass media—film, television, radio, and electronic sources—carried predictions of widespread power outages. Water systems, they reported, would fail. Nuclear power plants would malfunction, the stock exchange would close, airline operations would stop, payrolls would be frozen, and individuals would be unable to use ATMs, and their banking transactions and records would be lost. Popular magazines reported the possibility of gasoline shortages and shortages of automobile parts, with loss of service shop records.

Retailers reported significant increases in purchases of canned goods, gas lamps, military meals, water filters, and other survival supplies, including weapons, as food riots and foreign invasions were projected by some extremists. A few economists predicted a long-term economic depression as a result of computer failures. Some newspapers prepared for saturation coverage during the week leading up to January 1, 2000, cutting back on press runs and projecting higher sales of newspapers that would become souvenir items. The Red Cross posted instructions for the public to prepare for the year 2000 as if preparing for a natural disaster such as an earthquake or hurricane.

Serious planning, however, had long been under way. Its purpose was to ensure that critical computer equipment was either replaced or upgraded to be Y2K compliant. In February, 1998, the President's Council on the Year 2000 Conversion was established through executive order. It was charged with monitoring compliance in critical services, including benefits payments, communications, electrical power, financial services, and water supplies. John A. Koskinen, previously deputy director for management at the government's Office of Management and Budget, was named chair.

Planning took two forms, in reaction both to the potential computer problems and to the threat of public hysteria. The Securities and Exchange Commission required Y2K compliance reports from the companies it regulated and posted them on its Web site, and federal regulators sent Y2K examiners to banks and other financial institutions. Their findings were reported to Congress. Financial institutions were 96 percent compliant by January, 1999. Nonetheless, the Federal Reserve made additional cash available to banks, to avoid a shortage caused by public fears. The Federal Aviation Administration reported that its systems were compliant by July 1, 1999. Corporations and other institutions, including the military, also made sure their systems were compliant. Despite these efforts, public fear remained. Some experts predicted that, no matter how much was spent in the United States, undeveloped counties, connected worldwide with other computers, could produce problems.

Consequences

By December 31, 1999, an estimated $600 billion had been spent worldwide to prevent Y2K problems. U.S. expenditures were estimated at approximately $100 billion (about $365 for each citizen), including the government's $50 million Information Coordination Center (ICC), which housed government monitoring efforts.

Midnight, December 31, arrived on a weekend. No serious problems were reported, but ex-

On December 31, 1999, Hewlett Packard employees at the company's headquarters in Palo Alto, California, prepare for the entry of their computers into the year 2000.

AP/Wide World Photos

perts said the worst disruptions were not expected until the following Monday when work resumed. On Monday, however, the expected domino effect did not take place. Those problems that did occur were quickly fixed. Among potentially serious problems was one at a Chicago bank, apparently Y2K related, which caused a delay in Medicare payments to health care providers. The Pentagon reported that a Y2K problem had interrupted the communication of spy satellite data for several hours. The Hong Kong Futures Exchange reported a breakdown, fixed within a day. Other problems generally were trivial and local.

Government and corporate experts claimed that the smooth transition into the new century was a result of planning and allocation of neces-sary funds. Critics questioned whether all the expenditures were necessary, noting that undeveloped countries, having few resources to commit to Y2K compliance, reported no more problems than did developed nations. Some critics argued that the various planning operations and vast expenditures were a reaction to the media, rather than to the actual problem. The International Data Corporation estimated that U.S. businesses and government agencies had spent $122 billion on the problem since 1995, of which about $20 billion to $30 billion was unnecessary. When no problems occurred on February 29, 2000, when noncompliant computers were expected not to recognize a leap year, public discussion gradually ended.

Betty Richardson

America Online Agrees to Buy Time Warner

America Online, the world's largest Internet service provider, and Time Warner, the huge media conglomerate, agreed to a merger of their two companies. The new company would be called AOL Time Warner.

What: Communications; Business

When: January 10, 2000

Where: Grand Ballroom, Waldorf-Astoria Hotel, New York City

Who:

STEVE CASE (1958-   ), chair of America Online

GERALD LEVIN (1939-   ), chair of Time Warner

Largest Corporate Merger in U.S. History

On January 10, 2000, Steve Case, chair of America Online, and Gerald Levin, chair of Time Warner, announced that America Online would buy Time Warner for approximately $165 billion. This stunning announcement, the largest merger in U.S. history, seemed to offer a symbolic meaning for the new millennium. America Online, an Internet company, only fifteen years old and with few tangible assets, was purchasing an established media giant, with tangible assets in television, publishing, films, and music. Time Warner had 1999 profits of more than $7 billion on revenues exceeding $27 billion, AOL's 1999 profits of $685 million were based on revenues of less than $5 billion. At the time of the announcement, however, AOL's share price, reflecting the high value assigned Internet companies, was twice that of Time Warner. The new company would be known as AOL Time Warner. Case would be chair, and Levin would be chief executive officer.

The Benefits of Togetherness

Although both America Online and Time Warner were profitable companies, each was concerned about developments threatening its future growth. Before the merger, America On-line's revenues were growing at about 30 percent per year. America Online, however, faced the loss of subscribers to providers offering free Internet service or faster cable service. Broadband cable connection to the Internet was one hundred times faster than America Online's dial-up connection. Time Warner was the second largest cable company in the United States. In addition, Time Warner's assets included *Time, Sports Illustrated, Fortune,* and *People* magazines, television outlets such as the Cable News Network (CNN) and Home Box Office (HBO), Warner Music, and the motion picture company, Warner Brothers. Time Warner, however, feared being bypassed in the digital age. Levin's $40 million attempt to establish the Pathfinder Web site had proved disappointing. By merging with America Online, the nation's largest Internet service provider, Time Warner gained access to the talent and experience of a successful Internet company.

Both companies recognized the great potential of a merger. America Online's 20 million subscribers would have access to Time Warner's broadband cable connection to the Internet. They would also have Internet access to Time Warner's content in magazines, books, films, music, and television programs. The merger, by uniting television and the Internet, laid the foundation for interactive television. Time Warner's 13 million cable subscribers would eventually be able to use their televisions to access the Internet to gain more information about a television program they were watching, to join chat groups discussing a program, or to directly order items they observed on television.

The merger also promised increased profits through cross advertisement of each other's services. America Online's subscribers would be exposed to Time Warner's products, and Time

3143

America Online chairman Steve Case (left) and Time Warner chairman Gerald Levin.

Warner's customers would receive America On-line advertisements. In addition, a vast database could be collected. Time Warner would know what America Online subscribers selected in their cable packages, and America Online would know what Time Warner customers read, watched, and listened to when they were not on-line. The commercial potential of being able to advertise specific products to targeted audiences was an obvious benefit of the merger.

Consequences

Consumer advocates and rivals raised concerns about the threat to competition. When the Federal Trade Commission (FTC) unanimously approved the merger on December 14, 2000, it did so with three conditions. First, the FTC would protect consumer choice by ensuring access to

Time Warner's content. Second, it would ensure competition by providing access to AOL's broadband cable for competing Internet service providers. Third, it would keep open a second distribution channel such as telephone lines or satellites. AOL was therefore required to open its cable systems to competitors. In those cities served by Time Warner before the merger, at least one AOL competitor would be offered high-speed Internet service before AOL could use those cable lines. Finally, AOL was prohibited from withholding Internet service from companies offering high-speed digital service through telephone lines or by satellites.

The Federal Communications Commission (FCC) approved the merger on January 11, 2001, a year and a day after Case and Levin had made their dramatic announcement. By this date, a

3144

stock market decline had reduced the value of the merger to $106 billion. FCC approval went beyond that of the FTC by reinforcing open access with specific stipulations. Competing Internet service providers would not only have open access to AOL's platform, but also they would also control the screen display on the computers of users and have direct billing rights with their customers. The FCC reserved the right to review any contracts between AOL and competing Internet service providers. To maintain an open and competitive market for "advanced instant messaging," AOL was required to either support an industry standard for "server-to-server interoperability" or enter into agreements with at least three other instant messaging competitors before AOL could offer "advanced instant messaging" such as video.

These FCC conditions, especially its reserved right to review all contracts between AOL and competing Internet service providers, could establish a powerful precedent. To many observers, the conditions set the precedent for FCC regulatory control over virtually every Internet product. This was a position that some found troubling and threatening to the growth of the Internet. It was not clear as to whether AOL Time Warner would be a standard for the future. Corporate mergers, especially those with different corporate cultures such as those of AOL and Time Warner, have often failed. The future, however, may very well be shaped by the merger of Internet companies with established companies, representing the merger of the new economy with the old economy.

Thomas W. Judd

Clinton Creates Eighteen New National Monuments

In the final thirteen months of his presidency, Bill Clinton created eighteen new national monuments and increased the size of three existing monuments, protecting nearly six million acres of public land with special biological or historical significance.

What: Environment; Government; National politics

When: January 11, 2000-January 19, 2001

Where: Washington, D.C.

Who:

BILL CLINTON (1946-   ), president of the United States from 1993 to 2001

BRUCE BABBITT (1938-   ), U.S. secretary of the interior

JAMES V. HANSEN (1932-   ), U.S. representative from Utah, chairman of the Resources Committee

History and Purpose of National Monuments

Near the beginning of the twentieth century, some scientists became concerned about the looting and destruction of early Indian pueblos and cliff dwellings. In response to those concerns, Congress passed the Antiquities Act of 1906, giving presidents the power to preserve "historic landmarks, historic and prehistoric structures, and other objects of historic or scientific interest" on public lands by declaring them national monuments.

Advocates of conservation argue that quick presidential action is often necessary to protect federal lands of scientific, historic, and scenic value from immediate threats of irreparable harm. Most Americans have welcomed presidential action to create monuments, and Congress has often given its blessing as well. Congress has created national monuments on its own authority and has signaled approval of many presidentially created national monuments by making them national parks. More than half of all national parks were protected first as national monuments; some of the better known include Arches, Badlands, Bryce, Carlsbad Caverns, Death Valley, Denali, Gates of the Arctic, Glacier Bay, Grand Canyon, Grand Teton, Lassen Volcanic, Olympic, Petrified Forest, and Zion.

The language of the Antiquities Act suggested that national monuments would cover small areas, but ever since the act's passage, presidents have created controversy by interpreting their powers liberally. President Theodore Roosevelt used the power he had been given to save the Grand Canyon in Arizona from threatened mining. President Franklin D. Roosevelt created Jackson Hole National Monument to save the Yellowstone elk. In a successful effort to preserve areas for future national parks, President Jimmy Carter created seventeen national monuments in Alaska, totaling 56 million acres (23 million hectares). President Bill Clinton's 1996 proclamation of a 1.7-million-acre Grand Staircase-Escalante National Monument in Utah provoked protests within the state and from congressional critics.

Since 1906 presidents have created more than one hundred national monuments by executive action. Most national monuments are administered by the National Park Service, but presidents have sometimes—for economic or political reasons—left new national monuments in the care of the agency that had previously managed the lands in question.

The Clinton Monuments

President Clinton was unusually active in establishing national monuments during the final thirteen months of his presidency. In that relatively brief period, on the advice of Interior Sec-

3146

retary Bruce Babbitt, Clinton created or enlarged twenty-one national monuments, enhancing the level of protection for approximately 5.7 million acres (2.3 million hectares) of federal land and additional areas off shore along the California coast.

The monuments of fewer than one hundred acres protect historic sites. California Coastal and the two Virgin Island monuments protect rich marine environments. Hanford Reach and Upper Missouri River Breaks protect rare free-flowing sections of the Columbia and Missouri Rivers. The remainder protect a vast array of scenic landscapes containing mountains, deserts, canyons, and unusual rock formations. Within these monuments are exceptional plants, such as the giant sequoia and ironwood trees; important archaeological sites; and spectacular animal life, including a number of endangered species. Despite active lobbying from environmental groups, President Clinton declined to give national monument status to the Arctic National Wildlife Refuge in Alaska, which contains valuable wildlife resources as well as significant reserves of oil and natural gas.

Consequences

Clinton's declaration of twenty-two new or expanded national monuments was a record number for a single American president. The probable consequence of Clinton's proclamations is better protection and more environmentally sensitive management for the areas designated, but much will depend on the management plans that will eventually be adopted by the administration of George W. Bush. The number, size, and timing of Clinton's proclamations fueled controversy, and Congress retains the power to abolish, reduce, or alter the management of one of more of the monuments, and has occasionally done so in the past. Because Congress often defers to the wishes of local citizens, Clinton's new national monuments are most likely to be at risk wherever they are opposed by local residents.

A more remote possibility is that Clinton's proclamations may provoke a critical Congress to curtail the president's power to create monuments. In July, 2001, Chairman of the House Resources Committee James V. Hansen, a Republican from Utah and a vocal critic of Clinton's monument proclamations, reintroduced a national monument bill that would place very severe limitations on presidential power under the Antiquities Act. U.S. presidents have generally resisted legislation that curtailed presidential power, and the power to create national monuments has been frequently exercised. Every president in the twentieth century either declared national monuments on his own authority or signed bills to create them.

Craig W. Allin

Kasha-Katuwe Tent Rocks, near Cochiti Pueblo, New Mexico, one of the areas newly designated as a national monument.

Hague Court Convicts Bosnian Croats for 1993 Massacre

> *A war crimes tribunal convicted five Croatian soldiers for atrocities against civilians during the battles that followed the breakup of Yugoslavia in 1991-1992.*

What: Civil war; Human rights; Political independence; International law
When: January 14, 2000
Where: The Hague, Netherlands
Who:
VLADIMIR ŠANTIC, a Bosnian Croat soldier
FRANJO TUDJMAN (1922-1999), president of Croatia from 1990 to 1999
SLOBODAN MILOŠEVIĆ (1941-   ), president of Yugoslavia from 1997 to 2000

Aggression Against Bosnia

Yugoslavia began breaking up in 1991, when the republics (similar to U.S. states or Canadian provinces) of Slovenia and Croatia seceded from the federation that had existed since 1918. In 1992, the republic of Bosnia-Herzegovina also seceded. This sparked a multilateral war between Bosnia's three main ethnic groups (Croats, Serbs, and Bosnian Muslims, also known as Bosniaks) and two neighboring states, Croatia and the remainder of Yugoslavia, dominated by the republic of Serbia. The Serbs, who made up 32 percent of Bosnia, did not want to secede from Yugoslavia. They set up their own country, tied to Yugoslavia. The Bosnian Croats, who made up only about 17 percent of the republic's population, set up a mini-state called Herzeg-Bosna and linked it closely to the highly nationalistic government of Franjo Tudjman, the president of Croatia. Serb forces brutalized Muslims, as the world witnessed in the long siege of Sarajevo and the massacre at Srebrenica. At first, the Croats fought against Serbs, but in 1993, large-scale fighting broke out between the local Croatian army and the Bosnian Muslim army. By 1994, U.S. pressure and Serbian

success had pushed the Croats and Bosniaks into an alliance, which remained into the twenty-first century. However, in the meantime, the Croatian forces carried out several horrific massacres of Bosniak civilians.

The first of the infamous Croatian attacks took place in April, 1993, in the Lašva River Valley, northwest of Sarajevo. In one attack, Croats killed 103 Bosnian civilians by shelling the village of Ahmici and then going door to door, brutalizing people and destroying property, as part of a general effort to force all Muslims to flee. In 1997, eight Croats suspected of this attack gave themselves up or were captured by the international forces of the North Atlantic Treaty Organization (NATO) in Bosnia.

The Hague Tribunal

The International Criminal Tribunal for the Former Yugoslavia (ICTY), or Hague tribunal, as it is also called, is based in the city of The Hague, Netherlands. It was set up by the United Nations Security Council in 1992 to investigate and punish war crimes that took place during the breakup of Yugoslavia.

By spring, 2001, the ICTY had publicly indicted more than one hundred people from the former country of Yugoslavia and issued new indictments. It also made sealed indictments not open to the public. The most frequent charges involve crimes against humanity, as defined by the Geneva Conventions and United Nations resolutions. These crimes include executions of prisoners, torture, sexual slavery, rape, forced labor, plunder, and the destruction of cultural objects. The most serious charge the tribunal can bring is genocide (the killing of people because of their ethnic identity, religion, or language). The majority of those indicted were

3148

Serbs. The most well-known figures under indictment, such as the former Yugoslav president Slobodan Milošević and former Bosnian Serb leaders Radovan Karadzic and Ratko Mladic, are still at large. Indictee Milan Milutinovic remains in his post as president of Serbia.

In 1999 and 2000, NATO forces in Bosnia captured an increasing number of indictees through surprise raids. It was rumored, however, that NATO leaders were slow to arrest Bosnian indictees such as Karadzic because they remained popular among the Serbian community, and bringing them before the tribunal might provoke the Serbs into scuttling the Dayton Peace Accords, which ended the war in 1995. Current Yugoslav president Vojislav Koštunica has refused to extradite anyone to The Hague, claiming that the tribunal is a political tool being used unfairly against Yugoslavia, citing as proof the fact that there are more Serbian indictees than those of any other nationality. Other critics of the tribu-

nal claim that it is hypocritical because it has not brought charges against NATO leaders for collateral damage against Serbian civilians during the air war over Kosovo in 1999 or for the use of depleted-uranium weapons, which might pose a health threat. In response to some of these criticisms, the ICTY promised to investigate abuses against Serbs by Albanians in Kosovo. The large number of Serbs on the list of indictees reflects international recognition of the fact that Serbs overran 70 percent of Bosnia and that their leaders initiated the practice of ethnic cleansing.

Charges against two of the men in the Ahmici case were dropped, and one man was acquitted. The five others were sentenced on January 14, 2000. Three of them were enlisted men: Mirjan Kupreškic, Vlatko Kupreškic, and Drago Josipovic Another, Zoran Kupreškic (a relative of the other two men) was a local commander, while Vladimir Šantic was a military police commander who

Bosnian Croats—(front row, left to right) Mirjan Kupreškic and Zoran Kupreškic, (back row, right to left) Dragan Papic, Drago Josipovic, and Vlatko Kupreškic—before sentencing.

AP/Wide World Photos

headed a unit with the sinister nickname of "the Jokers." All were charged with violation of the laws and customs of war and with crimes against humanity. Šantic was sentenced to twenty-five years. The other four men received sentences of between six and fifteen years.

Consequences

This case is important because it highlights the fact that Croats would be held accountable by the international community for war crimes, although Croats were also frequently victims in the wars in former Yugoslavia. Another notorious case of Croatian war crimes involved Mladen Naletilic (also known as "Tuta") and Vinko Martinovic (also known as "Stela"), who, in spring, 2001, were in custody and on trial for their alleged torture and execution of Bosnian Muslims, as well as the destruction of a mosque, in 1993. The conviction of a prominent Bosnian Croat political leader, Dario Kordic, demonstrated the ICTY's willingness to hold politicians responsible for encouraging massacres. Many of the tribunal's individual cases highlight certain principles, as in the recent convictions of three Serbs for sexual slavery and rape, the first time that such crimes have been regarded as war crimes under international law.

At first, Croatia refused to cooperate with the Hague tribunal. After President Tudjman died in 1999, however, the new Croatian leaders began working actively with the ICTY in an effort to improve Croatia's international reputation. They also wanted the country to overcome its tendencies toward right-wing authoritarianism and expansionism. Under the new policy of cooperation, the government set out to capture a young general named Mirko Norac, accused of killing Serb civilians in Gospic, Croatia, in 1991. Norac surrendered in March, 2001. His trial was to be the first under the new Croatian government, which also planned to retry several soldiers who received light sentences under the Tudjman regime.

The most general impact of the Ahmici trials is in the reminder that most of the victims of the war in Bosnia were civilians. The tribunal hoped to demonstrate to the world that ethnic cleansing (forced expulsion of minorities) and the murder of religious or ethnic minorities would not go unpunished.

John K. Cox

Gene Therapy Test at University of Pennsylvania Halted After Man Dies

After the death of an eighteen-year-old test subject, the U.S. government decided to monitor human gene therapy experimentation more closely.

What: Medicine; Genetics
When: January 21, 2000
Where: University of Pennsylvania
Who:
JESSE GELSINGER (1981-1999), Arizona test subject
JAMES WILSON, director of the Human Gene Therapy Institute, University of Pennsylvania

Human Gene Therapy

On January 21, 2000, the Food and Drug Administration (FDA) halted all human gene therapy at the University of Pennsylvania because of the death of one of the test subjects in the Human Gene Therapy Institute's experimental program. Jesse Gelsinger, an eighteen-year-old Arizona man, had a toxic reaction within hours of receiving the gene therapy treatment and died within a few days at the University of Pennsylvania. He was enrolled in an experiment that was designed to correct the genetic liver disorder from which he suffered. Gelsinger died on September 17, 1999, from progressive organ failure, beginning with his liver. He was one of eighteen patients participating in this particular study, and he was the only one reported to have suffered any adverse reaction.

One of the charges against the Human Gene Therapy Institute and its researchers was that they failed to take the proper precautions or to report any serious adverse reactions before the test subject's death. There had been two serious adverse reactions previously reported, but James Wilson, the director of the institute, claimed that they were caused by other medical conditions, not by the gene therapy. The researchers had also failed to notify the FDA that two monkeys given gene therapy had died. Furthermore, the institute had changed the eligibility requirements for patients to enroll in the program without the FDA's knowledge or approval. Another charge was that Wilson's lab did not properly monitor the liver study and did not have a standard operating procedure for the program. According to the FDA guidelines, Gelsinger should have never been given the gene therapy treatment because he did not meet the proper health criteria for the program.

Gene Therapy in Action

A gene is a sequence of deoxyribonucleic acid (DNA) that codes for a single protein. These proteins are responsible for building organic structures and causing biochemical reactions. Human gene therapy is a very promising medical breakthrough that could aid in the cure, elimination, or at the very least, in the easement or remission of many serious and terminal human illnesses. The goal of gene therapy is to replace defective or missing genes with healthy ones. For example, gene therapy can be used to deliver genes that promote the destruction or reconstruction of cancer cells, rebuild damaged tissue, and vaccinate individuals against other illnesses. Gene therapy has been used for many serious diseases such as hemophilia, cystic fibrosis, cancer, heart disease, vascular disease, and metabolic diseases. It has also shown much promise in alleviating the aging process, and the outlook is very good for treating Alzheimer's disease.

The vector, or method through which these genes are carried into the subject's cells, is often a virus. Viral vectors take advantage of the natural ability of viruses to enter a cell and to deliver genetic material into the cell's nucleus, which contains its DNA. These viral vectors have been specially engineered in the laboratory so that they cannot reproduce. This makes the actual virus rather harmless and the test subject usually does not suffer from any symptoms of the virus. The virus used to deliver the gene therapy in Gelsinger's case was an adenovirus, which is a common cold virus. The cold virus was injected directly into Gelsinger's liver, but he suffered a toxic reaction within hours of treatment.

The investigation into Gelsinger's death prompted the government to convene the Recombinant DNA Advisory Committee (RAC) to set guidelines that must be strictly adhered to in the use of human gene therapy experimentation. Among the guidelines that the committee established was the requirement that any adverse reaction be reported to the National Institute of Health (NIH). The RAC sent out letters to all researchers involved in gene therapy programs, requiring them to report any adverse reactions. The committee then received reports of 970 adverse reactions in 1999 and 464 in the first half of 2000. Of those adverse reactions, less than 10 percent were considered serious. Although Gelsinger was the only death directly linked with the therapy, there were a few other deaths that were suspect upon review. These results raised many questions about the safety of human gene therapy.

Consequences

Human gene therapy is an exciting breakthrough in the medical frontier that can potentially cure or eliminate many serious human conditions and diseases. In June, 2001, at the University of Pennsylvania, researchers were able to cure dogs that were born with a genetic blindness. After treatment, these dogs were able to see, which once again has raised hopes in the area of curing human blindness. Although animal experimentation and human testing are quite different, there have been enough successes in human testing to keep hope alive for future cures. However, any future human testing must be strictly regulated to ensure the safety of the study participants.

Mary Ellen Campion

Intel Unveils 1.5 Gigahertz Microprocessor

Intel introduced its Pentium 4 microprocessor, with processing speeds of 1.5 gigahertz, a far cry from the Intel 8088, which it released in 1979 for the first IBM personal computer.

What: Business; Computer science; Technology
When: February 5, 2000
Where: Intel Corporation, Santa Clara, California
Who:
ANDY GROVE, chairman and founder of Intel
TED HOFF, developer of the first Intel microprocessor

Intel Processors Power Personal Computers

In 1971, Ted Hoff of Intel Corporation developed the Intel 4004 microprocessor. It was actually a programmable controller, but Intel scientists recognized that it could be the central processing unit (CPU) of a small computer. Intel continued developing microprocessors, and in 1979, the Intel 8088 became the first real CPU on a chip.

With the release of the 80486 in 1992, Intel enhanced its CPU on a chip to support a wide range of computing capabilities, including virtual memory, a floating-point unit, and some on-chip cache. The 80486 had more processing power than the mainframes of the 1960's. All of these CPUs were complex instruction set computer (CISC) processors, a type of processor that loads data from memory, executes an operation, and stores data in memory with a single instruction. Intel had also been developing a series of reduced instruction set computer (RISC) processors, a type of processor in which the loading of data, execution of operations, and storage of data are distinct operations.

In 1995, Intel released its Pentium Pro processor, which combined many of the best features of both RISC and CISC processors. As Intel developed more sophisticated Pentium processors, it added features to the floating-point units that had been developed for supercomputers. This included pipelined adders and multipliers, in which the arithmetic and logic operations were done in stages rather than as a single operation. With the Pentium 2, Intel decided to devote more chip surface to graphics and multimedia. This trend continued with the Pentium 3 and the Pentium 4. Rather than simply making processors that execute business and scientific applications more quickly, Intel began to concentrate on making processors that are optimized for multimedia and Internet applications.

Improvements in the density of transistors— an electronic unit that can be set to either zero or one—that could be placed on the chip facilitated advances in Intel processors. The original Pentium chip had 3.1 million transistors, and the Pentium 4 has 42 million. A hertz (clock cycle) is a unit of time used by a CPU to coordinate its actions.

The total number of hertz executed every second is called the CPU's clock speed. RISC processors usually execute about one instruction per clock cycle, so the number of instructions executed each second on RISC computers is close to the clock speed. CISC processors usually require several clock cycles to execute instructions so that the number of instructions executed each second is slightly less than the clock speed. Although all of Intel's Pentiums are CISC processors, the Intel micro-architectures have been designed to approximate executing one instruction per clock cycle, so the number of instructions executed per second of most Pentium processors is close to the clock speed. In 1999, many speculated that Intel would soon release a 1-gigahertz Pentium processor, and late in 2000, the first 1-gigahertz Intel-based computers appeared. Some wondered if faster processors were possible. They

3153

were, and Intel soon was shipping 1.5-gigahertz processors with an accompanying chipset on an Intel motherboard.

The Pentium 4 Processor

On November 20, 2000, Intel announced that it was delivering its long-awaited Pentium 4 processor. By February, 2001, Pentium 4 computer systems were actually being delivered to consumers. The Pentium 4 has a clock speed of from 1.3 gigahertz to 1.5 gigahertz, making it the fastest processor available in the first quarter of 2001.

With the Pentium 4, Intel made the first major changes to the Pentium micro-architecture since the introduction of the Pentium Pro. Intel refers to the new architecture as its NetBurst micro-architecture. Because of the unique design of the NetBurst micro-architecture, the Pentium 4 can actually do some integer operations in half a clock cycle. Floating-point operations are not as fast, but a 20-stage pipeline maximizes the speed of these operations as well. The long pipeline is especially good for Internet and multimedia applications. The Pentium 4 has a 12-kilobyte instruction cache and an 8-kilobyte data cache, specifically designed to support pipelining and to reduce the number of times that a branching instruction causes the pipeline to reload.

The Intel 850 chipset gives the Pentium 4 a 400-megahertz system bus—the computer bus connecting the processor and memory. With the required Direct Rambus DRAM (DRDRAM), an expensive type of memory, the Pentium 4 can transfer up to 3.2 gigabytes of data per second between the central processor and memory. Adding 256 kilobytes of high-quality cache mem-

ory to the memory subsystem means that the Pentium 4 also has the fastest memory subsystem of any processor available in the first quarter of 2001.

Consequences

The Pentium 4 micro-architecture was designed to support Intel's upgrading to processor speeds of 2 gigahertz by the end of 2001 and 5 gigahertz within four years. Although the major beneficiaries of these speed increases were Internet and multimedia applications, other applications would benefit as well. The work of Intel in attempting to predict branches to improve the pipelining of the Pentium 4 was advanced and promised to give Intel a major advantage over its competitors for several years. The Pentium 4 requires computer manufacturers to use DRDRAM memory. Intel announced that by 2002 it planned to have a new chipset that would support Double Data Rate DRAM (DDRDRAM), a less expensive type of memory.

By the end of 2001, Intel planned to market a low-voltage Pentium 4 for mobile computers. It also began developing chipsets for multiprocessor servers. Intel had a number of other projects in the works. These included improved transistor density, increased chip speed, and the development of a 64-bit processor so that the memory and the central processor would support 64-bit addresses. The Pentium 4 architecture should support all of these projects nicely, but by the release dates of these projects, Intel was expected to have developed a Pentium 5 architecture.

George Martin Whitson III

Hillary Clinton Announces Candidacy for U.S. Senate

> *Hillary Rodham Clinton's candidacy for one of New York's seats in the U.S. Senate made her the first First Lady to run for public office while her husband was still president of the United States.*

What: Politics
When: February 6, 2000
Where: Purchase, New York
Who:
HILLARY RODHAM CLINTON (1947-   ), First Lady of the United States from 1993 to 2001
RUDOLPH GIULIANI (1944-   ), mayor of New York City from 1994 to 2001

Testing the Waters

On February 6, 2000, First Lady Hillary Rodham Clinton declared that she would run for the U.S. Senate seat being vacated by the retiring Democratic incumbent, Daniel Patrick Moynihan of New York. The president of the United States, Bill Clinton, looked on as his wife told a packed audience of supporters and journalists of the unprecedented campaign that she intended to wage. No presidential wife before had sought public office from the White House. There had been talk of Eleanor Roosevelt running for a House or Senate seat after her husband died in 1945, but it never amounted to more than talk. Hillary Clinton opened a new chapter in the history of American first ladies. Because her Republican opponent was expected to be Rudolph Giuliani, the combative and controversial mayor of New York City, political observers anticipated one of the most hotly contested Senate races of the 2000 election year.

Clinton's decision to run grew out of two related developments. With Moynihan's retirement, Democrats needed a high-profile candidate to offset the well-known and popular Giuliani. None of the Democratic politicians mentioned as possible candidates seemed likely to excite the voters. If they hoped to retake the Senate in 2000, the Democrats could not afford to give up the New York seat.

For Clinton, the idea of running for the Senate made good political sense. She had spent her adult life in the shadow of her husband and had been subjected to the humiliation of having her husband's marital infidelity made public in the Monica Lewinsky scandal as well as to the scrutiny and censure surrounding the president's impending impeachment. Running for the Senate in her own right would be vindication and would demonstrate that she could be a political force on her own.

The possibility of a Senate race first arose in the fall of 1998 during the turmoil over Clinton's husband's impeachment. Representative Charles B. Rangel of New York told her she should think about being a Senate candidate. She asked him who in New York would want her to run. Even a hint that the First Lady might enter the contest galvanized New York Democrats. Clinton would be a candidate with instant name recognition and fund-raising ability. However, she also came with high negative ratings, but so too did Giuliani. His drive to make New York City safer had pleased many white voters but had alienated many in the black and Hispanic communities.

Until the impeachment proceedings were concluded in the early weeks of 1999, the New York Senate race took a backseat. Then Hillary Clinton's office released a statement that said she was carefully considering becoming a candidate for the senatorial seat. New York Democrats responded with enthusiasm, and the press began to look forward to a no-holds-barred campaign between the First Lady and the feisty mayor.

3155

Learning About New York

The race with Giuliani took shape quickly. The mayor began attacking Clinton as an outsider unfamiliar with the Empire State and its problems. He traveled to Arkansas to make fun of her, and the carpetbagger issue seemed likely to be a key point of Giuliani's strategy. To address that problem, the First Lady embarked on a listening tour of the entire state of New York. It began at Senator Moynihan's farm and proceeded through all sixty-two of the state's counties. No recent candidate for the Senate, or even the incumbent senator, had done that in recent years. The press scoffed at Clinton patiently hearing New Yorkers air their concerns, but it proved to be a key element in her eventual victory

Hillary Clinton and the president bought a house in Westchester County in August, 1999, and she continued her informal campaign into the autumn months. She encountered some dif-ficulties that threatened to derail her candidacy. At a public event, she kissed Suha Arafat, the wife of Yassir Arafat, leader of the Palestinians. The reaction in the Jewish community was strongly negative. After this and other missteps, political observers speculated that she might drop out of the race. In November, 1999, she reassured New York Democrats that she would formally announce for the Senate early in 2000.

Consequences

Once Clinton announced on her candidacy on February 6, 2000, the long-awaited Senate race with Giuliani got under way. As the spring of 2000 progressed, however, it was evident that personal difficulties were distracting the mayor, and Republicans questioned his enthusiasm for the contest. In April, marital problems emerged for Giuliani, and he withdrew from the race on May 19, citing his diagnosis for prostate cancer and the breakup of his marriage. Congressman Rick A. Lazio of Long Island immediately announced that he would oppose Clinton.

Although Lazio conducted a spirited race, he never found a compelling campaign to match the energy of the Clinton candidacy. She bested him in two televised debates and won a smashing victory by 12 percentage points in the general election. Clinton was sworn in as New York's junior senator in January, 2001. Her successful candidacy was a major innovation for presidential wives. She used the platform of the White House and her service as First Lady to establish herself as a national political figure. In turn, that standing enabled her to capitalize on the unique opportunity of the 2000 Senate race in New York state to win a national office.

After assuming office, Clinton suffered some embarrassment when her brother was suspected of accepting money from convicted criminals seeking presidential pardons from her husband. However, it was generally recognized that she had played no part in any wrongdoing. What other turns her political career and personal life would take remained uncertain, but with her announcement of candidacy in February, 2000, she opened a new chapter in one of the most fascinating public lives of any woman in the nation's history.

AP/Wide World Photos

Hillary Rodham Clinton announces her candidacy.

Lewis L. Gould

Hacker Cripples Major Web Sites and Reveals Internet Vulnerability

Spurious computer signals generated by a hacker known as "mafiaboy" clogged the network connections to Yahoo! and other commercial Internet sites, making it impossible for legitimate users to log on.

What: Business; Computer science
When: February 7, 2000
Where: Silicon Valley, California;
    Montreal, Quebec
Who:
RONALD DICK, Federal Bureau of
    Investigation computer investigations
    chief
"MAFIABOY," fifteen-year-old hacker from
    Montreal, Canada

The Coming of the Script Kiddies

February 7, 2000, marked a disturbing new chapter in the history of computer security, with the disruption of a number of important commercial Internet sites by a hacker. Web surfers visiting such important sites as Yahoo!, CNN.com, eBay, and Amazon.com were unable to complete their connections, sometimes for hours. They were facing the Internet equivalent of a traffic jam; however, it was one in which the crowds of users clogging the system were phantoms.

Before this point, hackers were generally perceived as bright young people who had mastered the inner workings of computer systems and generally were exploring others' equipment out of curiosity without any real understanding of the disruption they caused. A few more malicious individuals used their knowledge to damage data or to write viruses, destructive software that could replicate itself on vulnerable systems. Protective measures against hackers had concentrated primarily on maintaining the security of password files and logon procedures, as well as the creation of software to prevent the spread of viruses.

The February 7 attack used the connectivity of the Internet to enlist a large number of innocent computers, installing on them hostile software that sent requests for information to all the computers they could contact. Because the return addresses were forged to point back to Yahoo! or other commercial targets, the receiving computers were tricked into flooding the real commercial sites with messages, until the rush of traffic made it impossible for legitimate users to get through. Most disturbingly, the software that created this mess was available on the Internet and could be used without any understanding of the inner workings of computers. This new breed of hacker was commonly known by the contemptuous nickname of "script kiddies."

Denial-of-service problems are not that uncommon, particularly when so many people want to use a given resource at once that they overload its capacity. However, these problems are different from the sort of denial-of-service attack that took down Yahoo! because those lack malicious intent. They do not mean to do the system any harm, any more than the people using city streets and highways at rush hour mean to create a traffic jam. In the case of the denial-of-service attack on Yahoo!, the person who set up the requests had no interest in the answers but rather intended to tie up network resources.

A Game of Cat and Mouse

The first attack on Yahoo! started at 10:20 A.M. Pacific standard time and shut down the site for three hours before system programmers could devise filter software to shut out the false requests and let real users get through. Over the next two days, several other sites fell victim to similar attacks. One victim, Buy.com, was having its initial

3157

public offering (IPO) of stock shares on the day it was attacked, and the inability of potential investors to obtain access to its site had a definite negative affect.

Almost immediately, the Federal Bureau of Investigation (FBI) launched an investigation to locate and capture the culprits. Ronald Dick, the FBI's computer investigations chief, noted that the techniques behind this attack were so simple that a fifteen-year-old teenager could have accomplished it. He reasoned that even if the hackers were smart enough to hide or remove the traces their activity had left in the computer systems, they would probably give themselves away by bragging about their actions.

Analysis of traffic to the affected sites traced the attacks that had jammed CNN.com to a desktop machine at the University of California, Santa Barbara. Officials at Stanford University and the University of California, Los Angeles, reported finding that their computers had also been used. However, they believed that the hackers were probably not university students, but outsiders who had taken advantage of Internet connections to plant their software. Observation of activity in various Internet chat rooms and news groups frequented by hackers soon connected the attack with a hacker known by the alias of "Mafiaboy."

Consequences

Through cooperation between the FBI and the Royal Canadian Mounted Police (RCMP), the hacker was identified as a fifteen-year-old boy living in Montreal. On April 20, 2000, he was arrested by the RCMP. Canadian law prohibits the publication of the identities of juvenile crimi-

nals, so he is known to the public only by his on-line alias, "Mafiaboy." The fifteen-year-old was returned to his family and forbidden to use any computers except to do his schoolwork while he awaited trial in juvenile court.

Many of the targeted commercial sites lost business and the associated revenues. EBay offered to give credit to anyone whose online auction ended during the denial-of-service attack and who lost money as a result. Throughout the computer industry, these attacks increased awareness of security problems and of the ease with which this new generation of hackers can do damage without needing to understand what they are disrupting. As a result, many companies developed technical methods of controlling access to their sites and ensuring that only legitimate requests could get through. However, the software that filtered out falsified addresses had the side effect of slowing down many Internet service providers.

Among the public, there has been a strong move for legislation to clearly define such activities as crimes, so that the police and courts will no longer have to try to fit them into laws designed for a precomputer society. The U.S. Congress has drafted a number of bills that would create stiff federal penalties for hacking, in view of the growing economic importance of the Internet and the potential for companies and consumers to lose large amounts of money if key Web sites are shut down by malicious attacks. There has also been pressure to create a permanent body, either a subcommittee or a special committee, in the Senate to deal with issues of information security.

Leigh Husband Kimmel

Kurdish Rebels Lay Down Arms and Seek Self-Rule Without Violence

> *The Kurdish people, whose homeland once stretched across Turkey, Syria, Iraq, and Iran, ended sixteen years of fighting against Turkey, saying they would seek independence through peaceful means.*

What: Political independence; Civil war; Ethnic conflict
When: February 9, 2000
Where: Southeastern Turkey
Who:
ABDULLAH OCALAN (1948-   ), leader of the Workers Party of Kurdistan

An Ancient People

The Kurds, who number about 25 million, are a Mideastern people whose history goes back to the Persian Empire in the sixth century B.C.E. In medieval times, they converted to Islam and fell under the rule of many Mideastern conquerors. During World War I, Kurdish uprisings were put down by the Ottoman Empire (Turkey), but Turkey's loss in the war gave the Kurds hope for a homeland based on U.S. president Woodrow Wilson's self-determination of nations policy. Initially, the Treaty of Sevres, the peace treaty between the Ottoman Empire and the victorious Allies, called for a Kurdish government. However, the great Turkish leader Mustafa Kemal (later known as Atatürk) negotiated a new treaty in 1923 (the Treaty of Laussane), which eliminated the proposed Kurdistan. The Kurds revolted in 1925 and 1930, but Kemal's army defeated them. In 1937-1938, the Turks defeated another revolt with planes, bombardment, and poison gas. Thousand were killed. Kurds also revolted in Iraq and Iran, but they rejected offers for negotiated partial rights without full independence.

In the 1980's, the Iraqi government under Saddam Hussein and the fundamentalist Muslim Iranian government conducted a series of anti-Kurd attacks. In 1988, Iraq carried out a campaign against the Kurds in which it used poison gas and slaughtered male Kurdish captives, killing more than 200,000 Kurds. After the Persian Gulf War in 1991, the United Nations forced Iraq to leave the country of Kuwait, and Hussein began a new campaign against the Kurds. More than a million Kurds fled to Iran, and one-half million fled to Turkey. The United Nations established a protected autonomous region in northern Iraq for those Kurds who had not yet fled and those who returned, but fighting between Kurdish factions continued until 1999.

The Kurds who had fled to Turkey fought against the government of that country. Because of its historical tradition of grouping people according to religion, Turkey regarded the Muslim Kurds as the same ethnic group as the Turks and was willing to give extensive civil and cultural rights only to non-Muslim minorities. In 1984, the Kurds formed the Workers Party of Kurdistan (PKK), which engaged in terrorist attacks and guerrilla warfare against the Turkish government, which party members viewed as trying to stamp out Kurdish culture. Under international pressure, the government of Turkey gave the Kurds partial linguistic rights in 1991; however, the following year, Turkey resumed attacks against the Kurds, driving two million of them into refugee camps. In 1995, the Turkish army attacked the PKK in Iraq.

A Call for Peace

In 1980, Abdullah Ocalan, the leader of the PKK and principal commander of the Kurdish war against his native Turkey, fled his homeland to live primarily in Syria and Syria-controlled Lebanon, where he set up bases for his war against Turkey. In October, 1998, the Turkish government pressured the Syrians to close the

About one thousand Kurds march in support of Abdullah Ocalan in front of the United Nations building in downtown Beirut.

base camps and send Ocalan on his way. During the next four months, Ocalan traveled around Europe and Africa looking for a safe haven, winding up in February, 1999, in the Greek embassy in Nairobi, Kenya. The Greeks, longtime adversaries of the Turks in the Eastern Mediterranean, hid him in their embassy unbeknownst to the Kenyan government. After the Turkish successes against the PKK, Ocalan had lessened his demand from complete independence to a culturally autonomous region within Turkey. However, many people, especially the Turkish govern-

ment, criticized Ocalan, whom they called "the baby killer," for his terrorist methods, which included bombings, narcotics dealing, robbery, arson, extortion, blackmail, and money laundering.

Ocalan left the embassy, believing that he would be allowed to go to Amsterdam. Instead, the Kenyans arrested him and sent him to Turkey. Immediately, Kurds throughout the world demonstrated and rioted, protesting his arrest. Ocalan was put on trial in June, 1999, and sentenced to death. After his arrest, Ocalan called for an end to the war, and the PKK supported him and pulled its troops out of Turkey. On February 9, 2000, the PKK officially announced that it was ending the war.

Consequences

A faction of the PKK, the PKK Fighters for the Revolutionary Line, rejected the peace and vowed to continue fighting. The Turkish military claimed that the peace overture was a ploy by Ocalan to avoid the gallows and continued tracking down Kurdish rebels. The Turks used Kurdish guards as auxiliary forces in the war against the rebels. Of the 60,000 Kurdish guards, 1,300 were killed in the fighting, but 20,000 deserted to the rebels. Some 200,000 Turkish soldiers took part in the fighting; 4,000 were killed, and 10,000 were wounded. About 27,000 Kurdish rebels and more then 5,000 Kurdish civilians died.

In order for Turkey to join the European Union, it needed to guarantee various cultural rights, including the right for minorities to use and be taught in their own language. Therefore, in August, 2000, Ankara signed two international agreements granting this right but refused to sign others that would give the Kurds the full rights of other minorities. In February, 2001, two years after the capture of Ocalan, while he was appealing the death sentence, Kurdish leaders called for a campaign of civil disobedience to obtain full civil rights.

Frederick B. Chary

Shuttle *Endeavour* Gathers Radar Images for Three-Dimensional Maps of Earth

During an eleven-day mission, the space shuttle Endeavour *used imaging radar to capture data for assembling Earth's most comprehensive topographic map, covering 80 percent of its land surface.*

What: Earth science; Space and aviation
When: February 11-22, 2000
Where: Orbit around Earth
Who:
KEVIN KREGEL, shuttle commander
DOM GORIE, shuttle pilot

The World in Three Dimensions

On February 11, 2000, the space shuttle *Endeavour*, commanded by Kevin Kregel and piloted by Dom Gorie, lifted off from Kennedy Space Center to begin an eleven-day mission aimed at compiling the world's most comprehensive topographic map. Spearheaded by the National Imagery and Mapping Agency (NIMA) and the National Aeronautics and Space Administration (NASA), the Shuttle Radar Topography Mission (SRTM) was an international project utilizing new methods for mapping Earth's land surface area using imaging radar. The advantages of radar over other types of imagery such as photography are its ability to penetrate atmospheric obstructions such as clouds and to be used at night and in all weather conditions. Imaging radar was previously used aboard NASA's Magellan spacecraft between 1989 and 1994 to create surface maps of the cloud-covered planet Venus.

The impetus for SRTM was a 1995 decision by the U.S. Armed Forces Joint Chiefs to acquire a worldwide topographic map. Unlike the United States, which is relatively well-mapped through a mosaic of fifteen hundred sets of aerial measurements, the world has not been mapped comprehensively. In some countries, maps are inaccurate or nonexistent, especially for difficult-to-travel places such as deserts, mountain areas, and dense tropical rain forests.

Military benefits of having improved worldwide topographic data include their use in missile and weapons guidance systems, battlefield management applications, and flight simulators. More readily available topographic data are also of importance to a wide range of scientific and civilian uses, ranging from climate change modeling to identifying optimal locations for cellular telephone towers. For the earth sciences, three-dimensional land surface data is useful for studies of erosion, natural hazards, and vegetation as well as for evaluations of subsurface geologic structures and tectonic activities. Commercial aircraft applications include simulator systems and use in route planning and navigation. Digital elevation maps such as those collected by SRTM can be also combined with LANDSAT satellite images, aerial photographs, or other spatial information within a digital geographic information system.

U.S./Canadian RADARSAT and European remote sensing satellites were initially considered for the project. However, it was determined that commercial satellites were not well suited to collecting three-dimensional data because of problems in obtaining two clear images. Scientists at NASA's Jet Propulsion Laboratory (JPL) proposed using the space shuttle equipped with radar hardware that had previously been aboard shuttle missions. An earlier shuttle mission in 1994 used imaging radar to map approximately 30 percent of Earth's land surface.

Scientists at JPL designed SRTM's three-dimensional data collection effort to take advantage of a process known as interferometry. Interferometry involves capturing return radar

3161

signals from two slightly different locations to measure their patterns of interference. In contrast to a 1994 SRTM mission that captured radar data using two passes over Earth, the February, 2000, mission used two antennae to simultaneously collect radar data. An analogy representing interferometry can be seen in the ripples formed when two pebbles are dropped into a pool of water. The tiny waves travel outward from the center in the form of concentric circles that meet and interfere with each other. Like the interference between water ripples, SRTM measured interference between the two radar signals to generate topographic data. The separate antennae generate a "stereo" view in the same way that offset between a person's eyes enables depth of field and three-dimensional object visualization.

Endeavour's radar antenna contained panels for receiving return pulses that bounce off Earth's surface. One of these was positioned in the *Endeavour*'s payload bay, and the other was located at the end of a mast that telescoped 200 feet (60 meters) away from the spacecraft. Measuring the length of five school buses when fully extended, this lightweight boom was the largest rigid structure ever flown in space. At the conclusion of data collection, the boom could be retracted accordion-style, back into the payload bay.

The Mission

To concentrate on Earth's land surface areas, *Endeavour* maintained an altitude of 139 miles (223 kilometers) within an orbit that had a 57-degree inclination relative to Earth's equatorial plane. This facilitated data collection within a 145-mile (225-kilometer) zone in an area between 60 degrees north and 56 degrees south latitude. Traveling at a speed of 17,000 miles per hour (7.5 kilometers/second), the spacecraft completed 16 orbits each day for a total of 176 orbits during the eleven-day mission.

Endeavour's initial mapping effort began over the Maldives in the Indian Ocean. Astronauts formed two mapping teams working alternating twelve-hour shifts to monitor data collection equipment. During each ninety-second period, SRTM was capable of mapping an area about the size of Florida. In total, 46 million squares miles (119.05 million square kilometers) of Earth's land surface area were mapped at least one time (80 percent of Earth's land area). An additional 43 million square miles (112.6 million square kilometers) were mapped twice (75.3 percent of Earth's land area). Just more than 11.7 terabytes of data were captured, the equivalent of 20,600 CD-ROMs. This information was later analyzed at a ground facility to produce data sets for three-dimensional mapping.

Consequences

SRTM's products included 295-foot (90-meter) resolution (referring to the horizontal distance separating sampling points) topographic data for most of the land surface areas of the world. This data was vastly superior to the 0.6-mile (1-kilometer) data previously available for some portions of the world. Some of the first SRTM images released showed regions of New Mexico, the United States, and Rio Sao Francisco, Brazil. In addition to lower resolution data, a higher resolution 98-foot (30-meter) data set would be available for scientific research in some parts of the world. The U.S. Geological Survey planned to make this 98-foot topographic data available for the entire United States. SRTM data has already assisted scientists in developing an improved understanding of terrain patterns in remote areas. For example, although existing two-dimensional radar imagery of the Amazon Basin shows little relief, SRTM three-dimensional imagery reveals previously undetected drainage systems.

Thomas A. Wikle

NEAR Is First Satellite to Orbit an Asteroid

The Near Earth Asteroid Rendezvous (NEAR) spacecraft went into orbit around the asteroid Eros on February 14, 2000. It orbited Eros for a year, gathering data on the chemical composition, mineralogy, shape, and structure of the asteroid.

What: Astronomy; Space and aviation
When: February 14, 2000
Where: Asteroid Eros
Who:
ROBERT W. FARQUHAR, NEAR project manager
ANDREW CHENG (1951-   ), NEAR project scientist
JOSEPH VEVERKA (1941-   ), NEAR project imaging team leader

The Mission

The Near Earth Asteroid Rendezvous (NEAR) spacecraft was launched from Cape Kennedy, Florida, on a Delta II rocket on February 17, 1996. NEAR was the first spacecraft launched in the National Aeronautics and Space Administration (NASA) Discovery program. The Discovery program reflected NASA's move away from infrequent, large space exploration missions and toward more frequent, less expensive, highly focused missions. The NEAR spacecraft, designed by the Applied Physics Laboratory at Johns Hopkins University, incorporated several instruments developed by the laboratory for earlier Department of Defense spacecraft, in an effort to emphasize simplicity and reliability. The instruments included an X-ray/gamma-ray spectrometer to determine the chemical composition of Eros, a CCD Imaging Detector to photograph the surface of Eros with a resolution down to 3.3 feet (1 meter), and a magnetometer, to determine if Eros has any magnetic field.

NEAR was scheduled to go into orbit around Eros on January 10, 1999. However, on December 20, 1998, the NEAR spacecraft's engine shut down just seconds into a scheduled 20-minute burn designed to put the spacecraft into position to orbit Eros. The computer on the NEAR spacecraft detected accelerations exceeding the programmed limits, so the computer commanded the engine to shut down and put the spacecraft into safe mode, in which it does nothing except wait for radio instructions from Earth. On the evening of December 21, 1998, the NASA deep space network, which tracks and communicates with spacecraft, picked up a weak signal from NEAR. Two-way communications with the NEAR spacecraft were reestablished on December 22, 1998, but it was too late to get the spacecraft into position to orbit Eros in 1999.

The NEAR spacecraft made a fast flyby, passing within 2,600 miles (4,182 kilometers) of Eros at a speed of about 0.62 mile per second (1 kilometer per second) on December 23, 1998. The NEAR science and engineering teams, including NEAR project manager Robert W. Farquhar and project scientist Andrew Cheng, quickly devised plans to salvage as much science from the flyby as possible, while also trying to devise a plan to put NEAR into orbit around Eros a year later than originally planned. During the flyby, the NEAR spacecraft took several hundred photographs of Eros, providing information on the size and shape of the asteroid.

On January 3, 1999, NEAR's engine was programmed to fire for twenty-four minutes, increasing the spacecraft's speed by 2,085 miles per hour (3,355 kilometers per hour). A smaller engine firing on January 20, 1999, increased NEAR's speed by another 31 miles per hour (50 kilometers per hour), putting NEAR in an orbit around the Sun that closely matched the orbit of Eros and allowing NEAR to make a second try at orbiting Eros in February, 2000.

3163

NEAR Orbits Eros

The NEAR spacecraft went into safe mode again on February 2, 2000, when preparing for an engine firing that was to place it into position to orbit Eros on February 14. The engineering team quickly designed an alternative sequence of engine firings to maneuver the spacecraft into the right position to go into orbit around Eros. On February 3, the engine was turned on to slow the spacecraft from a speed of 43 miles per hour (69 kilometers per hour) to 18 miles per hour (29 kilometers per hour) relative to Eros. A second engine burn, lasting twenty-three seconds, was conducted on February 8, increasing the speed of the spacecraft to about 22 miles per hour (35 kilometers per hour) relative to Eros. At 10:33 A.M. Eastern standard time on February 14, the NEAR spacecraft fired its engine for fifty-seven seconds, slowing its approach by about 20 miles per hour (32 kilometers per hour). This allowed the spacecraft to be captured by the weak gravity of Eros, in an orbit about 124 miles (200 kilometers) above the center of Eros.

After becoming the first spacecraft to orbit an asteroid, the spacecraft was renamed the NEAR Shoemaker to recognize the achievements of planetary scientist Eugene Shoemaker. On April 11, 2000, the Near Shoemaker spacecraft fired its engine for five seconds to lower its orbit to 62 miles (100 kilometers) above Eros. Later in April, a series of engine firings reduced the orbit to about 31 miles (50 kilometers) from Eros, to allow the scientific instruments to achieve better sensitivity. NEAR Shoemaker continued to orbit Eros until February 12, 2001, when scientists, including NEAR imaging team leader Joseph Veverka, programmed the engine to fire in a series of maneuvers designed to provide closeup images of the surface. The NEAR Shoemaker spacecraft conducted the first landing on an asteroid at 3:02 P.M. Eastern standard time on February 12 and continued to transmit signals to Earth for several days.

Consequences

During the year that the NEAR Shoemaker spacecraft orbited Eros, its scientific instruments provided the first closeup measurements of an asteroid. The camera showed that Eros is about 21 miles (34 kilometers) in length and is shaped

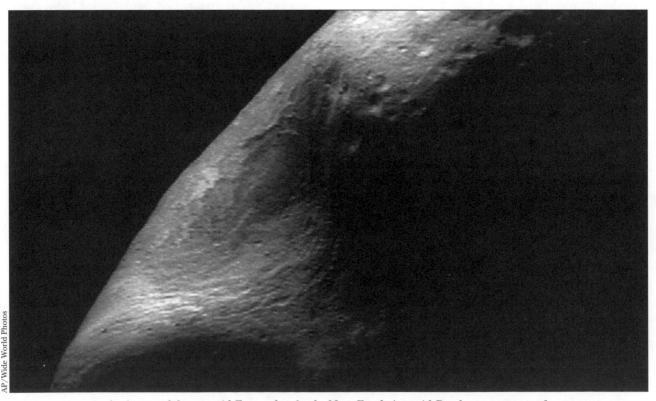

An image of the asteroid Eros, taken by the Near Earth Asteroid Rendezvous spacecraft.

AP/Wide World Photos

like a peanut, with two large ends and a narrower "saddle" at the middle.

Because an asteroid is exposed to continual bombardment by smaller objects, the number of craters it has accumulated can be used to determine the age of its surface. Some regions on Eros are saturated with craters, the largest one measuring 3.4 miles (5.5 kilometers) across. However, the saddle region of Eros has very few craters, indicating that the saddle has a relatively "young" surface. Photographs of Eros show areas with numerous boulders, some as large as 330 feet (101 meters) across. This observation was unexpected, as most other solar system bodies are dominated by craters on their surfaces with few obvious boulders.

Measurements of the chemical composition of Eros by the X-ray/gamma-ray spectrometer showed it to be a primitive object that has undergone little or no melting. Nickel and iron are present in abundance on the surface of Eros. The only major difference between the composition of Eros and that of primitive meteorites is Eros's low content of sulfur, one of the first elements lost when a stony body is heated. Therefore, Eros may have experienced some minimal degree of heating. Earth, in contrast, has experienced differentiation, a process of melting that separates an object into a metallic core and a stony outer layer.

The NEAR project demonstrated that a small, inexpensive spacecraft can return significant scientific results and can be designed with sufficient redundancy to accomplish its mission even after unexpected setbacks.

George J. Flynn

Investigation Finds Misconduct in Los Angeles Police Department

Members of an antigang police task force of the Los Angeles Police Department (LAPD) were discovered to have interrogated and brutalized residents without probable cause and then to have fabricated evidence to wrongfully convict more than one hundred Hispanics during the 1990's.

What: Civil rights and liberties; Crime; Law
When: March 1, 2000
Where: Los Angeles, California
Who:
BERNARD C. PARKS (1943-), Los Angeles chief of police
RAFAEL PEREZ (1967-), former Los Angeles police officer

The Fight Against Gangs

The Los Angeles Police Department (LAPD) has several geographic divisions and often establishes special units in which officers are given autonomous authority, developing their own headquarters, jargon, logos, radio frequencies, rituals, and slogans, with no direct supervision. The LAPD pioneered scientific policing, in which officers are evaluated by the number of arrests that they make, but that practice has been criticized by those who believe it leads to officers making unnecessary arrests in order to improve their records.

Beginning in the 1980's, the densely populated Rampart division, located halfway between Hollywood and downtown Los Angeles, experienced an influx of immigrants from Central America. Within the division, which had the highest crime rate in the city, the Community Resources Against Street Hoodlums (CRASH) team, consisting of twelve to twenty officers, was established to crack down on Hispanic gangs, which were selling drugs and collecting protection money. In 1994, the LAPD estimated that in Los Angeles, there were 403 gangs, which had committed almost 11,000 crimes, including 408 homicides. Rampart accounted for 10 percent of all gang crimes.

Misconduct Emerges

Evidence of police misconduct throughout the LAPD has tended to be eclipsed by the gigantic task of fighting crime. The CRASH unit's mission was to eradicate the gangs, but often the charges brought against gang members did not result in convictions because witnesses feared gang retaliation if they testified. To get around this problem, CRASH officers committed perjury and planted evidence to frame gang members to get them off the street and into prison. A deputy district attorney prosecuting a case resulting from one of the arrests questioned the credibility of a particular officer, Rafael Perez, and he asked the judge to dismiss the charges against the defendant in June, 1997. Later that month, when vital evidence (cocaine) in another case involving Perez was missing, Perez mysteriously produced the evidence. Accordingly, the prosecutor informed his superiors about the possibility of a "dirty cop"; however, his efforts did not result in notification of the LAPD.

Then three extraordinary incidents occurred that brought into question the integrity of the CRASH unit. In November, 1997, one officer robbed a bank. In February, 1998, a suspect in custody was severely beaten. In March, 1998, eight pounds of cocaine valued at about $1 million were missing from LAPD evidence lockers. Perez was arrested for the cocaine theft in August, 1998.

Perez's trial ended in a hung jury. In 1999, a deputy district attorney who was a member of the

Rampart special prosecution team negotiated a plea bargain with Perez. In exchange for Perez's testimony about other officers who abused police authority, Perez was given a five-year sentence for the cocaine theft and immunity from prosecution for all other offenses.

On September 21, 1999, Chief of Police Bernard Parks convened an internal board of inquiry within the LAPD to determine whether police procedures were being violated. On March 1, 2000, the board released a 355-page report, alleging that CRASH engaged in serious misconduct. In response, the LAPD disbanded all antigang units, fired five officers, and disciplined thirty others; nine officers resigned. Three Rampart division officers were tried, and two were convicted of felonies (one was acquitted), all based in part on Perez's testimony, though their convictions were reversed on appeal one month later, and new trials were pending in early 2001.

The board of police commissioners then formed a Rampart independent review panel. The panel later concluded that civilian control over the LAPD was inadequate and that LAPD supervisors were not doing their jobs to monitor subordinates. Like the LAPD investigation, the inquiry had a narrow scope; therefore, some people charged that the actual extent of corruption, which some believed might also involve the district attorney's office, remained unknown.

Victims, meanwhile, brought civil suits for harassment, false convictions, and wrongful death. More than one hundred convictions have been reversed, and the LAPD has been sued for millions of dollars in damages in more than two hundred separate cases. In one case, in which Perez admitted shooting an unarmed man, leaving him for dead, then arresting him and providing false testimony to send him to jail, the court awarded $15 million in damages, the largest LAPD payout in history.

Consequences

The fact that most of the victims of police abuse were Hispanic was of particular interest to the U.S. Department of Justice, which filed a civil

The Rampart section of Los Angeles, an eight-square-mile district west of the city's downtown skyline.

rights lawsuit against the LAPD in 2000, claiming that the latter had demonstrated a "pattern or practice" of violating the federal civil rights of Hispanics. In November, 2000, the Department of Justice negotiated a settlement with the Los Angeles City Council, approved by a federal district court judge in Los Angeles.

Under the settlement, an independent monitor was placed in charge of a computerized early warning system to identify problem officers, a new LAPD unit was to investigate officer-involved shootings, and more power was given to the civilian police commission and its inspector general. The monitor was to be jointly selected by the Justice Department and the Los Angeles mayor; but if they did not agree, the federal judge was to make the selection. Some members of the city council, however, wanted to abort the consent decree.

Michael Haas

Pope Apologizes for Catholic Church's Historical Errors

Pope John Paul II led observances for a Day of Pardon, acknowledging instances when members of the Roman Catholic Church acted in a manner that harmed people of different cultures or religions during the last one thousand years and asking for forgiveness for sins committed by both the Church and its members.

What: Religion
When: March 12, 2000
Where: Vatican City
Who:
JOHN PAUL II (KAROL JÓZEF WOJTYŁA, 1920-), pope of Roman Catholic Church from 1978

Forgiveness and the Jubilee Year

On March 12, 2000, the First Sunday of Lent, in preparation for Easter, Pope John Paul II led a worship service in St. Peter's Basilica in Vatican City for the express purpose of publicizing the need for Roman Catholics and other Christians to acknowledge that there had been instances in the last one thousand years of Christian history when Christians fundamentally violated the human rights of other people. The pope also asked Christians to acknowledge present-day behaviors that have led to a decline in respect for explicit religious values, lack of compassion for the poor, and violation of the right to life of the unborn. During the mass, the pope led prayers asking for forgiveness of sins committed by members of the Church against people who follow other religions, particularly Jews; people whose indigenous cultures were damaged by missionaries who imposed their faith rather than offering instruction in it; and people, including women, who have traditionally suffered some form of discrimination by the Church.

International Theological Commission Document

During his homily on the Day of Pardon, Pope John Paul II made reference to a document titled "Memory and Reconciliation: The Church and the Faults of the Past," published in December, 1999, and authored by the Vatican's International Theological Commission. This document, written at the behest of the pope, is an in-depth investigation into the historical and theological foundations necessary to permit a process of acknowledgment of past sinful behavior against others by the Catholic Church as well as by baptized members acting on behalf of the Church. This acknowledgment was combined with a request for forgiveness as well as a prayer for reconciliation with those who continue to suffer oppression as a result of past historical actions.

The document consists of six major sections on topics such as the problems associated with having people feel responsible for actions committed in the past, Biblical precedents for asking for both personal and corporate forgiveness, and the theological foundations of the Church that are simultaneously holy and sinful. Other topics include how historical events must be interpreted in terms of an overall theological perspective, the necessity to refrain from judging past persons and events in terms of contemporary moral standards, the ethical dimensions of asking for and granting forgiveness, and the implications this document of apology has for the Church's future missionary activities.

"Memory and Reconciliation" is an unusual document in many respects. Although there had been instances in both the remote and recent past when popes publicly acknowledged abusive behavior by the Church as a social institution, "Memory and Reconciliation" makes a clear request for forgiveness, a request previous papal

3169

AP/Wide World Photos

Pope John Paul II embraces a wooden crucifix during the Day of Pardon Mass in St. Peter's Basilica at the Vatican.

documents did not include. The document draws explicit parallels between instances of behaviors that are sins against God and behaviors that are sins against other persons. Forgiveness must be sought for both categories of sin because reconciliation with God is not possible unless one has also been reconciled with one's neighbors. As a prerequisite for celebrating the beginning of the third millennium of Christianity on January 1, 2000, Pope John Paul II authorized the publication of "Memory and Reconciliation" in December, 1999, to ensure that the Jubilee Year of 2000 started off in a penitent state of mind for Christians.

Consequences

Many groups, both religious and secular, while praising Pope John Paul II for his sincerity and honesty in acknowledging past abuses of others by Christians, also criticized the document "Memory and Reconciliation" because of its lack of any specific accountability of individual Christians for their own violent and abusive actions. "Memory and Reconciliation" distinguishes between objective and subjective responsibility for past abuses, that is, between actions that an individual knew were wrong at the time and actions now seen in the light of history as having been abusive. No persons are listed by name as being responsible for past acts of oppression, and the Catholic Church as a corporate entity is not blamed for any specific policies that might have led members to act in the Church's name in an oppressive manner. Past historical mistakes such as the Inquisition and the Crusades are not mentioned by name. Jewish groups in particular criticized the document because it makes no mention of either the Holocaust or other examples of persecution suffered by Jews as a direct result of anti-Jewish ideology embedded in Christian theology.

Victoria Erhart

Tribune Company and Times Mirror Company to Merge

> *The Tribune Company, owner of the Chicago Tribune and other interests, acquired the Los Angeles Times and additional communication resources of the Los Angeles-based Times Mirror Company.*

What: Communications; Business
When: March 13, 2000
Where: Chicago and Los Angeles
Who:

OTIS CHANDLER (1927-   ), former publisher of the *Los Angeles Times*
JOHN W. MADIGAN (1937-   ), chairman and chief executive of the Tribune Company
MARK WILLES (1941-   ), chairman and chief executive officer of the Times Mirror Corporation

Surprise in Los Angeles

On March 13, 2000, it was announced that the Tribune Company, owner of the *Chicago Tribune* and numerous other media outlets, had acquired the *Los Angeles Times*, one of the premier newspapers in the United States. In addition to the major newspaper of Los Angeles, the Tribune Company took over the many properties owned by the Times Mirror Company, including several other daily newspapers and eighteen magazines.

The total price was reported to be $6.46 billion, and Times Mirror shares were to be purchased at almost twice the current value of the stock. In addition to purchasing stock, the Tribune Company would also assume the Times Mirror debt of $1.4 billion. By 2000, mergers between large corporations, including those in the communications industry, had become more and more common. A few months earlier, Time Warner and American Online had agreed to merge. The belief that larger was better and that only the giants would survive was undoubtedly a consideration for all parties in the Tribune and Times Mirror agreement.

The Tribune was already a multimedia giant, owning the *Chicago Tribune* and other newspapers, twenty-two television stations, several radio stations and cable stations, and the Chicago Cubs baseball team. Its acquisition of Times Mirror, the owner of New York's *Newsday* as well as the *Los Angeles Times*, would give the Tribune Company major newspapers in the nation's three most populous cities: Chicago, Los Angeles, and New York City.

Trouble in Chandler City

Since the 1880's, the Chandler family had been synonymous with the *Los Angeles Times* and with the city of Los Angeles. Harrison Gray Otis acquired the *Los Angeles Times* in 1884. He was succeeded as publisher by his son-in-law, Harry Chandler, who served from 1917 to 1944, and who was in turn followed by his son, Norman Chandler, from 1944 to 1960. Southern California politicians, including police chiefs, consulted the Chandlers on issues large and small. Major civic boosters of Los Angeles, the Chandlers were politically conservative and supporters of the Republican Party. The influence of the *Los Angeles Times* spread to the state capital of Sacramento and to Washington, D.C.

In 1960, Norman Chandler's son, Otis, only thirty-two years old, took over the *Los Angeles Times*. The move was controversial within the family; some members thought an older Chandler should have been chosen. The rift deepened as Otis Chandler abandoned his family's partisan political posture and moved the paper in a moderate, at times a liberal, direction. In the past, the paper had been prosperous but known for having a large number of advertisements, and its journalistic reputation was slight in com-

3171

parison with other major newspapers. However, Chandler turned the *Los Angeles Times* into one of country's most respected newspapers, and the parent company, Times Mirror, became a national media force.

Chandler retired in the 1980's. Subsequent publishers and editors were not family members, and some had little journalistic background. Critics complained that the paper and the corporation were drifting, both as purveyors of news and as a prosperous economic commodity. Although long retired, in 1999, Chandler harshly criticized the paper's editors for bad judgment and a conflict of interest. The winds of change began to blow, although secretly. The Chandler family trusts, which controlled about 65 percent of Times Mirror stock, privately negotiated the sale with Tribune representatives. Mark Willes, chairman and chief executive officer of the Times Mirror Company, was not informed about the pending sale until the very end, nor was the *Los Angeles Times*' publisher, Kathryn Downing.

Consequences

As a business venture, the merger could be readily defended. In an environment of communication giants, the Times Mirror Company was simply not large enough. Willes observed that for the Chandler family, the sale was "strategically compelling." In addition to the money received from the sale, which would allow the family to diversify its financial resources, it was agreed that the Chandlers would have four positions on the Tribune's sixteen-member board as well as 40 percent of a new board that was to be established to run the *Los Angeles Times*.

The Tribune Company, which already had considerable assets, strengthened its position with the Times Mirror acquisitions. John Madigan, Tribune chairperson, noted that the combined visits to the *Los Angeles Times* and *Chicago Tribune* Web sites totaled 3.4 million per month, while the *New York Times* Web site received only 1.8 million visits per month. The Tribune claimed that it would also be easier for advertis-

Tribune chairman John Madigan (left) and Tribune president Jack Fuller.

AP/Wide World Photos

ers to deal with a single business entity.

However, there was considerable anxiety and anger over the sale of the *Los Angeles Times*. Critics noted that Los Angeles would become the only major city in the United States without a locally owned daily newspaper. Others feared that Los Angeles might simply become a colony for a corporation headquartered more than halfway across the country. It was argued that the newspaper was one of the very few institutions that had provided a center or a focus for Los Angeles and its environs. The Chandlers had long been bene-

factors to Los Angeles, instrumental in bringing water from the Sierras in the 1920's and the Dodgers baseball team from Brooklyn, and in the construction of such urban landmarks as the city's Music Center. Some accused the Chandler family, so directly involved in the civic life of Los Angeles, of turning its collective back on its position of community leadership. Whatever the response to the sale of the *Los Angeles Times*, there was broad agreement that a new day had arrived for the City of Angels.

Eugene S. Larson

Smith and Wesson to Change Handguns to Avoid Lawsuits

> *Major firearms manufacturer Smith and Wesson signed an agreement with the U.S. government to install child safety locks on all firearms and make a number of other changes in its marketing, manufacturing, and design practices to avoid lawsuits.*

What: Social reform; National politics
When: March 17, 2000
Where: U.S. Department of Housing and Urban Development, Washington, D.C.
Who:
ED SHULTZ (1941-   ), president and chief executive officer, Smith and Wesson
ANDREW M. CUOMO (1957-   ), secretary, Department of Housing and Urban Development
BILL CLINTON (1946-   ), president of the United States from 1993 to 2001

Keeping Guns Away from Children

At a March 17, 2000, press conference, U.S. Department of Housing and Urban Development secretary Andrew M. Cuomo announced that Smith and Wesson, located in Springfield, Massachusetts, and one of the nation's largest and oldest gun manufacturers, had signed a landmark legal agreement to make a number of changes in the way it designed, made, and marketed its guns. Although critics from the National Rifle Association and other gun lobbies denounced the pact, President Bill Clinton praised it as part of a widespread effort to reduce gun violence and save lives, especially the lives of children. Secretary Cuomo called the agreement a framework for "a new enlightened gun policy for this nation."

Shootings in schools in Colorado, Arkansas, and Kentucky during the past several years had contributed to the growing pressure on gun manufacturers to take more responsibility for their products. For example, since October 30, 1998, more than two dozen cities, counties, and other groups filed lawsuits against gun manufacturers seeking to recover damages for the law enforcement and public health expenses incurred from gun injuries and deaths. Supporters of such lawsuits argued that they are a reasonable way of cutting homicides and accidental deaths from firearms. They claim that vital safety features such as trigger locks, making and keeping the "ballistic fingerprint" of every gun before it leaves the factory, and reasonable controls on distribution can keep guns out of the wrong hands. Opponents such as the National Rifle Association argue that such efforts merely make private gun ownership difficult and actually cost lives and dollars.

Safety Locks and Background Checks

Under the March 17, 2000, agreement, Smith and Wesson agreed to include child safety locks, require background checks on gun buyers both at retail stores and gun shows, take ballistic fingerprints of its guns, and work on designing and producing "smart" guns that can be fired only by their legal owners. A child safety lock is a device that prevents a gun from being fired accidentally or by someone without a key to remove the lock. Background checks are intended to prevent anyone with a criminal record or not of legal age from buying a gun.

Although background checks were already required at all licensed retail outlets, they had often been avoided at gun shows and swap meets. Each gun's ballistic characteristics, the specific and unique marks made by the gun's rifling, were to become part of a permanent database linked to the gun's serial number. Such a ballistic fingerprint could be used by law enforcement officials to connect a bullet recovered from a victim or a crime scene with a specific gun and the person who is listed as its purchaser. A "smart" gun would contain a device to prevent its being fired unless the device recognized the shooter as its legal owner.

Under the terms of the agreement, state and local governments would drop pending lawsuits against the company, and the federal government would not file suit against the company as it had said it would in December, 1999, if a settlement had not been reached. Other gun manufacturers could sign on as well. Another consequence of the agreement was that law enforcement officials in 190 U.S. communities agreed that Smith and Wesson would be their source for all gun purchases, an agreement of considerable economic value to the company. Therefore, other gun makers immediately filed suits in federal courts against what they viewed as an illegal conspiracy to restrict free trade.

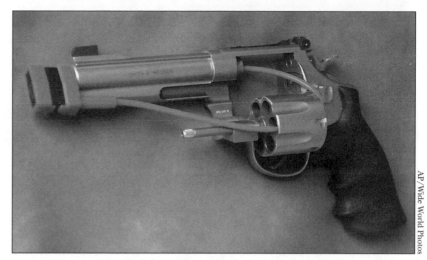

A cable-type gun lock on a Smith and Wesson revolver.

AP/Wide World Photos

Consequences

Immediately after the agreement was signed, the National Rifle Association and several other gun-rights groups called for a boycott of Smith and Wesson, filled Web pages and the airways with vigorous attacks on the agreement, and caused bills to be introduced in state legislatures prohibiting municipalities from filing lawsuits against gun makers. Various other groups opposed to any sort of gun controls attacked all the gun-control lobbies and the Clinton administration for the parts they played in getting the agreement. The ultimate outcome of these attacks on the agreement is uncertain. However, what is clear is that Ed Shultz, Smith and Wesson's president and chief executive officer, and at least one other of the company's chief executive officers are no longer with the company.

In addition, three months after signing the agreement, Smith and Wesson announced that it was shutting down two plants in response to what it called "normal summer softness," but which may have been influenced by the agreement and the resulting consumer boycott. Nearly twenty states passed legislation that prohibited cities and counties from suing gun manufacturers. Seven gun makers and an industry group sued the Department of Housing and Urban Development and sixteen cities on April 26, 2000, over the plan to give preferential treatment to Smith and Wesson when buying guns for law enforcement agencies. The suit, filed in an Atlanta federal court against a variety of federal and state officials as well as a number of cities, alleged that the plan constituted an illegal conspiracy in restraint of trade and was therefore illegal. The general counsel for Beretta USA, one of the seven suing companies, demanded that cities stop using purchasing power to control the design and distribution of firearms.

Theodore C. Humphrey

Taiwan's Nationalist Party Is Voted Out

Chen Shui-bian, leader of the Democratic Progressive Party (DPP), was elected president of the Republic of China (Taiwan) with 39 percent of the popular vote, defeating an independent and a candidate from the Nationalist Party.

What: National politics
When: March 18, 2000
Where: Taiwan (also known as Republic of China)
Who:
CHEN SHUI-BIAN (1950-), leader of the Democratic Progressive Party and president of the Republic of China from 2000
LEE TENG-HUI (1923-), president of the Republic of China from 1988 to 2000
LIEN CHAN (1936-), vice president of the Republic of China
JAMES SOONG (1942-), former governor of Taiwan
JIANG ZEMIN (1926-), president of the People's Republic of China from 1993

The Taiwanese Presidency

In the constitution adopted in 1947, the president and vice president of the Republic of China (Taiwan) were to be elected for six-year terms by the national assembly. The first person elected president was Chiang Kai-shek, leader of the Nationalist Party, or Kuomintang. When the Nationalists suffered defeat by the Chinese Communist Party (CCP) on mainland China and fled to Taiwan in 1949, the Nationalist Party declared martial law on the island and banned those opposition parties of which it disapproved. Protected by a Mutual Defense Treaty with the United States between 1952 and 1979, the Nationalist Party made educational, social, and economic reforms that resulted in rapid economic growth. Political reforms granting greater freedom and democratization followed in the 1980's under President Chiang Ching-kuo. They culminated in the lifting of martial law and constitutional amendments to allow direct elections for president and vice president. The first popular election for president was held in 1996; Lee Teng-hui of the Nationalist Party won.

The 2000 Presidential Election

Because both the original and amended constitution limited the office of president to two terms, and Lee was already serving his second term, he was not eligible to compete in the 2000 election. Three candidates emerged, Vice President Lien Chan, who was tarnished by the same corruption that had made the Lee presidency unpopular; James Soong, the popular former governor of Taiwan, whom the Nationalist Party did not endorse and who therefore ran as an independent; and Chen Shui-bian of the Democratic Progressive Party (DPP).

Chen was a lawyer and had been imprisoned in the 1980's for activities in the then banned DPP. He had served for one term as mayor of Taipei but had not won reelection. Chen had a flamboyant campaign style and credentials as a human rights activist. He was a native of Taiwan (as was Lien) and chose lawyer Annette Lu as his running mate, the Republic of China's first female vice presidential candidate. The DPP advocated independence for Taiwan, while the Nationalists stood for eventual unification with a democratic mainland China.

During the campaign, Jiang Zemin, president of the People's Republic of China, and other leaders from the mainland made repeated statements opposing the elections on Taiwan and the candidacy of Chen in particular. They warned that they would use force against Taiwan if Chen won and if he moved to declare Taiwan an independent nation. They mobilized troops across the Taiwan Strait to back up their threat. In the 1996 elections, the People's Republic of China had made similar threats against Lee Teng-hui, and had fired missiles that landed close to Taiwan's ports. However, although no one will ever know how many people voted for Chen to show defiance to the People's Republic of China, experts believe the threats backfired. With the Nationalist Party vote split between Soong (37 percent) and Lien (23 percent), Chen narrowly won with 39 percent of the vote.

Consequences

Chen was the first elected president not from the Nationalist Party. His election was a distinct triumph of democracy in practice on Taiwan. The smooth transition of power was also a testimony of the political maturity of the people of Taiwan. As president, Chen modified some of his earlier rhetoric. He assured his nervous compatriots and the world that he would not declare Taiwan an independent nation unless the Peo-

ple's Republic of China invaded, and he asserted his readiness to resume talks with the mainland leaders.

The United States warned China not to provoke military action and maintained its Seventh Fleet in the region. Although the People's Republic of China continued its extremely hostile stance against Chen, it did not make overtly hostile moves. Experts believed that China was not likely to initiate military action against Taiwan because although its army was five times the size of Taiwan's, it was undertrained and poorly armed. Similarly, China's larger air force consisted of mostly older planes, and it did not have the naval capacity to launch and sustain an invasion. Taiwan's army and air force, though smaller, were well trained and well equipped. In addition, China and Taiwan had built close economic ties, with Taiwanese businesses investing more than $30 billion dollars in mainland ventures. Any hostilities would seriously set back the economies of both sides.

Chen faced formidable obstacles domestically. The DPP had no experience in governing

Taiwanese lawmakers show their support for a possible impeachment of President Chen Shui-bian.

3177

and no trained cadre of people to staff important positions. In 2000, Chen narrowly avoided a recall election for high-handed executive orders without proper consultation with the legislature. The DPP held only about one-third of the seats in the legislature, where its members' roles had been to be disruptive, rather than to lead. Long accustomed to being the opposition, Chen and his DPP colleagues needed to develop expertise in governing and consensus formation to reverse the disruptions to Taiwan's economy caused by his election.

Most of the citizens of Taiwan are economically prosperous, well educated, and politically mature. They demonstrated their commitment to democracy in legislative and presidential elections, most notably in electing Chen of the DPP president in 2000. With this election, democracy appeared to have become well established in Taiwan.

Jiu-Hwa Lo Upshur

Highest-Ranking Army Woman Charges Sexual Harassment

> *Lieutenant General Claudia J. Kennedy, the highest-ranking woman in the U.S. Army, charged a fellow three-star general with sexual harassment.*

What: Military; Gender issues
When: March 30, 2000
Where: Washington, D.C.
Who:
CLAUDIA J. KENNEDY (1947-),
   lieutenant general in the U.S. Army
LARRY G. SMITH (1944-), major
   general in the U.S. Army

A Distinguished Career

On March 30, 2000, the Pentagon confirmed that the U.S. Army's highest-ranking woman, Lieutenant General Claudia G. Kennedy, filed a sexual harassment complaint against Major General Larry G. Smith, a married three-star general with a previously unblemished record. Kennedy, at the time one of only three women serving as three-star officers, accused Smith of groping her in her Pentagon office in 1996. Smith denied any wrongdoing, and the Army's office of inspector general initiated an investigation into the charges filed by the highly respected Kennedy.

Kennedy was born into and grew up in an Army family. Upon her graduation in 1969 from Southwestern University at Memphis with a bachelor's degree in philosophy, her father commissioned her into the U.S. Army as a second lieutenant. At that time, women served in a separate Women's Army Corps and were not allowed to be pregnant, command male soldiers, or advance beyond the rank of colonel. She began her career with a staff assignment at Fort Devens, Massachusetts, and then as a recruiting officer in Concord, New Hampshire. She continued with positions as strategic intelligence officer in the 501st Military Intelligence Group in Korea; as director of intelligence, Forces Command, in Fort

McPherson, Georgia; and as deputy commander, U.S. Army Intelligence School at Fort Huachuca, Arizona.

Kennedy also served stints as commander of the Third Operations Battalion in Augsburg, Germany; commander of the San Antonio Recruiting Battalion; and commander of the 703rd Military Intelligence Brigade, Field Station Kunia. At the time that Kennedy filed the sexual harassment complaint, she was the Army's deputy chief of staff for intelligence at the Pentagon. She was the recipient of many awards and honors in her long career, including the Legion of Merit, the Defense Meritorious Service Medal, the Army Meritorious Service Medal, and the Army Commendation Medal. Her career has often been cited as an excellent example of growing opportunities for women in the military.

The Charges

In the sexual harassment complaint filed by Kennedy on March 30, 2000, she told investigators that Smith grabbed her, held her against her will, and kissed her in her Pentagon office in 1996. Kennedy had informally reported the incident to her superiors at the time it occurred but declined to file formal charges until it was announced that Smith was being considered for the post of Army deputy inspector general. In this position, Smith would be responsible for overseeing investigations of sexual harassment charges.

On May 11, 2000, after an investigation headed by Army vice chief of staff General John Keane, the Army substantiated Kennedy's allegations. Smith denied the charges, stating, "I did not commit these allegations, and I am deeply disappointed with the decision to substantiate them." According to the Army inspector gen-

AP/Wide World Photos

Lieutenant General Claudia Kennedy.

uty inspector general was rescinded, and he was assigned as a special assistant to the Army Material Command's commanding general. Smith, who began his military career in 1966 and served three tours of duty in Vietnam, admitted to nothing, stating that "for the good of my family and the Army, we have elected to put it behind us and move on with our lives." He received no demotion in rank, grade, or pay; however, he requested his retirement on the same day that the charges were substantiated. Kennedy was satisfied with the outcome of the case, stating that in her mind, it was closed.

On June 2, 2000, Kennedy retired from the Army at a ceremony held thirty-one years to the day after she entered military service. The ceremony was held in the Pentagon courtyard and presided over by Army Secretary Louis Caldera, who declared that Kennedy was "not only a role model for service women of our armed forces but to girls and women across our nation, no matter what their professions and aspirations." Major General John G. Meyer, Jr., said that Kennedy "has set a standard and has stuck to it." In her own farewell speech, Kennedy noted that coming into the military during the Vietnam era, she had felt that it was as important for women to serve in the Army as for men. She did not believe that "women should be exempted when men were not exempted." Kennedy's plans for the future included writing a book and just enjoying life.

Kennedy's career was emblematic of the progress made by women in the military from the 1960's to the start of the new millennium. After she entered the Army, thousands of positions once held exclusively by men have opened up to women. She characterized this change as being measured and steady, but even she admitted that she never dreamed that there would one day be "stars on my shoulders."

Mary Virginia Davis

eral's report, there was complete disagreement as to what happened in Kennedy's office during the 1996 meeting. Although Kennedy claims Smith tried to kiss and grope her, Smith told investigators that he had hugged her and possibly given her a kiss on the cheek, denying any impropriety whatsoever. Smith provided fifty-two character witnesses in his defense, of whom nine were interviewed by the inspector general's office. Six others who had served with Smith were interviewed as well. None of these witnesses were aware of any misconduct on Smith's part, and the inspector general found that Smith was a highly respected officer with a good record.

Kennedy's credibility, however, was pivotal to the case. She had no apparent motive to falsely accuse Smith nor were the two in competition for the same position. Friends and acquaintances of Kennedy verified that she had told them of the incident at the time it had occurred.

Consequences

General Keane issued an "administrative memorandum of reprimand" for conduct unbecoming an officer and for sexually harassing an officer. However, no basis for criminal proceedings was found. Smith's proposed promotion to dep-

Controversy Arises over Estrogen and Heart Attacks

A major scientific study revealed that hormone-replacement therapy might not prevent coronary disease in postmenopausal women.

What: Biology; Gender issues; Health; Medicine
When: April 4, 2000
Where: Washington, D.C.
Who:
CLAUDE LENFANT (1928-   ), French-born director of the National Heart, Lung and Blood Institute at the U.S. National Institutes of Health

Hormone Replacement and Aging

To some, it seemed inevitable that the women's movement that emerged during the turbulent 1960's would eventually affect virtually every aspect of women's lives, especially issues related to their health and well-being. Countless activists sought ways of empowering women with information and technology designed to improve their physical and mental health. For the most part, earlier studies dealing with women's health had often identified normal processes as disorders or diseases relative to the female condition. In the case of menopause, considerable literature had ranged from folklore and old wives' tales to cursory studies lacking adequate research and information on an important biological function.

The term menopause has been applied to both a woman's last menstrual period and to the time when a woman's menstrual cycle begins to disappear. Often referred to as "the change" or the female climacteric, menopause is a normal process common to all women who reach the age in which it normally sets in. In the United States, for example, menopause typically occurs between the ages of forty-five and fifty-five. Once

believed to cause insanity, debility and in a sense, a signal of woman's uselessness as a human being, menopause does nothing more than to hail the end of a woman's capacity to reproduce offspring. It is neither a disease nor a disorder.

Menopause is triggered by decreasing levels of the female hormones estrogen and progesterone. Estrogen is produced in the ovary, the adrenal gland and in the placenta—the organ that provides nutrition for the developing fetus in a woman's uterus. Whereas progesterone is secreted by the corpus luteum—a follicle in the ovary—and is essential in preparing the uterus for the fertilized egg and in sustaining the pregnancy. If conception does not occur, the corpus luteum disintegrates and menstruation occurs.

For some time, researchers have known that the female hormones played a role in more than the reproductive processes. In subtle ways estrogen affects the skin, hair, blood, liver and bones. Moreover, scientific data exist to indicate that estrogen therapy, that is the replacement of the hormone, has relieved women of symptoms associated with menopause, for example hot flashes. Described as feelings of excessive warmth sweeping over the face, chest and often the entire body; lasting for a few seconds; and eventually yielding to perspiration, these sensations can cause temporary discomfort and even interrupt a woman's sleep.

Physicians have also prescribed estrogen therapy to relieve painful sexual relations which result from the thinning of the vaginal walls—another symptom of menopause. In addition, osteoporosis—a thinning of the bones, placing older women at greater risks for fractures has also responded positively to hormone-replacement therapy. Earlier findings, however, suggested that

3181

hormone-replacement therapy had a risk factor. For example, some studies indicated an increased risk of uterine cancer with the use of estrogen therapy. It was the increased incidence of uterine and breast cancer that prompted researchers to develop hormone-replacement therapy, combining both estrogen and progesterone to lessen the risk of cancer thought to be associated with using only estrogen.

Perhaps, the most important effect of hormone-replacement therapy was its apparent ability to lessen the incidence of heart disease which has become the leading cause of death of American women over the age of fifty. Evidence pointing to the relationship of estrogen replacement and the decreased incidence of heart disease resulted from a ten-year study of more than 48,000 nurses. The findings of the Harvard University Nurses' Health Project indicated that women who were taking estrogen were much more likely not to experience heart disease than those taking a harmless placebo. In that study, researchers attempted to show that hormone-replacement therapy could guard against heart attacks.

Based on the favorable results of the Harvard study, Premarin—an estrogen-replacement drug produced by Wyeth-Ayerst Laboratories—became the best-selling prescription drug in the United States. Nevertheless, there were some persons, especially women's health activists, who expressed some doubts about its effects. For example, some argued that the incidence of heart disease had also decreased among American men during this same period even though they were not taking estrogens. These groups called for more scientific tests, but the drug's popularity continued.

Estrogen: Risks and Results

In April of 2000, the favorable findings of the earlier studies on heart disease and hormone-replacement therapy were challenged. In another study known as the Hormone Replacement Therapy Trial of the Women Health's Initiative, which included about 25,000 healthy post menopausal women who were taking hormone-replacement therapy or a placebo. In March, 2000, researchers felt obliged to notify each of the participants that earlier investigators might have arrived at erroneous conclusions. It then appeared as though women taking Premarin might be experiencing slightly more strokes, heart attacks, and blood clots in their legs and lungs. Researchers and sponsors of the study believed that participants had to be made aware of these findings and they notified them in writing.

Not surprisingly, these letters were disturbing to women and the physicians who had prescribed hormone-replacement therapy. One principal investigator in the Women's Health Initiative Study admitted that her office had been flooded with phone calls from persons associated with the study as well as others who were taking the drug. This researcher's response was similar to others who felt that research must continue and that women should be informed of recent findings. Moreover, from a scientific point of view, a search for possible benefits from hormone-replacement therapy must continue. More important, research designed to decrease the incidence of heart disease among older females must also continue.

Consequences

There was little doubt that women's health activists would continue to study the "estrogen cure," as it represented another example of the consumer versus the pharmaceutical establishment. Some activists have argued that automatically prescribing estrogens as a cure-all is a risky business. It is therefore important to continue examining and advocating other ways of preventing heart disease in women, such as proper exercise and diet.

Betty L. Plummer

Petrified Dinosaur Heart Is Found in South Dakota

A team of researchers confirmed the discovery of a fossilized heart preserved within the skeleton of a sixty-five-million-year-old Thescelosaurus.

What: Biology
When: April 20, 2000
Where: Buffalo, South Dakota; Raleigh, North Carolina
Who:
MICHAEL HAMMER, fossil collector
ANDREW KUZMITZ, medical doctor
DALE A. RUSSELL (1937-   ), senior research curator for the North Carolina Museum of Natural Sciences

A Heart of Stone

The first dinosaur skeleton containing a fossilized heart was discovered near Buffalo, South Dakota, in 1993. On a fossil-hunting expedition with his son, collector Michael Hammer spotted the skeleton of a Thescelosaurus partially buried in sandstone. Hammer could see a large reddish-brown rock within the animal's rib cage and wondered if it could be an internal organ preserved along with the dinosaur's bones.

Hammer removed the entire block of sandstone containing the skeleton and asked Andrew Kuzmitz, a medical doctor and amateur paleontologist to perform a computed tomography (CT) scan of the animal's rib cage and the brown rock inside it. The CT scan showed that the rock was an internal organ and that it resembled a heart.

The dinosaur was nicknamed "Willo" after the wife of the man who owned the land where it was found. It might have been one of the species *Thescelosaurus neglectus*, so named because when it was first discovered in the 1920's, it was not considered worth serious study. Thescelosaurus ("marvelous lizard") was a plant-eating dinosaur, about 12 feet (nearly 4 meters) long and weighing more than 600 pounds (272 kilograms), with a small head and a long tail. It lived near the end of the Cretaceous period of the Mesozoic era, sixty-five million years ago.

A Closer Look Raises Questions

Willo was moved to the North Carolina Museum of Natural Sciences in 1996, still undisturbed in its sandstone block. Scientists at the museum used enhanced computer imaging software to change Kuzmitz's two-dimensional CT scan images into three-dimensional images. Once again, the images confirmed that the dinosaur fossil contained a heart, but even more surprising was the structure of the heart. Instead of the three-chambered heart typical of most modern-

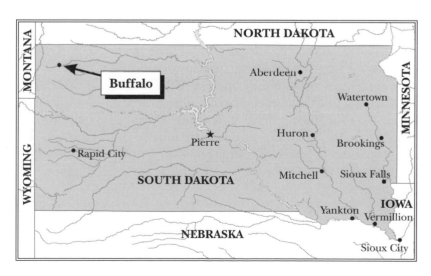

day reptiles, they saw the remains of a four-chambered heart, like that of a warm-blooded mammal or bird.

The pictures showed an organ with two lower chambers (the ventricles of the heart) and a single aorta, the large blood vessel that carries blood to and from the heart and lungs. The scientists believed the heart originally had four chambers. The upper chambers (or atria) one would expect to find in a four-chambered heart could not be seen in the computer images, but they could have decayed even while the two lower chambers were preserved as fossils.

The discovery of Willo's heart was reported in the April 21, 2000, issue of *Science*, and the dinosaur was then displayed at the North Carolina Museum of Natural Sciences. Some dinosaur specialists argued that the computer-generated images of Willo's heart did not really provide a clear picture of a four-chambered organ. They wondered if the appearance of a single aorta really meant there was never a second aorta. Some dinosaur experts believed the heart could have had three chambers and two aortas, and the computer images simply revealed a few parts of a typical cold-blooded animal's heart.

Various theories were proposed to explain how soft tissue such as a heart could have fossilized. Fossils are formed over long periods of time as minerals gradually replace teeth and bones, but internal organs decay too rapidly for fossilization to occur. Dale Russell, a dinosaur researcher and curator of the North Carolina Museum of Natural Sciences, believed that Willo might have been buried quickly under water, perhaps after a flash flood. If the dinosaur was immediately covered in mud, its internal organs might have been preserved as a soapy substance that gradually fossilized. Other scientists agreed that Willo probably died in a flood and remained underwater but suggested that Willo's heart hardened into a fossil as iron in the heart combined with minerals in the soil. Some researchers believed Willo's skeleton might contain other, as yet undiscovered, internal organs.

Consequences

Dinosaurs are thought to have evolved from an early ancestor into two branches, classified as either "lizard-hipped" (Saurischian) or "bird-hipped" (Ornithischian). It is actually the lizard-hipped, Saurischian dinosaurs that are believed to have resembled warm-blooded animals and evolved into modern-day birds. Thescelosaurus is part of the Ornithischian branch, "bird-hipped" dinosaurs that were supposed to have evolved into modern-day, cold-blooded reptiles. Finding the heart of a warm-blooded animal within the remains of a Thescelosaurus calls into question many scientists' ideas about different types of dinosaurs and how they evolved. It appears that either the lizard-hipped and bird-hipped branches of the dinosaur family both evolved in the same way or dinosaurs cannot be so easily classed within the two branches.

The discovery of a warm-blooded heart in a Thescelosaurus raises new questions about how dinosaurs lived, because warm-blooded animals move quickly and have more endurance than cold-blooded animals. Warm-blooded animals also consume more natural resources, ranging over more territory and eating more than less energetic creatures do. The environment might therefore have supported dinosaurs in smaller numbers than previously thought.

The methods used to examine Willo and to discover its heart could also change the way paleontologists approach fossil preservation and research. Willo was kept in its original state and transported and examined still lying in the chunk of earth in which it fell millions of years earlier. Traditional methods focused on extracting dinosaur bones from the surrounding environment and cleaning them; materials found within the bones or nearby were discarded. Some scientists wonder if, because of these methods, other fossilized soft tissues and organs were overlooked. In the future, dinosaur fossils are more likely to be preserved intact, subjected to CT scans, and examined through newly developed computer imaging.

Maureen J. Puffer-Rothenberg

Vermont Approves Same-Sex Civil Unions

As a result of successful legal action brought against the state of Vermont by three gay and lesbian couples, the state granted to same-sex unions legal status involving the benefits, protections, and responsibilities of traditional husband-and-wife unions.

What: Civil rights and liberties; Social reform

When: April 26, 2000

Where: Montpelier, Vermont

Who:

STAN BAKER, man in whose name three couples brought suit against Vermont

JEFFREY L. AMESTOY (1946-   ), chief justice of the Vermont Supreme Court

HOWARD DEAN (1948-   ), governor of Vermont

Homosexual Unions

Homosexuality has always existed among a significant minority of persons. However, because of society's disapproval, most gays and lesbians had been secretive about their sexual lives until the latter part of the twentieth century. Therefore, many people heard and saw little evidence of homosexuality and tended to regard it as strange and "unnatural" behavior.

In one important respect, however, homosexuals are just like those attracted to the opposite sex: They often form close attachments to one particular person. Among heterosexuals, that strong bond frequently takes the form of marriage, which is a union recognized and protected by law in the United States and in many other countries. In many faiths, marriage is celebrated by religious ceremonies, and the union is regarded as a sacred bond that is expected to last until death. However, until 2000, none of the fifty states had given legal recognition to homosexual partners. In Vermont, however, which takes pride in having been the first state to outlaw slavery, a series of events, starting in 1997, changed that situation.

A Stormy Process

On July 22, 1997, two lesbian couples and one gay couple, convinced that current law discriminated unfairly against them, brought a lawsuit against the state of Vermont for the purpose of obtaining marriage licenses. The three couples had been together for periods ranging from four to twenty-five years. The case, *Baker v. Vermont*, took its name from Stan Baker, who with his partner Peter Harrington formed one of the couples. The lesbian couples were Nina Beck and Stacy Jolles, and Lois Farnham and Holly Puterbaugh. After the Chittendon County Superior Court ruled against them, their attorneys appealed to the Vermont Supreme Court, which heard their arguments in January of 1999. Finally, on December 20 of that year, that court ruled unanimously in favor of the plaintiffs. Chief Justice Jeffrey L. Amestoy noted that the six men and women sought "nothing more, nor less, than legal protection and security for their avowed commitment to an intimate and lasting human relationship."

The lawmakers of Vermont were now faced with a problem. The court's decision obliged them to change the law, but a storm of protest arose against the idea of legal recognition of gay and lesbian unions. Polls indicated that roughly half the citizens of Vermont opposed such a measure; the other half supported it. As is usually the case in such controversies, the opposition proved more organized and determined. The Roman Catholic bishop of Burlington and fundamentalist Protestant organizations had been on record as opposing such a change even before the suit was brought, and more liberal groups, including the Unitarian Universalist Association, had supported the idea of recognizing same-sex unions. From these and many other organizations and

AP/Wide World Photos

Vermont governor Howard Dean addresses reporters after signing a law granting homosexual couples most of the benefits of male-female marriage.

individuals, state legislators heard much during their deliberations in the early part of 2000.

The legal process began in January with hearings conducted by the House Judiciary Committee. Many issues were raised, including fears that the legalization of homosexual unions would destabilize marriage and concerns that "separate but equal" protection of gay and lesbian couples might prove unconstitutional. Of several bills that were presented and debated, one of them, H.847, passed after often bitter and emotional debate lasting for two months. On March 17, the Vermont House passed H.847 by the narrow margin of 76 to 69 and, on April 25, agreed to a slightly different Senate version, 79 to 68. The next day, Governor Howard Dean signed into law the bill, which did not, as some of its supporters had hoped, refer to same-sex "marriage" but to "civil union." The law, recognizing the rights of such couples in regard to wills, state taxes, family

leave benefits, and many other matters, went into effect on July 1, 2000.

Consequences

On Monday, July 3, 2000, the first weekday after the law went into effect, an estimated two dozen same-sex couples obtained civil union licenses. The same division that marked public opinion generally affected town clerks. In seven communities, clerks kept their offices open late to accommodate gay and lesbian couples, and in five others, clerks at first refused to grant licenses or threatened to resign rather than comply with the law. Opponents paid for a black-bordered full-page ad in the *Burlington Free Press*. The furor calmed down as the weeks passed, and life went on normally for most Vermonters, although Governor Dean's approval rating dropped substantially in public opinion polls. In September, five Republican state legislators who had supported the civil union bill lost their bids for renomination, as did one Democrat who had opposed the bill. In November, however, Dean was reelected to his fifth two-year term, although by a narrower margin than he had enjoyed in 1998.

The main beneficiaries of the new law were the homosexual couples, including some from other states who came to Vermont to register. A number of civil ceremonies were performed at hotels, although the participants surely understood that a civil union accomplished in Vermont did not afford them protection elsewhere, for the federal Defense of Marriage Act (1996) in no way compels states to recognize same-sex unions legalized in other states. It is difficult to assess the effect of the Vermont initiative on other state governments. As of August, 2000, thirty-two states had laws banning same-sex unions, but legislatures in several states were exploring the possibility of following Vermont's example. Unless similarly motivated by court rulings, however, other states will probably be reluctant to enact similar laws in the face of strong opposition.

Robert P. Ellis

ILOVEYOU Virus Destroys Data on Many Thousands of Computers

A Filipino student created a computer virus that did an estimated $10 billion damage to computer systems worldwide but was not prosecuted for lack of applicable computer crime laws.

What: Computer science; Crime; Technology
When: May 4, 2000
Where: Manila, Philippines
Who:
ONEL DE GUZMAN (1976-   ), Filipino student
JOSEPH ESTRADA (1937-   ), president of the Philippines from 1998 to 2001

The Virus Strikes

On May 4, 2000, computer users around the world were greeted with an e-mail enticingly titled "ILOVEYOU." The text of the e-mail message read, "Please look at this love note that I've sent you," and contained an attachment labeled "LOVE-LETTER-FOR-YOU.TXT.vbs." Upon opening the attachment, computer users did not receive a love message but unwitting activated a program, or executable, that renamed or destroyed a variety of files on their hard drives—most notably, multimedia image and audio files. The rogue program, or virus, also created a backdoor on the user's computer, allowing hackers to access data (including passwords) surreptitiously. It then created further havoc by automatically forwarding the original e-mail message to every address in the user's address book. In addition to the file damage and breach of security to the user's computer, the almost endless replication of e-mail messages immediately overloaded and disabled network servers around the world. Institutions ranging from the Ford Motor Company to the British House of Commons were forced to shut down servers temporarily in response to the flood of electronic messages.

The ILOVEYOU virus (sometimes called the LOVE bug) operated by means of the attachment, written in Visual Basic Software (vbs) code. In some respects, it resembled the Melissa virus, which had been released in March, 1999, and had disabled e-mail servers around the world by a user's unwitting replication of e-mail messages. Both Melissa and ILOVEYOU are classified as a specific type of virus called a worm—a self-replicating program that surreptitiously alters one or more of a computer's primary operating systems.

The damage done by Melissa, however, paled in comparison to that done by ILOVEYOU. First, Melissa did little or no damage to files on the individual computer user's system. Second, Melissa forwarded its message to only the first fifty e-mail addresses in a user's address book, while ILOVEYOU forwarded its message to every available e-mail address in the user's system—exponentially increasing the flood of electronic messages and the threat to network servers. Finally, Melissa was aimed exclusively at Microsoft Outlook's message system while ILOVEYOU attacked a number of message systems and created an opening, or backdoor, on the computer's operating system for ease of future penetration. Indeed, the latter aspect of the virus initially led investigators to suspect electronic theft as the motive behind its release.

The Scene of the Crime

The pattern by which the ILOVEYOU virus spread pointed to an Asian origin. The code of the virus itself included encoded references to Manila, Philippines, and the hacker pseudonym "Spyder." Although the password theft aspect of the virus implied the possibility of a sophisticated criminal conspiracy involving electronic banking, the graffiti-like tags embedded in the code

3188

suggested a simple act of vandalism perpetrated by students or teenage hackers. The Philippines National Bureau of Investigation (NBI) initially suspected a criminal conspiracy when it traced the release of the virus to a computer located in the apartment of twenty-eight-year-old bank employee Reonel Ramones. However, their attention quickly transferred to twenty-four-year-old Onel de Guzman, the younger brother of Ramones's girlfriend and a frequent visitor to his apartment. De Guzman, a dropout from the Philippines' AMA Computer College, had recently submitted a master's thesis detailing a program capable of stealing Internet passwords and forwarding e-mail messages. The thesis, rejected by his professors, outlined a program that contained substantive similarities to the ILOVEYOU virus.

On the basis of this resemblance and physical evidence, the NBI arrested de Guzman on May 10, 2000. Although de Guzman admitted accidentally releasing the virus, he disavowed authoring it. The NBI suspected de Guzman's friend and fellow student Michael Buen to be "Spyder"; however, it had insufficient evidence to bring charges against him. Because no computer crime laws existed, the Philippine justice department initially charged de Guzman with violation of credit card laws; however, they subsequently dropped these charges as inapplicable to the case and were forced to release de Guzman on August 22, 2000.

In the meantime, Joseph Estrada, president of the Philippines, signed a law that criminalized computer hacking in that nation. The measure, however, could not be applied retroactively to de Guzman. Although de Guzman failed to capitalize on his act—he was unsuccessful in attempts to find work as a computer security consultant in the United States—he did bring a measure of international recognition to the Philippines as an environment capable of producing sophisticated computer programming and was elevated by some in that country to the status of folk hero. However, to most computer users, de Guzman's act simply brought a painful awareness of the vulnerability of their networked systems to hacking. It also reinforced the need for users to follow security protocols in configuring message systems and opening e-mail attachments.

On June 9, 2000, Microsoft released a "patch," or auxiliary code, for its Outlook message system to prevent future viruses such as ILOVEYOU from infiltrating its programming. However, by this time, many users had already taken responsibility for their own antivirus protection; antivirus programs such as McAfee Virus Scan became the most-downloaded programs in the wake of the ILOVEYOU virus. The success of such antivirus measures was evident in the relative ineffectuality of the NEWLOVE virus, a clone of ILOVEYOU released by copycat hackers several days after the original. In the United States, Senators Orrin Hatch, a Republican from Utah, and Charles Schumer, a Democrat from New York, used the widespread destruction caused by the ILOVEYOU virus to bolster support for a bill to aid law enforcement authorities in combating cybercrime and cyberterrorism.

Luke A. Powers

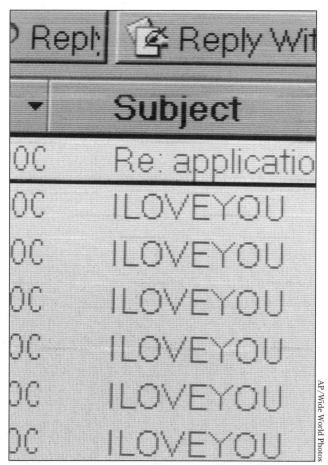

A computer screen shows an e-mail inbox to which the ILOVEYOU virus has been sent.

Unemployment Rate Reaches Lowest Level in Thirty Years

The drop in the unemployment rate to 3.9 percent represented good economic and financial conditions for American workers but caused alarm about potential inflation.

What: Business; Economics; Labor
When: May 4, 2000
Where: Federal Bureau of Labor Statistics, Washington, D.C.
Who:
BILL CLINTON (1946-　　), president of the United States from 1993 to 2001
GEORGE W. BUSH (1946-　　), Republican presidential candidate

Improving Economy

Estimates of the percentage of workers unemployed are released every month by the Federal Bureau of Labor Statistics. In 1992, as a result of a brief economic recession, unemployment rose to 7.6 percent. From that point, however, the rate gradually declined through the 1990's during the presidency of Democrat Bill Clinton. The announcement that unemployment in spring, 2000, was down to 3.9 percent caused a mixture of elation and alarm. Unemployment levels very close to 4 percent were subsequently reported for the remaining months of 2000.

Low unemployment is a sign of a booming economy, one in which workers are benefiting from an abundance of job opportunities. The unemployment rate had not been as low as 4 percent since 1970, when the military draft was still in effect, removing from the measured labor force many young men who might otherwise have been unemployed. This meant that the low rate was more significant and that the economy in 2000 was performing even better than it had been thirty years earlier.

The decline in unemployment meant that job opportunities were growing more rapidly than the number of people seeking jobs. The number of workers employed outside of agriculture increased from 109 million in 1992 to 131 million in 2000. This vigorous growth in job opportunities reflected strong consumer demand for goods and services. Although capital growth and technological improvement provided workers with more and better tools, this process did not result in job declines for the economy as a whole. Good employment performance in the United States was in strong contrast with European countries such as France and Germany, where unemployment rates during the 1990's seemed stuck in the neighborhood of 10 percent.

Some economists noted that labor markets in the United States were more free and flexible, and those in Europe were subject to more trade union constraints, government restrictions, taxes, and transfer payments. In addition, the European countries pay much higher unemployment benefits; which means that an unemployed worker there might not be as anxious to find another job. European firms are severely limited in their freedom to dismiss workers; in consequence, they are also reluctant to expand their job roles. The European environment is especially unfavorable to young people just entering the labor force.

Possible Problems

During and after the Great Depression of the 1930's, much attention was directed toward unemployment and toward government measures to combat it. Often in these discussions an unemployment rate of 4 percent was seen as a desirable and attainable approximation of "full employment." However, some economists observed a tendency for low unemployment to be associated with an increased threat of inflation—sustained increases in prices over time. Low unemployment could mean a situation in which demand for labor exceeded the available supply. Such conditions would strengthen labor's bargaining power. Further, employers would bid against each other for the available workers. These processes would drive up the levels of money wages and prices.

So as U.S. unemployment declined in the 1990's, economists watched anxiously for signs of worsening inflation. There were some such signs. The annual rate of increase of the consumer price index rose from about 1.7 percent in 1998 to about 3.7 percent in spring, 2000. There were some special circumstances, notably the increase in international petroleum prices. During the campaign leading up to the presidential election of 2000, which pitted Democrat Al Gore against Republican George W. Bush, voters demonstrated little public concern about inflation, although elderly people did complain about the rising costs of prescription drugs.

Although low unemployment is clearly a good condition, news of falling unemployment has often provoked a drop in stock prices on Wall Street. This is because investors fear that low unemployment will be used by the Federal Reserve as a reason for implementing a restrictive monetary policy to ward off possible inflation. A restrictive monetary policy means higher interest rates. Because stocks are often purchased on borrowed money, higher interest rates reduce investors' potential profits.

Consequences

Whether 4 percent unemployment is a feasible goal for public policy, and how much it threatens to cause inflation, depends a lot on how rapidly labor productivity is increasing. Rapid productivity gains permit employers to pay higher wages without the need to raise product prices. Economists were sharply divided in the late 1990's on this question. Other fiscal experts who believed the inflationary threat was not large pointed to the intense international competition experienced by U.S. firms and to the relative decline in the power of American labor unions.

A weakening in the demand for current production by consumers, business firms, and government is sometimes revealed by a rise in unemployment rates. There are various government measures that can counter such weakening, often termed demand-management policies and consisting chiefly of monetary and fiscal policies. Monetary policy, conducted by the Federal Reserve, can try to counter rising unemployment by reducing interest rates and by purchasing government securities in the open market to bring about an increase in the supply of money and credit. Fiscal policy can work through an increase in government expenditures and through a reduction in tax rates. As the U.S. economy showed signs of possible recession in early 2001, president elect Bush strongly advocated a reduction in federal income-tax rates to stimulate higher spending by consumers and business firms. Federal Reserve policy shifted in the direction of more expansionary measures. By August, 2001, the unemployment rate had risen to 4.9 percent.

The downward trend in the economy was aggravated by the effects of the September 11, 2001, terrorist attacks on New York City and Washington, D.C. By the end of the year, unemployment reached a six-year high of 5.8 percent.

Unemployment is an imprecise concept. The unemployment figures are estimates based on interviews with a sample of households. People are counted as unemployed if they were looking for work and did not work. The interview takes no account of the person's qualifications or what wage they require. However, changes in the unemployment rate from month to month are apparently quite accurate, and each month's figures are avidly watched by investors, politicians, and business-cycle analysts.

Paul B. Trescott

Fijian Rebels Attempt Coup

On Friday, May 19, 2000, a group of armed gunmen stormed Fiji's parliament, seized the prime minister and several cabinet ministers, and declared a coup. The coup failed, but it demonstrated the seriousness of ethnic divisiveness in Fiji's national politics.

What: Civil strife; National politics; Coups
When: May 19, 2000
Where: Suva, Fiji
Who:
GEORGE SPEIGHT, rebel leader
MAHENDRA CHAUDHRY (1942-   ), prime minister of Fiji from 1999 to 2000

The Coup

On May 19, 2000, gunmen armed with M-16 rifles entered Fiji's parliament building and took thirty-one hostages, including Prime Minister Mahendra Chaudhry and seven other ministers, parliamentarians, and staff. Chaudhry was Fiji's first ethnic Indian prime minister. The attempted coup took place on the first anniversary of the election of his government. There were no casualties; the rebels fired only warning shots as they entered the building. There were seven rebels, six members of an army special forces unit and George Speight, a failed businessperson and the coup leader. Speight claimed that he and his gunmen were standing up for the economic and political interests of indigenous Fijians because Chaudhry's policies favored Fiji's immigrant Indian population. He demanded Chaudhry's resignation and an end to Indian participation in Fijian politics.

Ethnic Divisiveness

The May 19 coup shows how ethnic differences between indigenous Fijians and Indo-Fijians divide politics in this island country. Indigenous Fijians, who are mainly Christians, are a mixture of Melanesian and Polynesian peoples, resulting from the original migrations to the south and central Pacific centuries ago. About half of them are dark-skinned people of Melanesian origin; they predominate in Fiji's western islands. The people of the eastern islands are largely of Polynesian descent. Both groups speak Fijian, although English is Fiji's official language. Together, indigenous Fijians make up a majority—51 percent—of this island nation's 800,000 people.

Indo-Fijians are immigrants or descendants of immigrants from India. About 80 percent of them are Hindu, 15 percent Muslim, and the rest mostly Sikh, with a few Christians. Hindi is the main language spoken by Indo-Fijians. About 44 percent of Fiji's population belongs to this group. The remaining 5 percent of the population is mainly a mixture of Europeans, other Pacific Islanders, and overseas Chinese.

The ethnic divisiveness that plagues Fiji's indigenous and Indian populations developed during British colonial rule (1874-1970). Between 1876 and 1916, the British introduced sugarcane plantation agriculture and imported 60,000 laborers from India, another British colony, to work on the plantations. These laborers were indentured servants, meaning they were under contract to work to pay off debts to their British employers. Upon completion of their contracts, the British governor invited the Indian workers to stay and contribute to Fiji's economy.

At the beginning of the twenty-first century, descendants of Indian plantation workers produced 90 percent of the sugar crop, but they could not legally own the land that they farm. They had to lease it from indigenous Fijian clan chiefs, who controled 85 percent of the nation's land.

A second group of Indians immigrated to Fiji voluntarily during the 1920's and 1930's. This group consisted of several thousand Gujaratis, from near Bombay. Descendants of this group form the core of Fiji's urban shop keeping and business class. Therefore, today, unlike the indigenous Fijians, who live throughout the country, the Indo-Fijians reside primarily near urban centers and in cane-producing areas.

Indigenous Fijians did not benefit as much as Indo-Fijians under British rule. British laws requiring permits to travel to other villages prevented native Fijians from seeking work and selling products outside their villages. Such restrictive laws did not end until the 1930's, after Indo-Fijians had come to dominate businesses and industries in urban areas. Many indigenous Fijians are poor farmers relegated to areas with poor soil, whereas Indo-Fijians are successful sugar growers and business owners. Consequently, the majority of indigenous Fijians resent the Indo-Fijians' relatively high levels of education, occupational attainment, income, and standard of living.

Despite the minority status of Indo-Fijians, free and peaceful elections in 1999 resulted in the election of the pro-Indian government led by Chaudhry. His political strength came from a major voting bloc made up of Indo-Fijian sugarcane farmers. These farmers organized widespread boycotts of the sugar industry. The boycotts crippled the economy and pressured many indigenous Fijian voters to support Chaudhry's election. After the election, indigenous opposition to Chaudhry criticized his attempts to renew expiring leases on farmland held by the ethnic Indian tenants who were behind the damaging boycotts.

While Speight and his accomplices were holding the hostages, Speight tried to start an anti-Chaudhry uprising within the indigenous Fijian community. He was partly successful in and around Suva, the capital of Fiji, where thousands of indigenous Fijians rioted. They set up roadblocks; occupied a power station, airstrips, a military barracks, a police station, and other government buildings; damaged hundreds of Indian-owned shops; and seized private land and businesses, including four tourist resorts. The army and police calmed the area after a few days. Fortunately, no deaths occurred during the rioting.

Consequences

Speight negotiated for three months with the Fijian government officials who were not among the hostages. Although the rebels killed no one during the standoff, at one point, they dragged Chaudhry onto the lawn of the parliament building and held a gun to his head. Speight agreed to release all the hostages a month before negotiations ended. At the end of negotiations, he gained an essential concession. Chaudhry agreed to resign and a

3193

temporary, military-installed government was set up. After surrendering with his men, Speight expected to have a role in selecting leaders of the new government; instead, the temporary government arrested him and imprisoned him for treason. The government would later hunt down a small group of conspirators in the military. Most of them were arrested; a few were killed resisting arrest.

The May 19, 2000, coup was a failed attempt to assert indigenous Fijian control of government through violent means. It did not have the support of the military leadership although the rebels were low-level members of the military, nor did the coup have the support of important political leaders. Nevertheless, the Chaudhry government was forced to resign, and the coup signaled to the world that divisiveness between ethnic groups in Fiji is a serious problem.

Richard A. Crooker

Israeli Troops Leave Southern Lebanon

After eighteen years of occupying southern Lebanon, Israel finally withdrew its troops, assured by the United Nations and the government of Lebanon that their own forces would prevent future terrorist incursions across Israel's northern border.

What: International relations; Military conflict; Terrorism
When: May 24, 2000
Where: Southern Lebanon and Israel
Who:
YASIR ARAFAT (1929-   ), head of the Palestine Liberation Organization
MENACHEM BEGIN (1913-1992), prime minister of Israel from 1977 to 1983
ARIEL SHARON (1928-   ), Israeli defense minister
HAFEZ AL-ASSAD (1930-2000), president of Syria from 1971 to 2000
RAFAEL EITAN, Israeli army chief of staff
EHUD BARAK (1942-   ), prime minister of Israel from 1999 to 2001

Lebanon and Israel

Lebanon is an Arab nation located in southwest Asia on the eastern shore of the Mediterranean Sea. One of the world's smallest sovereign states, with an area of 4,015 square miles, it is surrounded on the north and east by Syria and on the south by Israel. In 1970, a civil war in nearby Jordan saw the Jordanian army, led by King Hussein, decisively defeat the Palestine Liberation Organization (PLO) of Yasir Arafat, forcing the relocation of PLO headquarters to Beirut. The Palestinian portion of Lebanon's population rapidly rose to approximately 10 percent, with these displaced Arabs feeling betrayed by other Arab nations.

As more displaced Palestinians settled in southern Lebanon, the area soon became the major base for PLO guerrilla operations against Israel. When Israel retaliated by raiding PLO bases across its northern border, Lebanon's Christian-dominated government attempted to restrict PLO activities within its jurisdiction. Palestinian military forces, which had previously refrained from taking sides in internal Lebanese conflicts, began giving support to antigovernment Muslims, and a bloody civil war broke out in April, 1975. Although the official policies of both the PLO and the Lebanese army forbade intervention, dissident groups from both sides joined the fighting and sought international support.

Sporadic fighting continued into the early 1980's between the PLO and Israel despite ceasefire agreements between Syrian, Christian, and Israeli leaders in July, 1981, brokered by United States special envoy Philip Habib. Israel bombed PLO headquarters in Beirut, but the PLO continued to attack Israel from regions of Lebanon that were occupied by Syria.

The Peace in Galilee Campaign

Menachem Begin, who became prime minister of Israel in 1977, determined to end the threat that PLO bases in southern Lebanon posed by invading Lebanon and driving PLO camps farther north, thereby placing Israel out of the firing range of its missiles. The Israeli attack, which Begin called the "Peace in Galilee Campaign," was launched on June 6, 1982. The first week of that invasion witnessed an easy Israeli victory that eliminated the PLO's Soviet-supplied surface-to-air missiles.

The Israeli navy closed off all sea approaches to Beirut, the capital of Lebanon, while the army seized the PLO's large weapons supply. However, it has been speculated that Ariel Sharon, Israel's defense minister at that time, may have exceeded Begin's instructions by launching an unplanned attempt to capture Beirut itself. In any event, what had seemed to be an easy victory was swallowed up in bloodshed. Syria, which had been willing to stand aside if Israel's only goal was to crush the PLO became more active in supporting native Lebanese forces when Israel lent active support to Lebanon's Maronite Christians, who were strongly opposed to Syrian domination of their country.

On the night of September 14, 1982, under suspicious circumstances, the Israeli Army allowed Maronite troops to enter the Palestine refugee camps at Sabra and Chatila, supposedly to eliminate pockets of PLO armed resistance. Instead, the Maronites slaughtered seven hundred unarmed civilians. The blame for that atrocity was directed against Israeli minister of defense Ariel Sharon and Israeli army chief of staff Rafael Eitan.

Israel's supreme court convened what became known as the Kahan Commission, which laid the blame at the doors of generals Sharon and Eitan. For the first time in Israel's embattled history, large-scale street demonstrations protested the consequences of an Israeli military action. Prior to the ill-fated Lebanese invasion, loyal Israelis had always supported their government's prosecution of what they regarded as defensive wars. Prime Minister Begin himself was spared any direct condemnation, but he went into seclusion and retired from office on September 15, 1983.

An international armed expedition replaced the Israelis as they retreated from Beirut. In the disorder that followed, hundreds of United States Marines died when their barracks in Beirut were blown up by Lebanese Shiite suicide bombers. In February, 1984, U.S. president Ronald Reagan summoned United States forces home. In this bloody disaster, Israel's only apparent advantage lay in that Yasir Arafat and thousands of his PLO fighters withdrew to distant Tunisia in North Africa.

A small Israeli army contingent remained within southern Lebanon, in alliance with a Maronite Christian force. Under constant pressure by Hezbollah terrorists, supported by Syria, the Israelis in that vulnerable post, were constantly burdened by casualties. The demand of Israel's civilian population was increasingly in favor of withdrawal. They were destined to remain there for eighteen years until Israel's Prime Minister Barak withdrew them on May 24, 2000.

Consequences

Despite assurances by the United Nations and the government of Lebanon that their forces

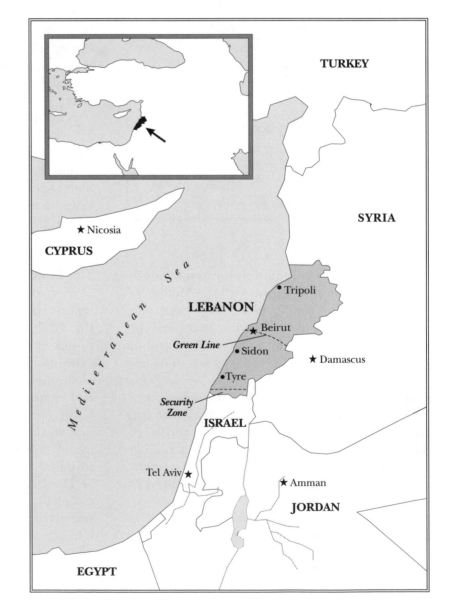

would prevent future terrorist against Israel, southern Lebanon was actually controlled by Syria. Hezbollah terrorists continued to operate there freely, unchecked by either Lebanon or the United Nations. Lebanon remains the last active front in the Arab-Israeli conflict.

Lebanon's internal conflicts have repeatedly been complicated by confrontations centered on the country's relationship to Israel, its Arab neighbors, and numerous Palestinian refugees who relocated there following the Arab-Israeli wars and the Jordanian civil war. The twentieth century saw Lebanon serve as a continual battlefield upon which both internal forces and foreign powers unrelentingly massacred militiamen and civilians. By the end of the twentieth century, Lebanon remained the last active front in the Arab-Israeli conflict. Lebanon's political instability continued to leave its southern neighbor, Israel, insecure.

Arnold Blumberg

South and North Korea Sign Peace and Unity Agreement

On June 15, 2000, the leaders of North and South Korea, which had been enemies since the Korean War began in 1950, signed a joint declaration calling for steps to be taken to work toward eventual unity between the two countries.

What: International relations
When: June 15, 2000
Where: Pyongyang, North Korea
Who:
KIM JONG IL (1942-   ), president of North Korea from 1994
KIM DAE JUNG (1925-   ), president of South Korea from 1998

Two Koreas

On June 15, 2000, the presidents of North and South Korea capped their historic three-day summit meeting by signing a declaration agreeing to work together toward peace and unity between the two nations. Korea has existed as a people and country for five thousand years, but hostility and division between parts of the Korean peninsula has also long been a reality.

In 1910, Japan invaded and annexed Korea, losing it with its defeat in World War II. In August, 1945, after the war ended, the Soviets almost immediately moved into northern Korea. In reaction, the United States asked that Korea be divided at the thirty-eighth parallel, with the northern part under Soviet control and the southern under U.S. influence. The Soviets agreed. The north became the Democratic People's Republic of Korea (known as North Korea) in 1948 and continued to be supported by the Soviets, and the south became the Republic of Korea (known as South Korea), which remained defended by United States troops.

On June 25, 1950, North Korea invaded South Korea in an effort to forcibly reunify the peninsula. The war lasted three years, leaving up to 5 million dead, wounded, or missing, half of them civilians. The United States became involved in the conflict, and other nations also fought alongside South Korea. The war ended in a truce but without a treaty, leaving the two countries still enemies, armed against each other, and technically at war. The truce called for a no-man's land called the Demilitarized Zone (DMZ), 2.5 miles (4 kilometers) wide and running the width of the peninsula at the thirty-eighth parallel. This narrow strip is filled with land mines and defended by thousands of troops, tanks, and artillery, including 37,000 U.S. troops. The war also left families separated on either side of the DMZ, unable to visit or communicate with each other.

Kim Il Sung became premier and supreme commander of North Korea in 1948, and he became dictator of the communist country. He was succeeded by his son, Kim Jong Il, in 1994. The communist nation was cut off from modern technology, trade, and business, and suffered food shortages. After the end of the Soviet bloc in 1994, the country experienced floods, crop failures, disease, and mass starvation. However, North Korea has continued to spend money on military advancement, including the development of nuclear missiles, which has been a cause of friction with the United States.

After the war, South Korea had a series of authoritarian leaders, two military coups, and several rebellions, although it eventually became a capitalist democracy and manufacturing and exporting country, the world's eleventh largest economy.

There have been several attempts at reconciliation between the two Koreas, beginning in 1972, but each has broken down without bearing much fruit. In 1998, Kim Dae Jung was elected president of South Korea. He had encouraged democratic principles even as an opposition leader before his election, and he urged engagement with North Korea, declaring his readiness to help the north economically, and his desire to meet and work together.

The Summit Meeting

Kim Jong Il invited Kim Dae Jung to Pyongyang, North Korea, for a summit meeting, which was held June 13-15, 2000. It was carefully planned to avoid the most controversial issues, including the south's concerns about northern missiles and the north's demand that U.S. troops leave the south.

The focus was to be on issues of family reunion for those separated by the war and economic and cultural exchange. Kim Jong Il unexpectedly met his South Korean counterpart at the airport, shaking his hand and guiding him through an inspection of North Korean troops, then they rode together to the state guest house

as huge, cheering crowds lined the way.

The meetings were warm and agreeable, and on June 15, they climaxed in the signing of an accord in which the two Koreas agreed to work toward peace and unity. The accord declared that the two countries would work together on unity issues, each using its own preferred formulas for reunification and acknowledging both the differences in the two methods and the common factors that would allow them to work together. They agreed to set up reunions between family members who had been separated and to allow Communist prisoners in the south who had served their jail terms to return to North Korea. They pledged to accelerate exchanges in social, cultural, sports, health, and environmental sectors and to meet again in the near future. Both countries' official names were included; this meant that the countries for the first time officially recognized each other's legitimacy.

Consequences

On June 19, 2000, the United States officially eased economic sanctions against the north for

South Korean president Kim Dae Jung (left) and North Korean leader Kim Jong Il, before signing the agreement.

3199

the first time in fifty years, thus providing hope for the north's economic survival, in exchange for a moratorium from North Korea on testing nuclear weapons.

A full reconciliation and reunification, if they were ever to occur, remained likely to take years. However, August 15 through 18, one hundred people each from the north and south traveled over the border to Seoul and Pyongyang to meet and visit with relatives they had not seen in fifty years. In addition, an agreement was reached to reconnect a railway line across the border, broken since the Korean War, and ministerial level talks and talks between defense chiefs took place. In the opening to the 2000 Summer Olympics in Sydney, Australia, athletes from the two Koreas marched behind a single banner, symbolizing their new era of cooperation and unity.

Eleanor B. Amico

Environmental Protection Agency Bans Pesticide Chlorpyrifos

The Environmental Protection Agency reached an agreement with pesticide manufacturers to end use of the organophosphate pesticide chlorpyrifos, known as Dursban, because of public health concerns.

What: Environment; Health
When: June 8, 2000
Where: Washington, D.C.
Who:

CAROL M. BROWNER (1955-   ), administrator of the Environmental Protection Agency

JAY FELDMAN (1953-   ), executive director, National Coalition Against the Misuse of Pesticides

Negotiating a Ban

On June 8, 2000, the U.S. Environmental Protection Agency (EPA) announced a ban on chlorpyrifos, a pesticide widely marketed as Dursban and also as Lorsban. The pesticide is one of a class of thirty-seven chemicals known as organophosphates. Concern about the effects of the pesticide on human health was expressed by the National Coalition Against the Misuse of Pesticides (NCAMP), the Environmental Working Group, the Natural Resources Defense Council (NRDC), and other organizations. The ban resulted from an agreement between the EPA and the manufacturer, Dow Chemical.

Dow Chemical developed chlorpyrifos in 1965 as a promising substitute for DDT and other pesticides that were already suspected of being harmful. After the EPA banned DDT in 1972 and chlordane (another major pesticide) in 1988, use of Dursban grew rapidly. However, reports of associated human illnesses also increased. Between 1993 and 1996, nearly 63,000 reports were made to U.S. poison control centers about organophosphates. Chlorpyrifos was used in more than eight hundred retail products such as Black Flag Liquid Roach and Ant Killer, Pest Control, and Hartz Yard and Kennel Flea Spray, in addition to products for agricultural pest control.

The Food Quality Protection Act of 1996 (FQPA) required the EPA to establish limits on pesticide residues in food. By August of 1999, the EPA had not met the deadline to reassess exposure limits for pesticides. Concern was expressed that the EPA was moving too slowly.

The Environmental Working Group was formed to press for a ban. Jay Feldman, director of NCAMP, helped lead a public information campaign on the dangers of Dursban. The Natural Resources Defense Council initiated lawsuits to force the EPA to take action. The EPA entered into extensive negotiations with Dow Chemical. Dow maintained that its chemicals were safe if used properly. However, it consented to an agreement in order to avoid long, costly legal battles. On June 8, 2000, EPA administrator Carol Browner announced the agreement, which also resulted in a revised risk assessment for chlorpyrifos issued on August 16, 2000.

The Basics of the Agreement

Factors to consider in crafting an agreement included the human health risks of chlorpyrifos and the extensive reliance on the product to kill insects. A white, crystal-like solid with a strong odor, chlorpyrifos is usually mixed with oily liquids. In the 1990's, the pesticide was being applied some 20 million times a year in homes, schools, and offices. Food and residential applications of the pesticide totaled between 15 million and 24 million pounds (5.5 million and 8.9 million kilograms) a year. Dow AgroSciences was earning around $100 million a year from sales of the pesticide. It was the most popular pesticide in use.

Its household uses included control of cockroaches, fleas, and termites. Agricultural uses included the control of ticks on cattle and as a spray for crop pests. Chlorpyrifos was found almost everywhere. Almost half the schools in some states were routinely using pesticides containing this chemical. A 1994 survey of a random sample of Americans found that 82 percent had chlorpyrifos residue in their urine. Research on fetal animals showed that even a single, low-level exposure can change brain chemistry and behavior. Breathing high levels of chlorpyrifos affects the nervous system and can cause headaches, blurred vision, salivation, unstable blood pressure, diarrhea, nausea, and muscle cramps. At very high concentrations, chlorpyrifos can cause paralysis, seizures, loss of consciousness, and death. The EPA tracks health effects and requires companies to report such information. In 1995, DowElanco, a Dow Chemical subsidiary company, was fined $732,000 for failure to report information on adverse health effects from Dursban.

The EPA conducted the most exhaustive scientific reviews ever done on a pesticide. Armed with this information, EPA administrator Browning announced an agreement that would "significantly minimize potential health risks from exposure to Dursban . . . for all Americans, especially children." The agreement ended higher-risk uses first and allowed eighteen more months of retail sales of products containing the pesticide.

Chlorpyrifos could no longer be used on tomatoes. The EPA cut the allowable amount that could be found on grapes and apples to 0.01 parts chlorpyrifos per million parts of food. These changes alone may produce a significant effect because the average one-year-old consumes twenty-one times more apple juice and five times more grape juice on a pound-for-pound basis than does the average adult.

The manufacture of home-use chlorpyrifos products ended on December 1, 2000. Retailers had until December 31, 2001, to sell their stock of the products. Some specialized uses, such as point applications to control termites in residential construction, were allowed until December 31, 2002, and, if applied before construction, could continue until December 31, 2005. Dursban could still be used in labeled container baits for ants if the packaging is child-resistant.

Nonresidential products were no longer made after December 1, 2000, and could not be sold after December 31, 2001. Applications of the pesticide on fire ant mounds ("drench treatment") could continue only if done by licensed professionals, who could also use the product on golf courses. Some other nonresidential and commercial uses could continue, but not in a setting that poses risks to children.

Consequences

The EPA/Dow agreement demonstrated the role of environmental organizations as well as individual citizens in bringing about a change in public policy. Moreover, it showed a process of negotiation by which an agreement was reached between a government agency and a corporation. It would have taken longer for the more traditional approach of a ruling issued by the EPA followed by an appeal or by court action. Most important, the agreement helped to make food and the environment safer.

The public lost a popular product in the control of pesticides, but there are alternative products that are not as harmful to humans. For example, products that can be used on tomatoes include pyriproxifen, buprofezin, imidacloprid, oil, soap, fenpropathrin, esfenvalerate, permethrin, endosulfan, Bt, methomyl, methamidophos, lamba-cyhalothrin, spinosad, tebufenozide, azadirachtin, and pheromone. Other products exist for different crop, home, commercial, and pet use.

Robert M. Sanford

Scientists Announce Deciphering of Human Genome

Two scientific research teams announced that they had finished the first complete reading of the human genetic code, fueling hopes that this breakthrough would lead to rapid medical advances but also raising ethical concerns as to how the information would be used.

What: Biology; Genetics; Medicine
When: June 26, 2000
Where: White House, Washington, D.C.
Who:
FRANCIS COLLINS (1950-   ), physician and head of the National Human Genome Research Initiative
J. CRAIG VENTER (1946-   ), scientist and president of Celera Genomics

The Hereditary Material

On June 26, 2000, U.S. president Bill Clinton hosted a ceremony at the White House at which a momentous announcement was made. The human genome—the genetic material in every person—had been sequenced. The announcement by Francis Collins, director of the National Human Genome Research Initiative and head of the publicly funded international effort to sequence the genome, and J. Craig Venter, president of the private company Celera Genomics, was regarded as one of the most significant advances ever made in biology.

The human genome is made of the chemical deoxyribonucleic acid (DNA) and is located inside the nucleus of nearly every cell of the body. DNA is the hereditary material passed from parent to child. DNA consists of four elements called bases, designated as A, T, C, or G. There are about three billion bases of DNA in the human genome, divided into twenty-three rod-shaped structures called chromosomes located in the nucleus of the cell. The sequence of the DNA bases contains information that is converted into proteins, molecules that have an impact on most human structures and biochemical processes.

Because human traits, as well as diseases, can be encoded in the DNA, it was believed that understanding the complete sequence would be the key to understanding human health and disease.

Deciphering the Sequence

The dedicated effort to sequence the entire human genome was begun in 1990 with the creation of the National Human Genome Research Initiative (NHGRI), a collective of many U.S. and five international research centers. This group of institutions expected to complete the sequencing in 2005. In May of 1998, however, J. Craig Venter and his company Celera Genomics announced that with newly developed technology they would complete the project in three years. Rather than work independently, the two groups became partners in sequencing the genome, sharing results and technology.

To sequence the genome, NHGRI researchers followed an orderly process of breaking the genome into large pieces, localizing them to particular sites on chromosomes, and then breaking the large pieces into smaller pieces. Automated sequencers were then used to establish the DNA sequence of the smaller pieces. Computer programs were used to arrange these short sequences so that they overlapped. As a final step, researchers called "annotators" located and identified genes, again aided by computers.

Celera used a different technique for sequencing the genome, one called "whole genome shotgun sequencing." The genome was shattered into thousands of short fragments. The ends of these fragments were sequenced and powerful computers were used to do literally trillions of computations to identify the overlapping regions of the fragments. These overlapping

3203

J. Craig Venter (left) and Francis Collins (right) join President Bill Clinton to announce the completion of an initial sequencing of the human genome.

fragments were then arranged, and the entire genome sequence was reconstructed.

Working together, the genome sequencing proceeded at a much faster rate than either initiative could have accomplished alone. The June, 2000, announcement of the nearly completed sequence occurred five years earlier than originally anticipated. At the time of the announcement, it was estimated that 99.9 percent of the human genome sequence was completed, but significant gaps and inaccuracies existed, which both groups continued to resolve.

Consequences

The news of the near completion of the deciphering of the human genome was met with both great excitement and caution. The medical community looked to the discovery for rapid advances in enhancing health and curing disease, and others regarded the announcement with ap-

prehension because of the possibility of misuse of genome information.

By comparing the sequence of genes in the human genome with the genomes of other organisms, scientists have localized suspected human disease genes in other animals. For example, the fruit fly has long been a research organism that has led to advances in understanding human genetics. In April, 2000, the sequencing of the fruit fly genome was completed. Scientists compared 289 human genes suspected to be involved in disease with the fly genome. More than 60 percent of these human genes had similar counterparts in the fly. Because genes are much easier to study in the fly, study of these human gene counterparts may lead to rapid understanding of disease processes in humans.

Scientists are also beginning to compare single nucleotide polymorphisms (SNP) between individual humans. Humans differ from one another in about one nucleotide per thousand bases, for example, an A instead of a T at a particular site. Although the vast majority of these differences are of no consequence, some may be involved in disease, or in the body's response to a particular medication. Scientists hope that by identifying these SNPs and establishing a database of these variations, researchers can use the information to identify disease genes, and drug companies can tailor medications to an individual patient.

On the darker side, access to the genome information of an individual could lead to genetic discrimination. Many genetically based diseases can be identified with diagnostic tests, and additional ones will be identified. Life and health insurance companies, as well as employers, may require genetic testing of their subscribers and employees. Individuals with a particular genetic

makeup could be denied or dropped from insurance or employment. Genetic discrimination is a reality that must be dealt with by society.

On February 10, 2001, Collins and Venter announced the completed sequencing of the human genome. Their initial analysis of the genome indicated that the human genome contained 30,000 genes—small segments of DNA that code for proteins—far fewer than expected by most scientists. The result has been interpreted to mean that the human complexity is caused by the modification of the protein products of genes and the interaction of proteins after they have been produced. The promise of sequencing the human genome was that it would help unravel the mysteries of health and disease. With the latest revelation, it appeared that much more research was needed to attain that goal.

Karen E. Kalumuck

Gore and Bush Compete for U.S. Presidency

One of the closest presidential campaigns in U.S. history pitted Democratic vice president Al Gore against Texas's Republican governor George W. Bush in a tightly contested race that remained undecided well after election day.

What: Government; National politics; Political reform
When: July-November, 2000
Where: United States
Who:
GEORGE W. BUSH (1946-   ), governor of Texas from 1995 to 2000
AL GORE (1948-   ), vice president of the United States from 1993 to 2001
DICK CHENEY (1941-   ), Republican vice presidential nominee
JOSEPH LIEBERMAN (1942-   ), Democratic vice presidential nominee
BILL CLINTON (1946-   ), president of the United States from 1993 to 2001

Primaries and Conventions

The 2000 presidential election in the United States featured two men whose families enjoyed long political histories. Governor George W. Bush was the eldest son of former president George Herbert Walker Bush and the popular First Lady Barbara Bush. Governor Bush's meteoric rise to the Republican nomination included a stint as an executive in the Texas Rangers baseball organization and the corporate oil industry. Bush's accomplishments included implementing educational reform, cutting property taxes, and getting tough on crime by presiding over a record number of Texas death penalty executions.

The Democratic nominee, incumbent vice president Al Gore, also came from a famous family. His father, Albert Gore, Sr., represented Tennessee for thirty-two years, fourteen in the House of Representatives and eighteen in the Senate. The younger Gore's record included assertions that he had played key roles in protecting the environment, leading congressional support in the growth of the Internet, and "reinventing government" by making the federal bureaucracy much more efficient. Some observers, however, felt that fund-raising activities occupied much of Gore's time, including a legally questionable visit to a Buddhist temple in California.

Both candidates easily defeated their opponents during the party primaries. Bush dispatched his chief competitor, Senator John McCain of Arizona, and Gore routed former New Jersey senator Bill Bradley. By March 7, 2000, Gore and Bush had enough delegates to capture their party conventions.

At first, the momentum seemed to favor Bush. Newspaper reports described Gore's campaign as "ebbing" or "failing." Analysts pointed to the "character issue" as one reason. Ironically, it surfaced when the vice president attempted to distance himself from President Bill Clinton by supporting asylum for the young Cuban refugee Elián González. Critics viewed it as a blatant act of pandering to Florida's Cuban Americans, who had fled from Fidel Castro's communist regime. The *Washington Post* polls reflected that such behaviors had damaged Gore's public image.

When the Republicans met in Philadelphia in their mid-summer national convention, Bush reinforced his theme of "compassionate conservativism." Democrats, who considered the Texas governor neither intelligent enough nor qualified to he president, conceded that Bush had walked a fine line between his party's moderate and conservative factions. This proved difficult for Bush because, although he pledged "strong civil rights enforcement" and most photographs depicted him with minorities, critics noticed that he carefully avoided issues such as affirmative action and racial profiling.

Other concerns arose within civil rights groups over Governor Bush's selection of his father's secretary of defense, Dick Cheney, as his vice presidential nominee. Feminists worried about the vote Cheney had cast as a young congressman against the Equal Rights Amendment to the Constitution. The media also questioned the wisdom of appointing a candidate with a long medical history of heart ailments. Bush's defenders, exhibiting great confidence, countered the attack by replying that although Cheney would add little charisma to the campaign, his considerable experience as a ten-term congressman and former president Gerald Ford's chief of staff would constitute a great asset to Bush once he became president.

The Democratic convention took an roundabout road to promoting its candidate. It devoted its first night to President Bill Clinton, who at length praised his administration's achievements but barely mentioned the vice president. Then, the convention dedicated the second night to celebrating the significant role the Kennedys had played in Democratic politics. Polls indicated that the first two sessions, while creating interesting television, scarcely promoted the presidential ticket.

Gore's prospects improved, however, with his selection of Joseph Lieberman as the Democratic vice presidential nominee. Lieberman, a Jewish senator from Connecticut, built a strong reputation for morality in public office and helped to offset the damage that President Clinton's scandals caused. Lieberman was one of the few Democrats who severely criticized Clinton's actions and statements—especially about the Monica Lewinsky affair—and his inclusion on the ticket lent a moral vitality to the campaign. Another factor behind the boost Gore received from the convention occurred as a result of his self-deprecating but highly effective acceptance speech in which he identified himself with the long economic boom the nation had enjoyed, pledged peace, prosperity, an extension of Social Security, and better public schools. Unfortunately for Gore, however, the media seemed to focus more on the prolonged kiss he gave to his wife, Mary Elizabeth "Tipper" Gore. The convention gave the vice president the "bounce" he needed in the fall campaign.

The Fall Campaign

Armed with $67.6 million in public money, Bush and Gore set out to clarify their stands on the issues and raise private contributions. No single issue dominated the campaign, but both candidates argued effectively on the major points. On education, Bush favored vouchers for private schools, while Gore wanted to improve the existing public system. Gore wanted to preserve Social Security in its present form, while Bush preferred semiprivatization with some funds being invested in the stock market. Throughout the fall, the vice president enjoyed a monetary advantage with $8.2 million in private contributions to Bush's $6.8 million. Late September polls accordingly gave Gore a slight lead among voters.

Then came the October debates. CBS News/ *New York Times* polls taken after each debate showed that, although Gore led Bush 45 percent to 41 percent following the first confrontation on October 3, Bush assumed the lead 44 percent to

Al Gore (right) shakes hands with George W. Bush before a debate at Wake Forest University in Winston-Salem, North Carolina.

AP/Wide World Photos

3207

42 percent after the last meeting on October 17. Surveys indicated that although Gore was more knowledgeable, Bush appeared the more likeable candidate and better leader. Two days before the election, a *Washington Post* poll revealed that Bush still held a slight lead, although the race was "too close to call," and a deadlocked campaign awaited voters who would be going to the ballot boxes on election day.

Consequences

On Tuesday, November 7, 2000, the television networks began their election night coverage. Their reports were marred by erroneous predictions that relied on analyses of exit interviews. Early in the evening, some networks reported that Gore had won Florida, a statement that they would later take back. The networks first called Gore, then Bush, the "new president." However, as the evening became the next morning, the outcome remained unclear. Although the popular vote favored Gore, the Electoral College count gave Bush a chance to win if he could carry Florida, where his brother, Jeb Bush, was governor. Several other states remained undecided. Gore placed a call to concede the election to the Texas governor. However, when the vice president heard rumors of voting irregularities in Florida, he revoked his withdrawal, demanding a recount and placing the campaign in a new stage in which the candidates looked to election boards and courts to determine who would become the next president of United States. The election was finally resolved by the U.S. Supreme Court on December 12.

J. Christopher Schnell

Fox's Election Ends Mexico's Institutional Revolutionary Party Rule

Charismatic self-made businessperson Vicente Fox, a member of the National Action Party (PAN), won Mexico's presidential election, defeating the candidate of the Institutional Revolutionary Party (PRI), which had dominated Mexican politics since 1929.

What: National politics; Political reform
When: July 2, 2000
Where: Republic of Mexico
Who:
VICENTE FOX (1942-   ), presidential candidate of the National Action Party
FRANCISCO LABASTIDA (1943-   ), presidential candidate of the Institutional Revolutionary Party
CUAUHTEMOC CÁRDENAS (1934-   ), presidential candidate of the Revolutionary Democratic Party
ERNESTO ZEDILLO PONCE DE LEÓN (1951-   ), incumbent president

An Upset Victory

In the July 2, 2000, Mexican presidential election, Vicente Fox, former governor of Guanajuato and candidate of the National Action Party (PAN), won 47.5 percent of the popular vote, while Francisco Labastida, the candidate of the ruling Institutional Revolutionary Party (PRI), received 31.8 percent. A third candidate, Cuauhtemoc Cárdenas, leader of the Revolutionary Democratic Party (PRD), got 15.8 percent of the vote. Fox's victory meant that the PRI, despite massive government support in terms of financing and manpower, lost control of the presidency for the first time since the party's founding during the administration of Emilio Portes Gil (1928-1930).

Observers judged the election to be perhaps the cleanest one in Mexico during the twentieth century. Ernesto Zedillo, the incumbent president, stated his determination to end the corrupt political practices of the past. In addition, the Federal Electoral Institute, an independent authority whose members were selected by the Mexican Congress, supervised the electoral process.

Fox, whose father operated a ranch in the state of Guanajuato, maintained ties to Mexico's rural areas throughout his career. He graduated from a Jesuit university in Mexico and studied business administration in the United States. A highly visible, charismatic figure, Fox is 6 feet, 4 inches (193 centimeters) tall, much taller than the average Mexican, and favors cowboy boots and buckles. Before entering politics, Fox was president of Coca-Cola Mexico. In fifteen years with that company, he rose from local distributor to become the company's chief executive. He used the marketing skills that he attained while with Coca-Cola in developing his political career. At the time of his election, Fox was divorced but maintained a friendly relationship with his former wife and remained close to his four daughters.

The Election

Fox waged a highly organized and aggressive presidential campaign. He did not rely solely on the facilities of PAN, his official political party, but formed a national group called Los Amigos de Fox ("the friends of Fox") that raised money in support of his candidacy. Fox instituted a television campaign stressing the corruption issue that had come to plague the PRI. He said that the government and the PRI had become part of the illegal drug business that had developed in Mexico. Fox described his plans to encourage foreign investment in the country by advocating an open

3209

Vicente Fox waves to supporters after his victory.

AP/Wide World Photos

tion. He was committed to free-market development, and the cabinet that he chose was filled with men trained at U.S. business schools. Fox intended to free the energy sector from governmental restrictions, encouraging foreign investment, which he believed would stimulate the Mexican economy, provide more domestic job opportunities, and decrease the need for Mexican citizens to leave for jobs north of the border.

Fox sought to reduce corruption within government as part of his aggressive war against illegal drug trafficking. He stated his intention to develop a team to fight drugs that would be able to resist all bribes and corruptive influences. He also planned to end the civil disorder in the southern state of Chiapas, beginning with the transfer of army units out of the territory. The heart of the problem, he felt, was in finding solutions to the complaints of the state's indigenous population. He planned to enter into negotiations with the leader of the Zapatista movement to further that end.

Finally, Fox sought to develop more effective relations with the United States and to improve the treatment of both Mexican citizens living in the United States and those working there temporarily. His ultimate goal, and one that was not likely to be achieved during his administration, was the freedom of movement for all Canadian, U.S., and Mexican citizens across each others' borders, similar to that which exists within the European Union.

Fox's presidential campaign and subsequent victory stimulated the Mexican people's participation in the country's politics. Although he gained the good will of the majority of the populace, questions remained about how much of his ambitious program he could accomplish.

Carl Henry Marcoux

market economy and to use oil revenues to improve the educational system. In a series of presidential debates, he easily bested his two opponents, Labastida and Cárdenas, in outlining his political goals.

One strategy failed. Representatives of the two opposition parties, PAN and PRD, attempted to form an alliance against the PRI, but the personalities and the ideologies of the two leaders, Fox and Cárdenas, prevented the combination. In the end, Fox did not need the Cárdenas compact in order to achieve his victory.

Consequences

President Fox faced numerous challenges in his six-year term. Because Mexican law precludes an incumbent president from seeking reelection, the reforms that Fox sought to accomplish had to be at least initiated during his administra-

Defense Missile Fails Test and Clinton Shelves System Development

The failure of a major test of the planned U.S. defensive missile shield raised doubts about whether the system would work and whether it should be developed.

What: Military technology; Military
defense
When: July 8, 2000
Where: Pacific Ocean; Washington, D.C.
Who:
BILL CLINTON (1946-   ), president of
the United States from 1993 to 2001
WILLIAM S. COHEN (1940-   ), U.S.
secretary of defense

A Controversial Program

The national missile defense (NMD) program is a new military technology in the process of development. Supporters of the NMD program say that it will provide the United States with a defensive shield that will protect it from incoming ballistic missiles. In the projected scenario, an enemy fires a missile at a city within the United States, but U.S. radar and communications systems, in combination with satellites in space, give early warning of the attack. U.S. interceptor weapons then track the missiles and shoot them down before they have a chance to do any damage.

The defensive shield is designed primarily to meet small-scale attacks by nations that the U.S. State Department refers to as "states of concern," such as North Korea, Iraq, and Iran. The United States is concerned about the rapid developments in missile technology in these and other potentially hostile countries. U.S. policymakers believe that such missiles could eventually carry chemical or nuclear warheads. U.S. intelligence also claims that North Korea is expected to be able to launch a missile attack on an American city by 2005. NMD would not, however, be able to deal with a major, sustained assault by a larger power such as Russia.

However, not everyone in the United States supports the development of NMD. Some experts point out that the technology is completely unproven and may not work. Others say it is too expensive—the estimated cost is $60 billion. A third factor highlighted by opponents is that most of the rest of the world, adversaries and allies alike, oppose NMD. They say it will harm international relations and lead to a new arms race. According to this view, Russia and China would feel compelled to build up their own missile arsenals so that in the event of a war with the United States, they would be able to overcome a missile defense by sheer force of numbers.

Test Is a Failure

On July 8, 2000, the Pentagon conducted a major test of an NMD weapon. To the consternation of its supporters, the weapon completely failed. The interceptor that was supposed to destroy a dummy target in space failed to separate from the booster rocket that would have directed it to the target. As a result, the interceptor never came close to doing the job for which it had been designed.

It was widely expected that had the test been successful, Defense Secretary William S. Cohen would have recommended to President Bill Clinton that the United States develop NMD. However, the failure made the situation more complicated. President Clinton was faced with a dilemma. Should he approve the development of an expensive military program when no one knew for sure that it could work? Should he cancel it, arguing that the technology was not yet ready? Or given that he had only another seven months in office, should he leave the decision to his successor, Vice President Al Gore or Governor George W. Bush?

On September 1, 2000, Clinton announced that he would not authorize the building of a missile defense system. He cited the fact that the necessary technology was not yet ready. "We have made progress, but we should not move forward until we have absolute confidence the system will work," the president said. Clinton's decision was influenced by strong Russian opposition to NMD. The Russian government feared that NMD would reduce the deterrent value of its own weapons. Russia thought NMD would provide the United States with a technology that could eventually blunt the force of a Russian missile attack. Russia also believed that despite U.S. statements that NMD was for protection from smaller potential adversaries such as North Korea, it was in fact designed to put the Russians at a military disadvantage.

One problem with NMD that directly concerned the Russians was that in 1972, the U.S. and the former Soviet Union signed the Treaty on the Limitation of Anti-Ballistic Missile (ABM) Systems. The ABM treaty banned the use of missile defense systems. In order to use NMD, the United States must either negotiate a modification of the treaty with the Russians or withdraw from it. Both the Russians and U.S. allies in West-ern Europe such as France, Britain, and Germany value the ABM treaty. They say it has been the cornerstone of arms control agreements for nearly thirty years. According to this view, the best way to maintain world peace is to work within the framework of the ABM treaty to reduce arms still further.

Consequences

Clinton's decision to put a hold on the development of NMD had little long-term effect because Bush, who repeatedly said during the election campaign that he strongly supported NMD, won the presidential election of 2000. After his inauguration in January, 2001, Bush made it clear that his administration would move ahead with all due speed on the development of NMD. In February, he announced a budget plan that called for $1 billion in new spending on research for missile defense. The Bush administration also said it intended to develop a more expanded form of NMD than had been sought by the Clinton administration. The more ambitious plan may include sea- and space-based systems as well as ground-based ones.

Bryan Aubrey

Florida Jury Orders Tobacco Companies to Pay Record Damages

Following a two-year trial, a jury decided that the tobacco industry should pay $145 billion in punitive damages to sick smokers in Florida. Although appeals were expected to take two or three years, observers warned that the judgment, if upheld, could force the tobacco firms into bankruptcy.

What: Business; Health; Law
When: July 14, 2000
Where: Miami-Dade Circuit Court, Florida
Who:
STANLEY ROSENBLATT and
SUSAN ROSENBLATT, lawyers for the plaintiffs

Liability of Tobacco Companies

The punitive award announced on July 14, 2000, constituted an unprecedented threat to the tobacco industry. In previous years, countless individual smokers had sued the companies for the harmful health effects of tobacco products, but the companies had almost always prevailed. Earlier juries had tended to blame the individuals for their inability to quit smoking, especially in view of the health warnings appearing on cigarette packages.

By the late 1990's, however, public opinion had turned against the tobacco industry. The change was a result of growing evidence that the companies had directed advertisements at young people, had manipulated nicotine levels to encourage addiction, and had withheld information about the harmful effects of their products. Based on this evidence, the state governments sued the companies in order to recover the public funds spent on the medical costs associated with tobacco use. In settlements during 1997-1998, the companies agreed to pay the states some $246 billion over twenty-five years.

Individual smokers continued to have the right to sue the companies. Although they could use the mountain of evidence uncovered in the state trials, individuals still had the problem of overcoming the tendency of juries to look upon smoking as a moral weakness. Some lawyers hoped that juries might focus less on the responsibility of the individual smoker in a class-action lawsuit, which meant that one plaintiff would bring a case in behalf of all persons sharing similar circumstances.

Florida's Class-Action Trial

In 1994, the small husband-and-wife firm of Stanley Rosenblatt and Susan Rosenblatt initiated a class-action suit on behalf of sick smokers in Florida. At the time, most lawyers believed that the plaintiffs had little chance of success. The Rosenblatts spent hundred of hours and at least $1 million of their own money to bring the case to trial. This class-action suit was the Rosenblatts' second major tobacco suit. In 1997, they won a $349 million settlement in a case dealing with flight attendants who had become sick as a result of secondhand smoke in airplane cabins.

Beginning in 1998, Judge Robert Kaye presided over the lawsuit, *Engle et al. v. R. J. Reynolds et al.*, which took place in the Miami-Dade Circuit Court. The six jurors included a smoker and three former smokers. The named plaintiffs in the case were Howard Engle, a retired pediatrician with emphysema, and Raymond Lacey, who had lost both legs as a result of a circulatory disorder associated with smoking. Engle and Lacey represented an estimated 500,000 smokers in the state, and they were asking the jury for at least $200 billion in damages.

The defendants in the case included Philip Morris, Brown and Williamson, Lorillard Tobacco Company, the Liggett Group, and Dorsal Tobacco of Miami. In addition, two industry

3213

Plaintiff Ralph Della Vecchia (right), whose wife died of cancer, hugs Margaret Amodeo, wife of plaintiff Frank Amodeo, after the jury announced its verdict.

groups were named as defendants, the Council for Tobacco Research and the Tobacco Institute.

The two-year trial featured thousands of exhibits, 157 witnesses, and more than 55,000 pages of transcript. The Rosenblatts were able to produce abundant evidence of company wrongdoing, including destruction of evidence, secret research in other countries, advertising directed at young people, and enhancement of the addictive nature of cigarettes. Ian Uydess, a former employee of Philip Morris, told jurors that he had been told to abandon promising research on a safer cigarette.

The trial was divided into three stages. In their initial verdict of July, 1999, the jurors ruled that smoking was a cause of twenty diseases and that the companies had a long history of deceiving the public. In April, 2000, they awarded compensatory damages of $12.7 million to three of the sick plaintiffs. Then on July 14, after deliberating only four hours, they ordered the companies to pay a damage award of $145 billion, which was the largest punitive judgment in the history of U.S. litigation to that date.

Consequences

Shortly after the punitive award was announced, the tobacco companies promised to appeal the amount. Analysts cautioned that the award would likely be overturned or greatly reduced. In any event, no payments would be made for many years, after long and complex smaller trials to determine which individuals were entitled to participate in the award. The companies were also expected to go to the U.S. Congress in an attempt to get immunity from future class-action suits.

Most informed observers agreed that the huge award was beyond the financial means of the tobacco companies, even though a state law stipulated that an award could not be high enough to bankrupt a company. Outside of Florida, lawyers in other states were initiating similar suits. A spokesperson for the Citizens for a Tobacco-Free Society expressed hopes that the combination of lawsuits would destroy the industry. Many of the state governments, in contrast, were concerned that the award could make it impossible for the states to collect their settlements with the companies.

Business analysts predicted that the first impact of the settlement would be higher prices. Because most smokers were addicted to nicotine, the companies had the ability to raise prices significantly. If the cost of tobacco products increased too much, however, there was the danger of the emergence of a black market and the illegal smuggling of cigarettes into the country. Also, there was the distinct possibility that new companies might be established, having no financial obligations for the health problems associated with smoking in the past.

During the months following the verdict, the stocks of Philip Morris and other tobacco companies did unexpectedly well. This appeared to indicate that investors expected that the award of $145 billion would not be upheld and that they considered it unlikely that juries in other states would follow Florida's example.

Thomas Tandy Lewis

Clinton Vetoes Bill Reversing Taxpayer "Marriage Penalty"

> *Millions of married couples pay more federal income tax than they would if they were not married. Congress passed a bill to address this, but President Bill Clinton vetoed it because it would take too much of the projected federal surplus and most of the benefits would go to wealthy families.*

What: Economics; Government
When: August 5, 2000
Where: Washington, D.C.
Who:
BILL CLINTON (1946-   ), president of the United States from 1993 to 2001

The Tax Code and Marriage

One of the most persistent problems in tax policy is the treatment of married couples relative to singles. The United States uses a progressive income tax, which means different amounts of income are taxed at different rates. Couples, because they often have two incomes, are more likely to reach the higher tax brackets. The amount sheltered from taxes, the standard deduction, also creates a higher tax load for couples. For 2001, married couples have a standard deduction of $7,600 annually, while single persons receive a $4,550 deduction. Because the amount for couples is not twice as great as for singles, many couples end up paying more tax. The current tax treatment of marriage is the result of several important changes over the years.

Before 1948, wealthy couples with lots of investment income could split their income and file separate tax returns so both husband and wife could take advantage of lower tax rates. Several states had laws that treated each spouse as the owner of half the family income, but most wage-earning couples had no way to divide the family income for tax purposes. In 1948, Congress changed the law so that each member of a married couple was considered to own exactly half the total income, and each was taxed on that income at the same rate as single people. Although this seemed fair to some, people who remained single argued that it penalized them. A single person earning $25,000 would pay more taxes than two married people with the same total income because more of the married couple's income (split into two $12,500 incomes) would fall in the lower tax rate. In response to these complaints, in 1969, a new tax rate table, more favorable to singles, was adopted.

For a majority of couples (those making up to about $60,000 a year), the most common marriage penalty stems from the fact that the standard deduction for a couple is less than the standard deduction for two unmarried individuals. In 2001, if two roommates each earned $20,000, they would have paid taxes on $30,900 of their combined income because each would have had a $4,550 deduction. A married couple earning $40,000 would have paid taxes on $32,400, only deducting $7,600. The marriage penalty caused by this increase in taxable income would have been $225.

Not all couples face the marriage penalty. A couple in which one spouse earns all the income is still better off married than single. Taxwise, a couple whose earnings are roughly equal is better off unmarried than married. In 2000, the Congressional Budget Office (CBO) estimated that 20 million married couples paid higher income taxes than they would have if they were single, and 25 million couples paid lower taxes than they would have if they were single. The CBO also projected that those couples hit by the marriage penalty paid an average of $1,480 more in taxes each year.

A Presidential Veto

On July 20, 2000, the House of Representatives approved a Republican-sponsored tax bill that eliminated the marriage penalty by reducing tax rates for married couples, doubling their standard deductions, and extending other tax credits. The Senate voted in favor of the measure on July 21, 2000, sending the bill to the White House for action.

During the debate on the legislation, there was disagreement on the financial impact of the proposal and the distribution of benefits. Supporters estimated that the legislation would help lower the tax bill for up to 50 million couples and would only decrease government tax receipts by $182 billion over the next decade. Opponents projected the impact of lost revenue at $292 billion over ten years and estimated that two-thirds of the total tax cut would go to the 20 percent of married couples with the highest incomes. Opponents argued that this group, with average incomes of $184,000 per couple, did not need additional tax relief.

President Clinton had proposed a more limited relief from the marriage penalty but had offered to sign an earlier version of the final bill if Congress also passed a prescription drug benefit for Medicare recipients. Congressional Republicans refused to accept that deal and sent the legislation forward even though President Clinton had said he would veto it. President Clinton described the bill as "the first installment of a fiscally reckless tax strategy" that would deplete projected budget surpluses and vetoed it on August 5.

Consequences

For most people, the economic impact of the presidential veto was limited. Although the average couple paid an extra $1,480 more in taxes, this was only one part of the total tax picture. Married couples also benefited from a variety of tax deductions that helped them more than single people. For example, married couples are more likely than singles to be buying a home. Interest payments on the home loan are tax deductible.

House majority whip Tom DeLay talks to reporters after the House sustained President Bill Clinton's veto of the "marriage penalty" bill.

3217

The political impact may well have been greater than the economic impact. Democratic presidential candidate Al Gore supported Clinton's veto, while Republican presidential candidate George Bush supported the bill as passed by Congress. Given the very close presidential race in 2000, if a small percentage of voters changed preference based on this issue, it might have helped Bush gain the presidency.

Although taxes might be unpopular, they are inescapable and necessary. The structure of the progressive tax code makes it impossible to tax singles and married couples equally. Elimination of the marriage penalty would shift the tax burden onto someone else. Under a progressive tax system, the only way to eliminate the marriage penalty is to go back to relatively higher taxes on singles. However lawmakers change the rates, there will always be a perceived penalty on somebody.

Allan Jenkins

Nine Extrasolar Planets Are Found

Scholars from two research institutions announced the discovery of at least nine planets around stars other than the Sun, bringing the total number of known extrasolar planets to at least fifty if these discoveries were confirmed.

What: Astronomy; Physics; Space and aviation

When: August 7, 2000

Where: International Astronomical Union assembly, Manchester, England

Who:

WILLIAM D. COCHRAN, astronomer at University of Texas McDonald Observatory

ARTIE P. HATZES (1957-), astronomer at McDonald Observatory.

SALLIE BALIUNAS (1953-), astrophysicist at Harvard-Smithsonian Center for Astrophysics

DEBRA FISCHER, postdoctoral fellow and astronomer at University of California, Berkeley

A Bundle of Nine

On August 7, 2000, a team of astronomers from McDonald Observatory and the Harvard-Smithsonian Center for Astrophysics announced their discovery of at least nine extrasolar planets (sometimes called exoplanets) at the International Astronomical Union assembly in Manchester, England. Included in the collection of newly discovered objects was the smallest extrasolar planet yet detected—a gas giant roughly half the size of Saturn. In addition, astronomers brought to light new evidence that several nearby stars may actually have multiple-planet systems. Previously, only one star, Upsilon Andromedae, had been found with multiple planets surrounding it. If confirmed, these discoveries could bring the total number of planets detected beyond the solar system to fifty. Many of this batch of detected planets seemed to have companion objects similar to the moons of planets in the solar system orbiting near them. Some astronomers, like Debra Fischer of the University of California, Berkeley, noted that these recent findings encourage further exploration.

Looking for the Wobble and Planets Like Earth

Because the stars these planetary bodies orbit emit so much light, no telescope yet invented can view these planets directly. Therefore, the question arises as to how astronomers can be so certain that they have discovered extrasolar planets. The answer lies in looking for and measuring the effects of the gravity of extrasolar planets on their stars. A specialized part of astronomy, astrometry, measures the proper motion of stars as a function of time. In the case of the study of one particular star named Barnard's Star, astrometricists examined photographic plates over many decades to determine its motion. Another part of astronomical study, called perturbation, is a way of describing any abnormalities in the star's motion. Because scientists have learned that the Moon affects Earth and that Earth and its sister planets affect the Sun, when they detect a "wobble" in the mass of the star being observed, they can measure the slight shifts in a star's light and therefore reasonably conclude that the wobble signals the presence of some large object, such as a planet or a brown dwarf star that may be orbiting the star.

Using ground-based telescopes from such widely separated institutions as the University of California, Berkeley, the Geneva Observatory, and the McDonald Observatory of the University of Texas, astronomers collected and analyzed wobbles from new batches of stars. Their analyses of the evidence led them to announce a number of additional extrasolar planets and companion stars. They have been able to detect only so-called gas giant planets (like Jupiter and Saturn). However, their latest discoveries suggest that extrasolar terrestrial (rocky or iron-based) planets may outnumber the gas giants that astronomers have so far discovered.

Using these methods, a team of astronomers from the Geneva Observatory in Sauverny, Switzerland, announced on May 8, 2000, that they had discovered eight new extrasolar objects. The researchers believed the objects to be six planets and two brown dwarfs (objects that weigh less than a star and more than a planet). One of the six planets has the shortest year (the time it takes to orbit its star) as well as distance to its star of any extrasolar planet discovered before those announced on August 7, 2000. Lick Observatory astronomers report quite bizarre-behaving extrasolar systems.

The search for planets like Earth has a long history. *Other Suns, Other Worlds? The Search for Extrasolar Planetary Systems* (1996) by Dennis L. Mammana and Donald W. McCarthy, Jr., provides a useful background for the history of human speculation about the possibility of there being other planets like Earth in the cosmos and traces the history of efforts to discover them. The book summarizes what is known about the size of the universe, which is believed to contain at least 50 billion galaxies, the largest of which contains thousands of billions of stars visible to such telescopes as the Hubble Space Telescope. Scientists estimate that a million solar systems are formed in the universe each hour. Therefore, they reason that the laws of probability make it nearly certain that planets like Earth do indeed exist.

Consequences

Astronomers must rely on helpful but often limiting ground-based observatories to detect extrasolar planets. However, spurred by the discoveries such as those made by the astronomers at the Geneva Observatory, the McDonald Observatory, and the Harvard-Smithsonian Center for Astrophysics, the National Aeronautic and Space Administration (NASA) decided to launch two new space-based telescopes dedicated to the pursuit of Earth-like planets beyond the solar system. NASA planned to launch the Space Interferometry Mission (SIM) in 2006 and the Terrestrial Planet Finder in 2013. The latter project was to be equipped with specialized instruments that can detect and photograph extrasolar planets and even analyze planetary atmospheres. These launches should provide astronomers with even better tools to detect wobbly stars.

Theodore C. Humphrey

Bridgestone Apologizes for Tires That Caused Numerous Deaths

Bridgestone/Firestone chairman and chief executive officer Masatoshi Ono, speaking before the U.S. Senate Appropriations Subcommittee for Transportation, apologized for rollover accidents associated with certain models of defective Firestone tires.

What: Business; Transportation
When: September 6, 2000
Where: Capitol Building, Washington, D.C.
Who:

MASATOSHI ONO (1937-   ), chairman and chief executive officer of Bridgestone/Firestone

JACQUES NASSER (1947-   ), chief executive officer of Ford Motor Company

SAM BOYDEN, officer of State Farm Insurance Company

ANNA WERNER, investigative reporter at KHOU-TV, Houston, Texas

Congressional Hearings

On September 6, 2000, Masatoshi Ono, chairman and chief executive officer of Bridgestone/Firestone, the U.S. unit of the Japanese company Bridgestone, appeared before the Senate Appropriations Subcommittee for Transportation. Ono offered his condolences to the American people in general and "especially to the families who have lost loved ones in these terrible rollover accidents." At the time of his appearance, eighty-eight people had been killed in accidents involving certain models of Firestone tires. On August 9, 2000, Firestone had announced a recall of 6.5 million tires.

Many of those killed had been riding in Ford Explorers. In his testimony, Ono mentioned the tendency of Ford Explorers to roll over. Jacques Nasser, chief executive officer of Ford Motor Company, responded by asserting, "This is a tire issue, not a vehicle issue." He explained that a large number of Explorers were involved because Firestone was the main tire supplier for this vehicle.

The Senate subcommittee hearings, as well as those held on the same day by the House of Representative's Subcommittee on Consumer Protection and Oversight, wanted to know what each company knew and when they knew it. Legislators rejected attempts by Ford and Firestone to blame each other.

Discovering a Problem

The Ford Explorer was introduced in March, 1990, as a sports utility vehicle. Although Ford selected Firestone to provide tires for the Explorer, Goodyear provided approximately half the tires used on Explorers between 1995 and 1997. The first recorded accident of a Ford Explorer involving Firestone tire separation occurred in 1992. The first Explorer Firestone tire lawsuit was filed in 1993. It was not until the late 1990's, however, that either Firestone or Ford acted on reports of tire failure with Explorers.

In October, 1998, Ford sent to Firestone examples of failed Firestone tires used on Explorers in Venezuela. The tire company blamed the tread separation on local conditions of heat and bad roads. In early 1999, Ford dealers in Saudi Arabia noted problems with Firestone's 16-inch (41-centimeter) Wilderness tires. Again Firestone blamed local conditions—a hot climate and fast drivers. Nevertheless, in August, 1999, Ford unilaterally began replacing Firestone tires on Explorers in foreign countries. By spring, 2000, Ford had replaced Firestone tires on 46,912 vehicles in Saudi Arabia, Thailand, Malaysia, and South America. The National Highway Traffic Safety Administration (NHTSA) did not learn of this action until May, 2000. At the time, there was no federal law requiring companies to inform the NHTSA about overseas recalls of products being sold in the United States.

In July, 1998, in response to claims adjuster inquiries, Sam Boyden of State Farm Insurance's Strategic Resources Office investigated twenty-one cases of Firestone tread separation. He found that all involved Firestone ATX tires, and fourteen had been on Explorers. He e-mailed this information to the Office of Defects Investigation of the NHTSA. The NHTSA took no action at this time but did react to a February, 2000, telecast featuring Anna Werner, an investigative reporter for KHOU-TV, a CBS affiliate in Houston, Texas. Werner reported on twenty-four accidents involving Explorers and Firestone tire tread separation in which thirty people had been killed. Local viewers, who related additional stories to the station, were encouraged to notify the NHTSA. In May, 2000, federal regulators began their investigation. Ford Motor Company also began an investigation about the same time.

In its investigation of Firestone's data, Ford noted a problem with ATX tires produced at the Decatur, Illinois, plant. On August 9, 2000, Firestone announced a recall of all P235/75R15 ATX and ATX II tires and P235/75R15 Wilderness AT tires made at the Decatur plant.

Consequences

On September 12, 2000, Firestone and Ford executives again testified, this time before the U.S. Senate Commerce Committee. Firestone executives suggested that Ford's recommended tire pressure may have played a key role in tread separation. Whereas Firestone recommended 30 pounds (13.6 kilograms) per square inch for its 15-inch (38-centimeter) tires, Ford had recommend 26 pounds (11.8 kilograms) per square inch. At 26 pounds per square inch, more rubber is on the road, creating greater friction and heat,

Bridgestone's Masatoshi Ono.

and is particularly dangerous at high speeds in high temperature areas. On September 22, 2000, Ford adopted 30 pounds per square inch as its recommended pressure for 15-inch Firestone tires on Explorers.

Firestone's internal investigation, concluded in December, 2000, suggested tire design and lubricants, especially those used at the Decatur plant, as possible explanations for tire failure. Firestone also noted that since the early 1990's, Ford had increased the weight of the Explorer by more than 300 hundred pounds (136 kilograms) without increasing the recommended pounds per square inch tire pressure for 15-inch tires. An independent investigator hired by Firestone, Sanjay Govindjee, an associate professor of civil engineering at the University of California, Berkeley, in his February, 2001 report, cited vehicle load as a critical factor in tread separation. By the time of Govindjee's report, 148 people in the United States and 48 people in foreign countries had been killed in accidents related to Firestone tires, mainly in Explorers.

On October 19, 2000, House Resolution 643 of the 106th Congress, second session, expressed the "sense of the House of Representatives with regard to Sam Boyden, who admirably notified transportation safety officials of a serious threat to American consumers." In November, 2000,

President Bill Clinton signed a bill requiring car and tire companies to report to the federal government when they recall products overseas for safety reasons. The law also required these companies to turn over data on warranty service and legal claims. In addition, the bill provided money to hire NHTSA staff to interpret this data. The new law also provided civil and criminal penalties for failure to report defects.

Thomas W. Judd

Unjustly Accused Scientist Lee Is Released

Wen Ho Lee, a nuclear scientist at Los Alamos National Laboratory, was unjustly accused of spying for the People's Republic of China, fired from his job, and jailed for nine months without a trial before he was finally released.

What: Human rights; International relations; Law

When: September 13, 2000

Where: Santa Fe County Detention Center, New Mexico

Who:

WEN HO LEE (1939-　　　), nuclear physicist at Los Alamos National Laboratory

BILL RICHARDSON (1947-　　　), secretary of Department of Energy

JAMES A. PARKER (1937-　　　), U.S. Albuquerque district judge

The Release

On September 13, 2000, Wen Ho Lee, a nuclear scientist who had been working at Los Alamos National Laboratory, was released from the Santa Fe County Detention Center, New Mexico, after nine months of solitary confinement since his arrest on December 10, 1999. The release of Lee was gained as a result of his pleading guilty to one count of unlawfully gathering national defense information out of fifty-nine counts of violating national security. In exchange, Lee agreed to cooperate with Federal Bureau of Investigation agents to reveal all he knew about seven computer tapes onto which he was accused of downloading sensitive information on the design of W-88, a thermonuclear warhead built for U.S. missile subs, and giving the information to China.

During his confinement, Lee, a diminutive, soft-spoken, sixty-year-old Taiwan-born physicist, stayed in a 13-by-7-foot (4-by-2-meter) jail cell with his legs shackled, his hands manacled, and the handcuffs chained to his waist. To pass the time, Lee read books, listened to classical music, and wrote a mathematics textbook. His family members, including his wife, Sylvia Lee; daughter Alberta Lee, a student at the University of California, Los Angeles; and son Chung Lee, a medical student in Ohio, were horrified by the treatment he received. U.S. Albuquerque district judge James A. Parker, who was responsible for Lee's release, termed Lee's imprisonment "draconian" and "unfair."

The release of Lee delighted the Asian American community, which had been petitioning the U.S. government to free him unconditionally and drop all fifty-nine charges against him. However, because Lee pled guilty to one count, he had become a felon, costing him some of his civil rights, including his right to vote. Consequently, the Asian American community continued to fight for his civil rights by petitioning for a presidential pardon.

Racial Profiling

The Asian American community viewed Lee's case as an example of the use of racial profiling. Lee was born in Taiwan, where he graduated from Cheng Gong University. He came to the United States to pursue graduate study and later worked at the Los Alamos National Laboratory in Albuquerque as a contract employee from the University of California. In 1995, a U.S. agent in Asia was approached by a Chinese defector with a sensational document that contained the blueprint of China's nuclear weapon program and that suggested a leak of classified information from Los Alamos. In May, 1996, federal investigators identified Lee as a possible spy for China, and Lee soon became the main target of a criminal investigation. On March 6, 1999, a *New York Times* article revealed the case.

Two days later, Los Alamos fired Lee because he had breached workplace rules, including downloading classified files to an unsecured computer. The case became international news. On August 1, 1999, Lee, after five months of silence, denied spying for China on the television news program *60 Minutes*, saying that downloading of classified files to unsecured computers was a common practice at Loa Alamos and that he was singled out because he was an ethnic Chinese. Bill Richardson, the secretary of the Energy Department, responded to Lee's statement by calling the practice "wrong."

On December 10, 1999, Lee was indicted on fifty-nine counts and was held without bail in Santa Fe County Detention Center; however, investigators conceded that they did not have the evidence to convict Lee as a spy. On August 24, 2000, U.S. district judge Parker announced that Lee was eligible for bail. On September 13, 2000, after 279 days of imprisonment, Lee pleaded guilty to one felony count of mishandling nuclear secrets and was free. The gov-

ernment dropped other fifty-eight charges. Judge Parker apologized to Lee and declared that the actions of the top decision-makers at the Department of Justice and the Energy Department had "embarrassed our nation."

Consequences

The case of Lee outraged the Asian American community, which felt that a great injustice had been done to Lee and viewed the case as an example of racial profiling and discrimination against Asians and Asian Americans. Asian American scientists in particular expressed fears of racial profiling and discrimination. After the Lee case began, the number of Asian American scientists applying for jobs at national defense and nuclear weapons labs dramatically declined. After Lee's arrest and imprisonment, some national Asian American organizations with direct interest in higher education passed resolutions demanding that all fifty-nine charges against Lee be unconditionally dropped, that Lee be unconditionally set free and reinstated to his job with back pay, that the University of California president and secretary of the Department of Energy publicly apologize to Lee and his family, and that all racial profiling and discrimination against Asian Americans be ended. Asian American communities in Los Angeles, Silicon Valley, San Francisco, Albuquerque, Seattle, New York, and Detroit, Michigan, also rallied for Lee's freedom and justice.

After Lee's release, the Asian American community continued to fight to obtain justice for Lee and to restore complete confidence in the national defense labs and in the government. The Asian American community believed that the case against Lee should not have been brought. It was tainted with racial prejudice and it was brought not on the basis of legal merit but under great political pressure from a group of influential figures in the U.S. Congress and from a prominent newspaper, the *New York Times*, which falsely accused Lee of spying for and stealing nuclear weapons secrets for China. The case deeply wounded Lee, his family, and Asian Americans in general.

Wen Ho Lee, after his release.

AP/Wide World Photos

Huping Ling

Whitewater Inquiry Clears Clintons of Wrongdoing

> *After a six-year investigation, independent counsel Robert W. Ray announced that there was not enough evidence to prove that President Bill Clinton or his wife, Hillary Clinton, had knowingly participated in crimes relating to the Whitewater real estate venture.*

What: Law; National politics
When: September 20, 2000
Where: Washington, D.C.
Who:

KENNETH STARR (1946-　　), former independent counsel for investigation of the Clintons

ROBERT W. RAY (1961?-　　), independent counsel for investigation of the Clintons

BILL CLINTON (1946-　　), president of the United States from 1993 to 2001

HILLARY CLINTON (1947-　　), First Lady of the United States from 1993 to 2001

The Clinton Scandals

The criminal investigation of the Whitewater scandal officially ended with Robert W. Ray's lengthy statement of September 20, 2000. The statement was especially good news for Hillary Clinton, as it occurred just seven weeks before the election in which she was a candidate for senator of New York. President Clinton's legal problems were not over, however, because Ray indicated that he had not yet decided whether to prosecute the president for lying under oath about his relationship with White House intern Monica Lewinsky.

Whitewater was the shorthand name for an Arkansas waterfront development that failed in the 1980's. When Clinton was governor, he and his wife invested money in the Whitewater Development Corporation, and they had several business relationships with James McDougal, the banker who headed the failed venture. McDougal hired Hillary Clinton to do some legal paperwork on the project, and he later claimed to have provided the governor with some "spending money."

When Clinton was running for president in 1992, a few journalists in Arkansas reported stories about the criminal use of federal funds in the Watergate venture. Both of the Clintons denied that they had knowingly done anything illegal. After Clinton became president, more information about Whitewater emerged. Meanwhile, the Clintons were accused of being involved in other irregularities, such as improper campaign contributions, the strange disappearance of legal records, and special favors for their friends. The suicide of a White House counsel from Arkansas, Vincent Foster, produced additional allegations and rumors.

The Independent Counsel's Inquiry

The independent counsel law of 1978 was designed to prevent abuses of power by high-ranking government officials. Inspired by the Watergate scandal that had forced President Richard Nixon to resign, the controversial law provided for the attorney general to request a special three-judge panel to appoint an independent counsel whenever there was "credible and specific" evidence of wrongdoing by a president or another official. The counsel, once appointed, was given almost an unlimited budget to conduct investigations, with great prosecutorial powers and very little oversight.

In 1994, Attorney General Janet Reno decided that there was sufficient evidence to warrant appointment of an independent counsel to investigate the Whitewater venture. On August 5, the judicial panel appointed Kenneth Starr, a prominent attorney with many connections to the Republican Party, to direct the investigation.

At first, the public had little interest in or knowledge about the matter.

After organizing a large staff, Starr launched broad investigations into almost anything and everyone remotely related to Whitewater. Eventually some fourteen people in Arkansas, including McDougal and the governor of Arkansas, received jail sentences for criminal acts. In one dramatic incident, Hillary Clinton was subpoenaed to testify under oath about her involvement in Whitewater-related matters. Critics of the inquiry charged that Starr was obsessed with trying to find evidence against the Clintons and that he was using the threat of prosecution to force witnesses to say that the Clintons had participated in illegal transactions.

In early 1998, Starr obtained authorization to expand his investigation to determine whether President Clinton had lied under oath about a sexual relationship with Lewinsky, who had worked as a White House intern. On September 9, Starr's 453-page report to the House of Representatives concluded that the Lewinsky scandal provided eleven possible grounds for impeachment, but he acknowledged that there was no evidence of impeachable offences relating to Whitewater. Several months after the Senate voted on acquittal in the impeachment trial, Starr resigned and was succeeded by Ray, who had served as a senior prosecutor in the independent counsel's office.

Consequences

When Ray announced an end to the Whitewater inquiry, polls indicated that a large majority of Americans believed that it should have ended much earlier. The public continued to identify the inquiry with Starr, who was often perceived as a partisan extremist. The public had paid less attention to Whitewater than the Lewinsky scandal, which was often seen as a matter of private morality rather than criminal wrongdoing. Many people believed that the independent counsel's entire project had been repudiated by the Senate's acquittal of Clinton in the impeachment trial of early 1999.

The many critics of Starr emphasized that the Whitewater inquiry, at a cost of more than $50 million, was the most expensive of the twenty operations conducted under the independent counsel law. At the same time, other indepen-

Hillary Rodham Clinton at Laborers' Local 79 in New York, after being cleared of wrongdoing in Whitewater.

dent counsel investigations, mostly directed at members of presidential cabinets, were also unpopular. Conservative Republicans, while often defending Starr's work, remained angry about earlier investigations that had been directed at Republican officials.

The independent counsel law, for all these reasons, had been allowed to expire on June 30, 1999. Although a minority of politicians and lawyers had wanted Congress to extend the law, there was a consensus in favor of returning the criminal investigations of high officials to the Department of Justice. Congress, as it had done in the Watergate affair, could always appoint a "special" prosecutor if considered necessary in a particular situation. Even Starr had stated publicly that he favored the law's demise.

When the law expired, a few independent counsel offices were concluding ongoing investigations. Ray's inquiry into the Lewinsky scandal finally ended with a compromise settlement in January, 2001. Clinton acknowledged that he had given misleading and evasive answers under oath, and he agreed to a five-year suspension of his Arkansas bar license. Ray agreed not to seek a criminal indictment in exchange for these two concessions.

Thomas Tandy Lewis

3227

Senate Approves Plan to Restore Everglades Ecosystem

South Florida's natural ecosystem was gradually deteriorating as a result of continuing human development. A complex plan was developed to restore the health of this ecosystem by reestablishing the natural flow of fresh water that supports the system.

What: Environment; Earth science
When: September 25, 2000
Where: South Florida
Who:
AL GORE (1948-   ), vice president of the United States from 1993 to 2001
LAWTON M. CHILES, JR. (1930-1998), governor of Florida from 1991 to 1998

The Human Impact

In the latter part of the nineteenth century, the interior of South Florida was a vast wetlands, covering an area of almost 9 million acres (3.5 million hectares). A mosaic of environments existed within this region, known as the Everglades, producing an extensive and complex ecosystem containing large areas of sawgrass sloughs, wet prairies, and cypress and mangrove swamps, bordered by coastal bays and lagoons. The Everglades supported an abundance of plant and animal life, including large populations of storks, alligators, and panthers.

Connecting the South Florida ecosystem was a broad flow of fresh water derived from the region's abundant rainfall. Water flowed southward from the headwaters of the Kissimmee River to Lake Okeechobee. The lake seasonally overflowed its southern shores and flowed slowly in sheets through the vast expanse of grassy marsh that made up much of extreme southern Florida toward Florida Bay, producing what has been termed a "river of grass."

During the twentieth century, the topography and hydrology of South Florida was continually modified to accommodate a rapidly growing human population's ever-increasing needs for water,

flood control, and agricultural land. From the 1950's to 2000, South Florida's population grew from 500,000 to more than 6 million, while the Everglades shrank by 50 percent. Canals and levees, built to drain lands for development and flood control, disrupted the natural sheet flow that nourished the Everglades.

In particular, the Central and Southern Florida (or C+SF) Project, begun in the early 1950's by the U.S. Army Corps of Engineers and completed in the mid-1960's, resulted in the construction of 1,000 miles (1,609 kilometers) of canals, 720 miles (1,158 kilometers) of levees, and almost two hundred water control structures. This plan accomplished its hydrologic goals but at a great cost to the environment. So much water was diverted that about 70 percent less flowed through the South Florida ecosystem than originally, and its quality was degraded by excess phosphorus, mercury, and pesticides.

The Everglades were also compartmentalized by the intersecting pattern of drainage canals, so that the sheet flow of fresh water was disrupted. As a result, the once prolific ecosystem deteriorated greatly. By 2000, wading bird populations had declined by 90 percent, sixty-eight native plant and animal species had been classified as threatened or endangered, and populations of commercial and recreational fish species had been drastically reduced. The decline in water levels also led to soil deterioration, frequent drought, and periodic outbreaks of large-scale smoky fires that sometimes burned for weeks.

The Everglades Restoration Plan

In the 1990's, the U.S. Army Corps of Engineers was authorized to restudy the C+SF Project. The goal was to develop a comprehensive plan to restore South Florida's natural ecosystem while enhancing water supplies and maintaining flood control. As a result of the second study, Vice President Al Gore announced in February, 1996, that the administration of President Bill Clinton had approved an initial $1.5 billion program to restore the Everglades. At the same time, Florida governor Lawton Chiles appointed a state commission to work with the federal task force. Work continued on the development of a more complete restoration plan, and on July 1, 1999, the Comprehensive Everglades Restoration Plan (CERP) was presented to the U.S. Congress. After a lengthy period of examination and debate, the plan was approved by Congress on September 25, 2000.

The goal of the CERP was to allow the South Florida ecosystem to heal itself, as far as possible, by reconfiguring water delivery systems to mimic the original natural flow pattern. The project had more than sixty components designed to favorably influence the quantity, quality, timing, and distribution of water flow. In order to improve the region's water connectivity and en-

hance sheet flow, more than 240 miles (386 kilometers) of levees and canals were to be removed within the Everglades.

Much of the 1.7 billion gallons (6.4 billion liters) of runoff that were diverted to the Atlantic or Gulf of Mexico were to be captured in surface and underground storage areas. Water was to be stored in more than 217,000 acres (87,885 hectares) of new reservoirs and wetlands, as well as in three hundred underground aquifer storage and recovery wells. Of the new water captured, 80 percent was to go to the environment and 20 percent was to be used to enhance urban and agricultural water supplies. Hydroperiods (alternating periods of wetting and drying) were vital to the natural functioning of the Everglades ecosystem. These were to be mimicked by timed water releases from water storage areas. Water quality was to be improved by directing water to storm-water treatment areas.

Large amounts of land in the Kissimmee River basin were to be purchased to help restore the river and its surrounding wetlands. In addition, some 61,000 acres (24,705 hectares) of land were

Park ranger Robert DeGross at the Shark Valley Park in the Everglades National Park.

AP/Wide World Photos

added to Everglades National Park, and further additions were planned.

Consequences

Proponents of the Comprehensive Everglades Protection Plan believed that it was possible to restore much of South Florida's original ecosystem while meeting the demands of a large and growing population. The degree to which the goals of the plan have been realized will not be known for decades or longer. By 2001, funding had been provided for only the first package of projects. Based on an analysis of the outcome of this initial group, authorization and funding of the remaining features of the plan were to be requested in

subsequent proposals beginning in 2002, and it was believed that the entire project would take more than twenty years to complete. The plan's estimated cost of $7.6 billion was to be split evenly between the federal government and the state of Florida.

The CERP was unprecedented in both its scale and complexity, and detailed planning for many of its components remains incomplete. If it proves successful, however, it will serve as a model for the development of plans to restore ecosystems in other regions that have been damaged by human activities.

Ralph C. Scott

Scientists Find Interferon, Drug That Fights Multiple Sclerosis

Medical researchers announced their discovery that early treatment with a drug known as beta interferon dramatically slowed the onset of multiple sclerosis (MS) and might halt the disease in some cases.

What: Biology; Medicine
When: September 25, 2000
Where: Buffalo, New York
Who:
LAWRENCE D. JACOBS, professor of neurology at the State University of New York (SUNY) School of Medicine at Buffalo
NANCY A. SIMONIAN, director of Medical Research at Biogen in Cambridge, Massachusetts
ROY W. BECK, medical researcher and consultant at Jaeb Center for Health Research in Tampa, Florida
R. PHILLIP KINKEL, neurologist at SUNY School of Medicine at Buffalo

Multiple Sclerosis

Multiple sclerosis (MS) is a chronic autoimmune disease of the central nervous system that can cause severe disability. Normally, an individual's immune system attacks foreign elements that enter the body. In the case of autoimmune diseases such as MS, the immune system attacks the body's own tissue instead. MS affects nerves in the brain, eyes, and spinal cord. More than one million people worldwide have MS, with more than 400,000 cases reported in the United States.

Typically, nerves in the body are coated with an insulating substance known as myelin, which allows the nerves to communicate rapidly with each other. This fatty covering protects the axons, elongated extensions of nerve cells, that send information to target cells in the brain and spinal cord. In MS, the myelin coating of some nerve fibers breaks down, interrupting communication between the nerves. Damaged myelin is eventually replaced by scarlike tissue that further interferes with nerve signals. The damage builds up over time and can cause muscle weakness or paralysis, loss of balance, fatigue, numbness, dim or blurred vision, memory loss, and other problems with thinking. The symptoms can mysteriously disappear as rapidly as they appeared, so that the disease is often marked by episodes of disability and then remissions of the symptoms. Some victims suffer recurring attacks, with their recovery being less complete until the damage finally becomes irreversible. Other victims may only suffer mild weakness for decades.

Interferon Treatment

After extensive clinical studies, researchers at the State University of New York (SUNY) School of Medicine at Buffalo announced on September 25, 2000, that a drug called beta interferon can limit the development of multiple sclerosis. If the drug is given to patients soon after the first signs of MS appear, it cuts the likelihood of further development of MS symptoms by at least 44 percent.

The body's immune system releases interferons to fight viruses, cancer cells, and other types of disease. Because beta interferon appears to interfere with the immune system's attack on healthy tissue, scientists were led to test this protein as a treatment for MS. Researchers searched for and found the human gene responsible for creating beta interferon naturally in the body and inserted it into living mammalian cells to produce large quantities of the substance for treating MS.

Lead researcher, Lawrence D. Jacobs, a neurologist at SUNY, along with Biogen director of medical research Nancy A. Simonian, neurologist R. Philip Kinkel of SUNY, medical researcher and consultant Roy W. Beck of the Jaeb Center for Health Research, and other colleagues, decided to test a form of beta interferon known as *1a* in the treatment of MS. They carried out a study in fifty different clinical centers on 383 patients who had suffered from a single attack of MS-like symptoms. All the patients had abnormalities on brain scans resembling those typically seen in MS. The patients were randomly divided into two groups. The first group received weekly injections of beta interferon *1a*, known by the brand name of Avonex, and the other group received injections containing no active medication.

In addition to delaying the onset of MS, beta interferon *1a* cut the number of new or actively inflamed brain lesions in the first group of patients by 44 percent relative to the comparison group. It also cut the total volume of such lesions by 91 percent over the comparison patients. In some patients, the drug prevented further MS-type episodes from occurring, effectively halting the progression of the disease. Thomas Leist, MS expert at Thomas Jefferson University in Philadelphia, and William Stuart, MS expert at the Shepherd Center in Atlanta, reviewed the findings of Jacobs and his colleagues and gave their full support.

In addition to beta interferon *1a*, the Food and Drug Administration has also approved another form of beta interferon known as *1b* for treatment of MS. Both *1a* and *1b* appear to have the ability to inhibit naturally produced gamma interferon in the body, a substance that plays a key role in MS attacks. Betaseron is the brand name for beta interferon *1b*. It has a slightly different molecular structure than *1a*. While the *1a* variety reduces the frequency of MS episodes and slows down the progression of MS, the *1b* variety appears only to reduce the frequency of epi-

Actor David L. Lander, a spokesperson for the Multiple Sclerosis Society, examines a model of interferon with scientist Laura Runkel of Biogen.

AP/Wide World Photos

sodes. A third form of interferon, alpha inter-feron, is also undergoing tests to determine if it is effective in treating MS. Common side effects of the interferon drugs include fever, chills, sweating, muscle aches, fatigue, depression, and injection site reactions.

Consequences

MS is typically not diagnosed until patients have more than one MS-type episode. More than two hundred new cases are diagnosed each week in the United States alone. According to Leist and Stuart, the findings of Jacobs and his colleagues are likely to change the way the disease is diagnosed and treated. Early treatment of MS symptoms with beta interferon *1a* shows that the fate of the patient can be changed, delaying or even preventing the onset of the disease. Clinical data show that the sooner patients receive the in-

terferon treatment, the better the outcome. When attacks do occur, they tend to be shorter and less severe. Magnetic resonance imaging (MRI) scans indicate that beta interferon 1a decreases the destruction of myelin.

One of the remaining questions about using beta interferon *1a* to treat MS is just how long the benefits will last. Another question is if the drug might prevent the disease altogether. Jacobs and his colleagues suggested that more research would answer these questions. They also stated that in order to determine how long patients will need interferon treatments, the patients would need to be monitored for longer than the three-year period designated for the clinical tests. They also noted that something would need to be done to reduce the price tag of about ten thousand dollars per patient per year.

Alvin K. Benson

FDA Approves Sale of Mifepristone for Nonsurgical Abortion

> *The Food and Drug Administration approved medical abortions using mifepristone (RU-486) as an alternative to surgical abortion.*

What: Gender issues; Health; Medicine
When: September 28, 2000
Where: Food and Drug Administration, Washington, D.C.
Who:
JANE HENNEY (1947-), Food and Drug Administration commissioner
RON WYDEN (1949-), Democratic representative from Oregon

Medical Abortion

On September 28, 2000, the Food and Drug Administration (FDA) announced the approval of RU-486 for use in the United States as an alternative to surgical abortion. The ruling mandated that in order to prescribe the drug, a physician must be able to date the pregnancy conclusively and, if anything unusual occurred, to provide surgical intervention, either to complete the abortion or to stop heavy bleeding. Three visits to the physician are required. The first is to ensure that the pregnancy is early enough for the pill to be used safely (forty-nine days after the last menstrual period).

Two sets of pills are required: The first, mifepristone, prevents the production of progesterone, which the uterus needs to nourish the fertilized egg and keep it alive. Without progesterone, the embryo is dislodged from the uterus and is expelled from the body. The second drug, misoprostol (Cytotec), to be ingested two days later, triggers contractions and expels the fetal tissue. Normally used to prevent gastric ulcers, misoprostol also makes the uterus contract so that it expels the fertilized egg, mimicking a miscarriage. A few days after taking it, there is some cramping and bleeding. After twelve days, the woman must return to her physician to confirm the success of the abortion and to check for any complications. In 1 percent of women, the bleeding may be excessive and require uterine curettage.

RU-486 has a lengthy chronology, having been developed in France and approved for use there in 1988. The FDA responded to pressure from antiabortion congressional representatives by banning the import of RU-486 for personal use. Having launched a large public education drive on RU-486 (the Campaign for RU-486 and Contraceptive Research), in July, 1990, a ten-member Feminist Majority Foundation delegation consisting of feminist leaders and prominent scientists, armed with more than 115,000 petitions from American citizens supporting RU-486, traveled to Europe to meet with officials of the drug companies Roussel Uclaf and Hoechst AG to urge introduction of RU-486 in the United States.

The Road to Approval

In November, 1990, Congressman Ron Wyden of Oregon held hearings on RU-486 before the House Small Business Committee. Scientists testified that the then-existing import alert (ban) hindered research on possible use of RU-486 as an alternative treatment for breast cancer. Following these hearings, Congressman Wyden introduced legislation to remove the import alert. The American Association for Advancement of Science endorsed the testing and use of RU-486 in February, 1991, and in May, 1991, New Hampshire became the first state to pass a resolution urging the commencement of clinical trials of RU-486 in that state. Subsequent resolutions were passed in other state legislatures.

In the first direct challenge to the FDA import alert on RU-486, Leona Benten, pregnant

3234

Eric Schaff of the University of Rochester Medical Center was one of the chief investigators in a study to determine the effectiveness of mifepristone.

and seeking abortion services, agreed to fly to Europe and attempted to openly bring a personal supply of RU-486 into the United States. She was arrested at U.S. Customs at the airport in New York City. A federal judge ruled that RU-486 was to be released for use as an abortifacient, but that decision was blocked by the Second Circuit Court of Appeals. The Supreme Court refused to overturn the import ban or order customs to return the drug to Benten. President Bill Clinton lifted the import ban during his first term.

In October, 1992, a study in the prestigious *New England Journal of Medicine* concluded that RU-486 was a safe and effective drug. Again in 1993, the *New England Journal of Medicine* reported that RU-486, in combination with misoprostol, was 99 percent effective in terminating pregnancy during the first nine weeks. In December, 1992, then FDA commissioner, David Kessler, wrote to Roussel Uclaf encouraging the company to submit an application to license RU-

486 in the United States. A clone of RU-486 was tested and made available in China in 1993, and in February, 1994, British health authorities allowed the Marie Stopes Clinic in London to administer RU-486 to American women who traveled there for early abortions.

From October, 1994, until December, 1995, the Population Council, a New York-based scientific research organization, conducted clinical trials on mifepristone involving 2,100 women at more than a dozen sites in the United States. In July, 1996, at public hearings, the FDA Advisory Committee on Reproductive Health Drugs determined that mifepristone was safe and effective. In September, 1996, the FDA issued an "approvable" letter to the Population Council in response to the latter's application. In October, 1997, Hoechst AG turned over worldwide (non-U.S.) patent rights for mifepristone to Edouard Sakiz, whose new company, Exelgyn, would distribute the compound as a method of early abortion. In December, 1999, the Feminist Majority

3235

Foundation was awarded sole responsibility for distributing the drug, which was finally approved by the FDA, under Commissioner Jane Henney, on September 28, 2000, and was marketed by Danco Laboratories.

Consequences

Antiabortion forces launched a major offensive in Congress against medical abortions. They labeled the drug "baby poison" and threatened to boycott its manufacturer. They proposed restrictions that would affect the drug's availability in some parts of the country, drastically reducing the pool of physicians prescribing the drug. Proponents of RU-486, approved as a safe and effective method of nonsurgical abortion that can be used during the first nine weeks of pregnancy, say that it involves a less invasive procedure than surgical abortion, thus permitting abortion to take place in the privacy of a doctor's office rather than at an abortion clinic where the women may encounter protesters and earlier in the pregnancy, at a time when fewer people have moral qualms. It was determined that RU-486 may also have possible use as treatment for fibroid tumors, ovarian cancer, endometriosis, meningioma, and some types of breast cancer.

Marcia J. Weiss

Protesters Topple Yugoslavian President Milošević

> *After two weeks of demonstrations throughout Yugoslavia's Serbia Republic, prodemocracy protesters seized control of the national parliament, forcing the resignation of President Slobodan Milošević, who had attempted to retain power after losing an election to Vojislav Kostunica two weeks earlier.*

What: Civil strife; Coups; Human rights; Political reform
When: October 5, 2000
Where: Belgrade, Yugoslavia
SLOBODAN MILOŠEVIĆ (1941-   ), president of Yugoslavia from 1997 to 2000
VOJISLAV KOSTUNICA (1944-   ), Milošević's opposition in the national election

Twelve Years and Three Wars

After being elected president of Serbia in 1989, when it was still one of six republics within Yugoslavia, Slobodan Milošević began to implement his vision of a Serb-dominated Yugoslavia, beginning with the stripping of autonomy from Kosovo, a province within Serbia with an ethnic Albanian majority. His efforts, along with those of nationalists in other Yugoslav republics, helped break up Yugoslavia and generate three wars fought over ethnic differences in Croatia (1991-1992), Bosnia-Herzegovina (1992-1995), and Kosovo (1999). More than 250,000 people died, and almost three million became refugees in those wars. The Kosovo conflict resulted in a massive bombing campaign by North Atlantic Treaty Organization (NATO) against Serbia and the entry of 25,000 NATO-led peacekeeping troops into Kosovo.

Restricted by law from continuing as Serbian president, Milošević was elected president of what was left of Yugoslavia in 1997. Meanwhile, the stresses of nearly a decade of war and international economic sanctions were leaving the Serbian economy severely weakened, with massive unemployment and occasional labor unrest over unpaid wages. In late 1996 and early 1997, demonstrations over fraudulent elections failed to topple the government, as Milošević was able to intimidate or co-opt large elements of the divided opposition.

In mid-2000, Milošević called for an early national election to get the people to ratify a constitutional change permitting him to remain in office longer. However, when the election was held on September 24, Vojislav Kostunica, a constitutional scholar and lawyer, defeated him for the presidency by winning an outright majority. Milošević's supporters claimed that Kostunica had fallen narrowly short of an absolute majority in the first round and called for a runoff election on October 8, with Milošević remaining in office until the final resolution. Popular protests began almost immediately. On September 27, more than 200,000 people took to the streets in Belgrade to reject the runoff plan and demand Milošević's immediate resignation.

At the same time, the leaders of other nations questioned the legality and necessity of a runoff election, and indicated that economic sanctions against Yugoslavia would stay in place so long as Milošević—an indicted war criminal—remained president. The United States, United Nations, and European Union—all of which had observers in Serbia—declared Kostunica the victor in the first round. On September 30, Russia offered to mediate the dispute, further indicating the isolation of Milošević, who had long considered Russia's leaders as his strongest allies.

On October 2, Kostunica and other opposition leaders called for a general strike to force Milošević from power. On that day much of Serbia's economy ground to a halt, as hundreds of thousands of protesters left their factories and

3237

stores idle and marched in Belgrade. On October 3, the Serbian police began to arrest hundreds of coal miners, truck drivers, and other workers who had begun work stoppages the previous day. By October 4, power was failing in major cities, as generating plants lay idle. On that same date, Milošević appeared on television to vow that the runoff would go forward, and that he would not yield to the demands of the protesters, who, he claimed, were acting on behalf of foreign powers.

Also on October 4, the Yugoslav Constitutional Court annulled the results of the election, handing Milošević a victory by insisting on a runoff. Some opposition leaders claimed that this was the first step toward canceling the election entirely, allowing the president to remain in power indefinitely.

Storming the Barricades

Events in Belgrade reached crisis proportions on October 5. The day began with an opposition march on the official pro-Milošević television station, which the protesters set ablaze. The most

critical events occurred around the parliament building near the center of town, which had seen many previous demonstrations. In contrast to earlier protests against Milošević and the stolen election, this time the police were not unified in their defense of the regime. Sensing, perhaps, that the public mood had shifted, or perhaps not wanting to start a bloodbath, several police units broke ranks with the state hierarchy and allowed small groups of Kostunica supporters to enter the parliament. The trickles soon grew into a flood, as more than 100,000 people gathered in and around the parliament building.

From the upper floors of the building, protesters began throwing pictures of Milošević out windows, waving Serbian flags, and otherwise showing their control over the events. Faced with this mass uprising, nearly all remaining police fled the city center, and the regime collapsed.

Milošević initially refused to resign, but after the United States, European Union, and Russia recognized Kostunica as president on October 6, the crisis ended, and Milošević congratulated Kostunica on his victory.

Protesters hold signs saying "He has to go," referring to Slobodan Milošević, in a rally in Belgrade.

Consequences

The next day, on October 7, Kostunica was officially sworn in as president of Yugoslavia. He received immediate international recognition and promises of aid from the United States and European Union. Two days later, the European Union lifted all economic sanctions and restored normal diplomatic relations with Yugoslavia. Within three months, Kostunica received the resignations of most of Milošević's allies in the government. In December, he held new parliamentary elections, leading to a democratic and reformist majority for the first time in Serbia's history.

Meanwhile, Milošević remained under international indictment for war crimes he allegedly committed during the wars in Croatia, Bosnia-Herzegovina, and Kosovo and faced the prospected trial for domestic abuses as well. In late June, 2001, the Yugoslav government took the unprecedented step of turning former president Milošević over to the International Court for trial on war crimes in The Hague.

Yugoslavia was the last of the former communist states of Eastern Europe to become democratic, ushering in the hope that a measure of peace might come to the troubled region.

Wayne H. Bowen

Explosion Tears Hole in U.S. Destroyer *Cole*, Killing Seventeen

Terrorists on a suicide mission aboard a small boat approached the U.S. Navy destroyer Cole, *setting off an explosion that created a large hole in the hull of the ship.*

What: International relations; Terrorism
When: October 12, 2000
Where: Aden, Yemen
Who:
KIRK S. LIPPOLD, captain of the *Cole*
JAMAL AL-BEDAWI, chief suspect

The Explosion

On October 12, 2000, while the U.S. Navy destroyer *Cole* was refueling in the port of Aden in Yemen, it was attacked by two men who pulled alongside the ship in a small boat and caused an explosion that blew a large hole in the hull of the ship, killing seventeen people. Although Yemen was officially a friend of the United States and had been trying to promote good relations, it was also home to terrorists and Islamic extremists who considered the United States a major enemy.

The *Cole* entered the harbor on the morning of October 12, planning to stay only a few hours. The destroyer was on threat condition bravo, a moderate level of security alert. The mooring of the *Cole* was complete about 9:30 A.M. local time, and refueling began soon thereafter. Shortly after 11:00, a small boat with two men in it approached the ship. No one recognized the small boat that pulled alongside as a danger because it blended in with the boats that were servicing the ship. At 11:18 as the two men in the small boat stood at attention, about 400 to 500 pounds (181 to 226 kilograms) of military type explosives went off, blowing a 40-by-40-foot (12-by-12-meter) hole in the hull of the *Cole*. The attack was a suicide mission; the men on the small boat knew they would die in the explosion but believed it was their duty to do so in order to carry out their mission.

Sailors were thrown into the air, and thick, black smoke was everywhere. Hatches were blown open, doorways bent, and parts of the upper deck buckled. Entire lower compartments were blown upward, trapping some crew members. The floor of the mess galley was pushed up against the ceiling. The *Cole* lost electrical power, and all onboard communication equipment was disabled.

The hole was near engine rooms and eating and living quarters, and if it had come minutes earlier, it might have caught many more crew members in the ship's mess area just above where the explosion took place. The ship listed at a four-degree angle, but the keel was not damaged, no fires started, and the ship did not take on water. It was later found that seventeen crew members had been killed in the blast, and thirty-nine were injured, but the captain, Commodore Kirk S. Lippold, was not hurt, and he and the remaining crew acted decisively and quickly to help the injured and keep the ship from sinking.

The Aftermath

The injured were taken to a local hospital, then to a military hospital in Germany, and finally home to the United States. The crew and U.S. investigators had to cut through the wreckage to retrieve all the bodies of the victims, but they tried to disturb as little as possible so as to preserve evidence. It took a week to recover all the bodies.

The crew then worked to make the ship seaworthy enough to be towed into deeper water, where it could be loaded on a special Norwegian ship built to transport offshore oil rigs. This ship, the *Blue Marlin*, was a floating dock, and once the *Cole* had been towed out of the harbor into deeper water, the *Blue Marlin* took on ballast, or extra water, which sank its upper deck under the sea. The *Cole* was towed onto its deck, and the ballast was removed, lifting the deck back above water. The *Cole* was then secured for the trip home. The ship arrived in the United States on December 13, where it was to be repaired and returned to service.

Consequences

In the United States, two investigations were carried out soon after the incident, one by the U.S. Navy into what happened aboard the ship, and the other by the Pentagon into what could have been done overall to prevent such an attack. Neither investigation cast blame on any individual, although the Navy investigation found that the captain and crew of the *Cole* were lax about following a number of security procedures. The Pentagon report recommended better methods all along the chain of command for preventing terrorist attacks.

In May, 2001, congressional hearings were held regarding the incident, and lawmakers criticized the apparent relaxation of the accountability standards to which the *Cole*'s commander was held. Admiral Vern Clark, chief of naval operations, said the Navy declined to punish Lippold because even if Lippold had carried out the security procedures he ignored, he would have been unable to prevent the attack. William S. Cohen, defense secretary at the time of the attack, supported Clark's decision.

The Federal Bureau of Investigation (FBI) sent investigators to Yemen, but initially, they were not allowed to directly question suspects or witnesses. Later they were given permission to question suspects, but Yemen refused to allow the suspects to be taken to the United States for trial. Finally it was decided that suspects found outside Yemen could be tried in the United States, although those found and charged by the Yemeni government were to be tried in Yemen.

The United States suspected from the beginning that Osama bin Laden was behind the attack. Bin Laden is a fugitive from Saudi Arabia, a millionaire who fought in the Afghan War against the Soviets in 1980, and who has since organized a worldwide terrorist network focused on the United States. There is no direct evidence

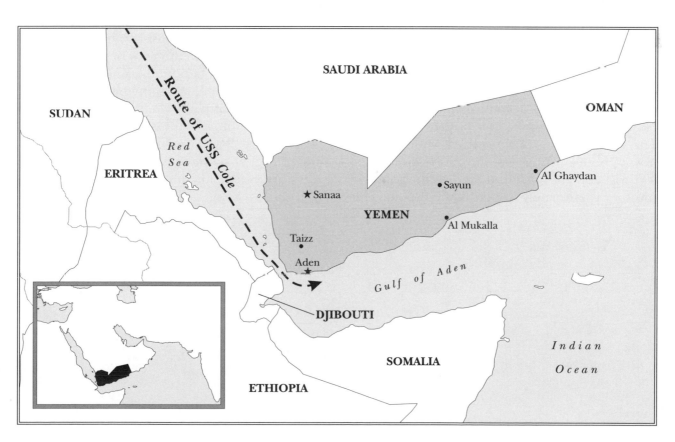

AP/Wide World Photos

Investigators examine the hole in the side of the destroyer Cole.

linking him to the attack, but reports from suspects show he could have been behind it.

Six men were arrested and prepared for trial in Yemen; other suspects are still at large. The chief suspect is Jamal al-Bedawi, who has said he received his orders from Muhammad Omar al-Hazari, a man who may have links to bin Laden. The United States is concerned with deciding how to better protect its interests from further terrorist acts and with finding out and stopping whoever is ultimately responsible for them. The first goal must be met by improving intelligence and communication measures, but the second requires working with the Yemeni government to find out the truth behind the suicide attack.

In the summer of 2001, the probe into the attack ran into difficulty when the United States sought to continue its investigation and widen its scope, as it believed the Yemeni government was not providing access to prominent Yemenis who were suspected of involvement in the attack.

Eleanor B. Amico

Post-Election Violence Rocks Ivory Coast

For the first time in the forty-year history of the Ivory Coast's independence, political violence reached the streets of the nation's capital.

What: Civil strife; National politics
When: October 26, 2000
Where: Ivory Coast
Who:
FÉLIX HOUPHOUËT-BOIGNY (1905-1993), president of the Ivory Coast from 1960 to 1993
HENRI KONAN BÉDIÉ (1934-   ), Houphouët-Boigny's successor as president until 1998
ROBERT GUEI, military leader who overthrew Bédié and claimed victory in the 2000 national election
LAURENT GBAGBO, Popular Front (FPI) candidate and eventual winner of the election
ALASSANE OUATTARA, disqualified candidate for president

Demonstrators Bring Down a Government

On October 25, 2000, thousands of the inhabitants of Abidjan, the capital of the West African country of Cote d'Ivoire (also known as Ivory Coast) demonstrated against the government to protest General Robert Guei's refusal to accept his defeat in an election that had been held three days earlier and step down from power as head of the government. Remarkably, the protest was joined by members of the police and gendarmerie, a paramilitary police force. Eventually that day, Guei gave in, demonstrating that ordinary citizens—like the Serbs of Yugoslavia—could compel their leaders to honor the results of general elections.

The following day, however, some members of the victorious crowd were turning on others, and perhaps one hundred persons were killed. The violence of October 26 was triggered by the fact that although opponents to Guei had presented a united front, once their victory was secured, a new split emerged between those who wanted new elections and those who thought that the winner of the earlier elections, Laurent Gbagbo, should automatically become president. The split mirrored broader splits in Ivorian society that were based on religious and ethnic loyalties.

Background to the Elections

For thirty-three years following its independence in 1960, the former French colony of the Ivory Coast was ruled by one man, Félix Houphouët-Boigny. During this period, the country was frequently cited as an island of stability in an otherwise turbulent continent. Four reasons were generally given for its stability. First and foremost was Houphouët-Boigny's generally wise leadership; he did a masterful job of balancing the country's various group interests, and stitching together an effective heterogenous national society. Though he headed a one-party government, he drew ministers and government officials from all parts of the country.

Second, although the country lacked the rich mineral riches of some of its neighbors, it enjoyed remarkable prosperity from its agricultural products, notably cocoa and palm products. Moreover, the country was unusually in not trying to restrict foreign participation in its economy. It allowed Lebanese and French nationals in sectors of the economy such as the retail trade, and in such professions as secondary school teaching. Finally, Houphouët-Boigny, retained close ties to France; with its help, he maintained close control of his army and avoided the common African malady of being overthrown by the military.

One thing that Houphouët-Boigny did not do well was build political institutions that would survive his death. In 1993, Henri Konan Bédié, a member of his party, the Rally of the Republic,

3243

succeeded him as president. Under the existing constitution, each term of the presidency was to last no more than five years. In 1998, Bédié sought to amend the constitution to increase his term to seven years. He also sought to include in the constitution a provision that would bar any one whose parents were not born in the Ivory Coast from becoming president. The first provision was probably aimed at avoiding the need for him to run for reelection in the midst of the economic decline the country was then experiencing. The second provision was clearly intended to keep a potentially strong opponent, Alassane Ouattara, from running against him.

The effects of the second provision also would have been felt more strongly by the inhabitants in the Northern part of the country who predominantly were Muslims than those in the South who were mainly Christians. This is so because, until 1960, France had administered many of its colonies as a single unit (or federation) in which people moved freely between what was later to be the Ivory Coast and the northern territories that were later to be Burkina Faso and Mali. It was thus primarily the citizenship and rights of northerners that was put in question.

These provisions, together with the deteriorating economic situation caused some uproar. Nonetheless, most people were caught by surprise when, on Christmas Eve, 1998, a bloodless military coup d led by General Guei overthrew Bédié's government. Most Ivorians initially welcomed Guei's government, particularly because Guei promised to "restore democracy." By early 2000, however, Ivorians were becoming disillusioned with the military government, which was not adept at managing the economy, and with Guei's hints that he intended to remain in power indefinitely. Moreover, a new constitution that Guei submitted for public approval in July, 2000, included the controversial citizenship provision.

Presidential elections were eventually set for October 22, 2000, with Guei as a candidate. In September, the supreme court disqualified thirteen of the nineteen candidates who sought the presidency, including Ouattara. The election was consequently boycotted by many voters, including most of Ouattara's supporters. The election then became a contest between Guei and Laurent Gbagbo, a veteran politician that had earlier run against Houphouët-Boigny and Bédié. Early returns suggested that Gbagbo would win, but tabulations of the returns were suspended without explanation. Two days later, after dismissing the election board, Guei declared himself the victor.

The result was the street demonstrations that drove Guei from the capital. Gbagbo then seized power and became president. Although many observers argued that new elections should be held, Gbagbo resisted such demands. Indeed, his supporters, who were mainly from the country's largely Christian south, went into the streets and killed supporters of

Ivory Coast opposition protesters welcome defecting members of the security force.

AP/Wide World Photos

Alassane Ouattara, most of whom came from the Muslim north. Gbagbo afterward took measures to bring about reconciliation in the country, but it remained unclear whether political stability would return.

Consequences

The events of October 25-26 are notable for two reasons. First, they shattered the reputation of Ivory Coast as an island of peace and stability in Africa. That reputation had already been challenged by Guei's bloodless coup; however, the violence of October 25-26 confirmed its complete destruction. Second, and perhaps more important, the demonstrations seemed to be another display of people power—the willingness and capacity of ordinary individuals to put their lives on the line to defend the results of democratic elections in which they had participated. In this, the behavior of the citizens of the Ivory Coast may be comparable to the people of Yugoslavia who brought down Slobodan Milošević during the very same month.

Maxwell O. Chibundu

Napster Agrees to Charge Fees for Downloading Music over Internet

The digital music file-sharing pioneer Napster teamed with Bertelsmann AG to find a way to compensate musicians and recording companies for music files shared using its revolutionary and previously free file-sharing network.

What: Business; Communications;
   Computer science; Entertainment;
   Technology
When: October 31, 2000
Where: San Mateo, California
Who:
SHAWN FANNING (1980-   ), Napster
   founder and software programmer
MARILYN HALL PATEL (1938-   ), U.S.
   district judge

Napster Redefined

On October 31, 2000, the German media giant Bertelsmann AG agreed to team up with Napster, the young and popular Internet company whose software allowed users to trade music with each other online. Besieged throughout the year by lawsuits from the music industry, Napster continued to allow users to obtain music files from each other without payment. However, in its agreement with Bertelsmann, one of the companies suing Napster, it agreed to join with the German company and work out a way to charge users and pass on the money raised so that artists and recording companies could be paid for their work. In exchange for help in setting up the technology to allow this to occur and a loan to finance the work, Bertelsmann agreed to drop its part of the lawsuit once the new system was in place and to purchase part of the Napster company.

Napster permits users to connect with one another to download music files. What Napster provides is the directory in which specific music can be found and the software to log on to the other person's computer and download the files. The company, based in San Mateo, California, had been very popular, especially with college students, who used it so heavily that some colleges banned it, complaining that the downloading files slowed their Internet systems. Similar programs, such as Gnutella and Freenet, exist, but none is as easy to use or as popular as Napster. After six months in service, by February, 2000, Napster had five million users. That number grew, one year later, to fifty-five million.

Shawn Fanning, who invented Napster, wrote the code for the program in a few months in 1999 after dropping out of Northeastern University in Boston. After listening to his roommate complain about the difficulty of downloading music from the Internet, he decided there must be a way to make it easier and set out in a single-minded way to write the code that became the backbone of Napster. The company opened in May, 1999, and its beta version was introduced in August. By December, it was enough of a threat to the music industry that the lawsuits began.

An Ongoing Conflict

On December 7, 1999, the Recording Industry Association of America (RIAA) filed suit against Napster in the U.S. District Court in Northern California, charging Napster with violating federal and state laws about copyright infringement and arguing that Napster facilitates music piracy and harms musicians. RIAA represents five of the largest record companies: Warner Music, Sony Music, Universal Music, BMG (Bertelsmann's recording company), and EMI.

Napster argued that it was trying to cooperate with the recording industry but had been rebuffed by an industry that wants to gain control of the way music is distributed on the Internet by destroying Napster and taking over its technol-

3246

ogy for its own use. One of Napster's prime arguments is that it does not host any content and merely provides the technology for users to link up with each other; therefore, it cannot be held liable for piracy of copyrighted works. It has always warned users that downloading and then distributing copyrighted music is illegal, but Napster argues that it is perfectly legal to make copies for one's own personal use.

In April of 2000, heavy metal band Metallica and rapper Dr. Dre both filed suits against Napster, although other musicians, notably Limp Bizkit, supported the company, reasoning that allowing fans to download and listen to music would boost sales of music, encouraging fans to purchase compact discs. On May 5, Judge Marilyn Hall Patel refused Napster's request to dismiss the case, finding that Napster is not, as it argued, merely a conduit for information and thus not responsible for illegal activity occurring on its site.

On July 26, Judge Patel ruled that Napster contributes to widespread copyright infringement and issued an injunction requiring Napster to close down while the case was tried. Napster argued that this injunction would be a death sentence for the company and received a stay of the injunction by the Ninth Circuit Court of Appeals on July 28, allowing it to continue until further information could be presented.

The first hearing on the injunction was held October 2, and before that date, many organizations, including the U.S. Copyright Office, filed friend-of-the-court briefs both for and against Napster. The Copyright Office argued that the widespread sharing of files facilitated by services such as Napster does not fall under the protection users are allowed when they copy copyrighted material for their own personal use.

Consequences

After the Bertelsmann-Napster agreement became public, Bertelsmann announced that it was developing a subscription plan, and some ob-

Bertelsmann eCommerce Group president Andreas Schmidt (right) and Napster CEO Hank Barry announce a possible deal.

AP/Wide World Photos

3247

servers hoped that Napster would be able to turn its customers into subscription purchasers. In January, 2001, TVT Records, an independent record company that had sued Napster, dropped its lawsuit, and Napster hoped that others would also see that its attempt to compensate artists and the music industry would make it profitable to stop litigation and work with Napster to promote methods to make music downloading work for everyone concerned.

On February 12, 2001, the judges of the Ninth U.S. Circuit Court of Appeals in San Francisco sent Judge Patel's injunction back to the court for revision, saying it was too broad and should keep Napster from allowing piracy of copyrighted music without shutting the service down completely. The court directed Patel to modify the injunction in such a way that not only Napster but also the record companies must monitor violations, and upon being informed of them, Napster would have to rectify the situation.

On March 5, 2001, Judge Patel ordered Napster to remove all copyrighted material from its system within three days. The order required the recording companies that had sued Napster to provide the company with a list of songs to be removed from its system. A day earlier, Napster had installed song-blocking software meant to prevent users from downloading copyrighted material. This software, however, used specific song titles and was being circumvented by Napster users, who altered or deliberately misspelled song titles.

After Patel's order, analysts reported that membership in other Napster-like, free services such as Music City were increasing, and downloads from Napster were decreasing. However, some analysts saw a future for subscription services, as did some media companies. In June, 2001, Napster announced plans to join with AOL Time Warner, Bertelsmann AG, the EMI Group, and software maker Real Networks to create MusicNet, a subscription service that allowed its members to download music for a fee. MusicNet would compete with Duet, a subscription service formed by Sony Corporation and Vivendi Universal. Industry experts expressed doubt that these ventures would be successful. Their creation demonstrates that the advent of Napster has forced music companies to rethink the way they do business and to find new ways to defend intellectual property rights.

Eleanor B. Amico

Peruvian President Fujimori Resigns While in Japan

While visiting Tokyo, Peruvian president Alberto Fujimori shocked the political world by announcing his retirement. Earlier in the year, he had been under fire for questionable campaign tactics during his race for the presidency.

What: National politics
When: November, 2000
Where: Tokyo, Japan
Who:

ALBERTO FUJIMORI (1938-   ), president of Peru from 1990 to 2000

VLADIMIRO MONTESINOS, Peruvian intelligence chief

ALEJANDRO TOLEDO, presidential candidate for opposition party

ABIMAEL GUZMÁN REYNOSO (1934-   ), leader of the Shining Path guerrilla movement

Fujimori Gains Control

Alberto Fujimori, the first person of Japanese descent to head a South American government, won his first term as Peru's president in 1990 by securing the backing of the country's mestizo and Native American majority. Fujimori trained as an agronomist and had taught at the National Agrarian University in Lima. He hosted *Getting Together*, a local television program addressing the contemporary problems facing the country. By 1990, Peru's annual inflation rate had reached 8,000 percent, and the country's economy had virtually ground to a halt. Through the television program, he came to be recognized by the general population and gained their support.

Once in office, Fujimori introduced economic reforms designed to end the extremely high rate of inflation that plagued the country's growth and development. His political party, Change Ninety, gained the majority of the seats in the Peruvian congress. Through his party, he instituted legislation that gave the office of the president broad powers in combating internal opposition and unrest as well as in introducing economic changes. He opened many of the country's industries to privatization. In a massive reorganization of the country's bureaucratic structure, he dismissed more than seven hundred judges and demoted or discharged hundreds of police and military officers for corruption and fraud.

In September, 1992, during Fujimori's first term, Peru's military succeeded in capturing and imprisoning Abimael Guzmán Reynoso, a former academic from Ayacucho, who led the Shining Path, a Marxist guerrilla group. The group had allegedly used kidnappings and assassinations to eliminate its political enemies and had kept much of rural Peru in a state of constant turmoil. Fujimori gave the army a free hand in combating this group. Guzmán and his followers received life sentences. The government imprisoned them in a maximum security facility on an island off the country's coast. The defeat of the Shining Path ended internal fighting in Peru.

Using the legislature, Fujimori made himself eligible for two more terms as president, in effect waiving the two-term restriction imposed by the country's constitution. When certain members of the Peruvian supreme court stated that the move violated the constitution, Fujimori's congressional adherents replaced the dissenting judges and adopted a new constitution.

In 1995, the incumbent president easily defeated his opponent, Javier Pérez de Cuéllar, a diplomat who had served as secretary general of the United Nations. An attempt by Fujimori's former wife, Susana Higuchi, to run for the presidency also failed. In the 1995 election, Fujimori

3249

received 65 percent of the popular vote.

During Christmas time in 1996, another group of leftist guerrillas, called the Tupac Amaru, seized the Japanese embassy in Peru, taking many hostages, including foreign diplomats, government officials, and Fujimori's brother. After failing to end the hostage situation by negotiation, Fujimori authorized the recapture of the embassy by an army combat team. The army strike force killed all fourteen guerrillas in the process of retaking the building. The army lost two commandos, and one hostage lost his life during the battle. The majority of the Peruvian population applauded this dramatic action.

Fujimori added to his popularity by conducting a hands-on relief effort when the country suffered severe damage from a series of El Niño storms in late 1997 and early 1998. He toured the country widely by helicopter, directing the distribution of aid to remote communities.

The president continued to persecute members of the political opposition. The government authorized telephone wiretapping of opposition leaders. A scandal developed when a television station revealed that the intelligence service tortured members of its own service who had disclosed their knowledge of illegal government activities. Fujimori had the citizenship of the foreign-born owner of the station revoked, forcing him to relinquish control of the station.

Peruvian president Alberto Fujimori tells reporters he plans to stay in Japan indefinitely.

The 2000 Presidential Election

With the backing of his adherents in congress, Fujimori sought the Peruvian presidency for the third time. Fujimori's Peru 2000 party lost its overall majority in the April 9 election, and Fujimori failed to win a majority of the ballots for president. Amid charges of election fraud, the opposition called for a runoff between the second largest vote-getter, opposition candidate, Alejandro Toledo, a business school professor, and Fujimori. The president's supporters criticized the United States for interfering in the election. Despite street protests and domestic and international calls to delay the election until irregularities could be resolved, on May 28, the runoff election was held between Toledo and Fujimori. In the runoff, which Toledo boycotted, Fujimori received 74 percent of the vote.

Soon damning evidence of the dishonest nature of the Fujimori campaign was revealed. Television cameras caught one of the president's closest associates, intelligence chief Vladimiro Montesinos, paying off an opposition politician for switching his support to Fujimori. Montesinos escaped arrest although Fujimori publicly and personally sought to lead his capture. He later was taken into police custody where he awaited trial on 140 corruption-related charges.

Consequences

Although Fujimori agreed to schedule another presidential election the following year and to give up the presidency permanently at that time, the hue and cry against his besieged administration continued. Toledo insisted that the 2000 election was fraudulent. Finally, Fujimori left the country voluntarily, flying to Japan, where he resigned the presidency. Under Japanese law, because his parents had registered his birth in their home prefecture, Fujimori possessed dual citizenship. As a Japanese citizen, he could not be extradited to his birthplace, Peru. The former president indicated in a public statement that he did not intend to return to Peru in the foreseeable future. The Peruvian congress refused to accept his resignation and, in a congressional condemnation, fired him from office as being morally unfit.

The congressional president served as acting president until the April 8, 2001, elections could be held. That election resulted in a June 3 runoff, in which Toledo was victorious. In late Au-

gust, 2001, the congress of Peru voted to charge Fujimori with homicide and forced disappearance for massacres at Barrios Altos and La Cantuta, in which government forces killed fifteen and ten people, respectively, in the early 1990's. These killings occurred at the height of the government's crusade against the Shining Path guerrilla group. The Peruvian congress then voted to rescind Fujimori's immunity so that he could face these charges in Peru, but the Japanese government refused to return him.

Carl Henry Marcoux

International Space Station Welcomes First Crew

> *The establishment of a space station in permanent orbit marked the beginning of both a new era in space exploration and a new era in international cooperation.*

What: Engineering; Space and aviation; Technology; International relations
When: November 2, 2000
Where: Earth orbit
Who:
BILL SHEPHERD (1949-   ), an American astronaut and mission commander
YURI GIDZENKO (1962-   ) and
SERGEI KRIKALEV (1958-   ), Russian cosmonauts

Approaching the Final Frontier

Toward the beginning of the twentieth century, Russian educator Konstantin E. Tsilokovsky suggested the possibility of traveling in space. Soon afterward, science fiction writers spun tales of futures in which such travel had become everyday fact. With American inventor Robert Goddard's development of liquid-fueled rockets, space travel went from theory to achievable reality. During the 1960's, television viewers could watch space travel in both fiction and reality, as producer Gene Rodenbury's visions of humankind's spacefaring future in episodes of *Star Trek* jostled with news reporting of real-life U.S. and Soviet space missions. One could even see in *Star Trek*'s imaginary Klingons a coded reference to the rival Soviets in what was dubbed the "space race."

However, the early space missions were purely exploratory, brief forays into the skies comparable to Christopher Columbus's voyages to the New World. Even such programs as the U.S. Skylab and Soviet Mir were often vacant, although they did set duration records for humans remaining in space. If the human race was to become a truly spacefaring race, it had to begin establish-

ing a permanent presence in space. In 1984, U.S. president Ronald Reagan called for the creation of a permanent space station, which he tentatively named *Freedom*, as a way for the United States to gain the ultimate high ground.

However, the reality took fifteen more years to develop, and required more than American resources. Alongside the American contribution would be those of U.S. allies—Japan and the European Union. And in a fascinating case of life imitating art, Reagan's former Evil Empire had become a partner instead of a rival. Much as later incarnations of *Star Trek* television series featured the Klingons as allies, rather than enemies, of the Federation, the new post-Soviet Russia would have its important role in building and crewing the new International Space Station.

Putting the Station in Space

The first component of the International Space Station, the Russian unit *Zarya* (Sunrise), was launched in November of 1998 from Russia's launch site in now-independent Kazakhstan. Over the next two years, additional missions by Russian rockets and U.S. space shuttle flights would deliver and assemble more parts, until enough of the station was in place for a crew to take residence.

That point arrived at the end of October, 2000. A three-man crew was assembled for this first mission, called Expedition 1. The mission commander would be Bill Shepherd, a captain in the U.S. Navy. Under him would be two Russians, Yuri Gidzenko and Sergei Krikalev. These men were no strangers to hardship, especially Krikalev, who had spent ten consecutive months on Mir doing back-to-back missions that spanned the period when the Soviet Union broke up in 1991.

AP/Wide World Photos

From left to right, cosmonaut Sergei Krikalev, cosmonaut Yuri Gidzenko, and U.S. commander Bill Shepherd, inside the station.

This experience was important because conditions aboard the ISS would be spartan. Many of the comforts that later crews would enjoy were still waiting to be sent aloft. For example the 6.5-foot satellite dish that would enable high-speed, high-bandwidth communication with Earth was scheduled to go up on the same space shuttle mission that was to return the Expedition 1 crew home. Without the dish, the crew's contacts with Earth would be limited to what could be received through a small antenna, equivalent in capacity to a slow computer modem.

The Expedition 1 mission lifted off from Russia's Kazakhstan launch site on October 31, 2000. Two days later, on November 2, their Soyuz capsule docked with the ISS and the cosmonauts moved into the station. Their first priorities were turning on the various lights, alarms, and life support systems and getting their zero-gravity toilet working properly.

For the first several days after their arrival, mission controllers in Moscow kept the crew on a light duty schedule to give them time to recover from the stress of launch and to acclimate themselves to weightlessness. After that initial period, their duties were steadily expanded to full workdays, getting everything in working order for the scientists who would follow on later missions.

Consequences

Crewing the International Space Station for the first time marked an important change in the nature of human activities in outer space. No longer would humans be occasional visitors to the realm beyond their native world's atmosphere. Now that the human race had established a permanent base on the final frontier, which would be continuously occupied by a succession of crews. There they would be able to perform experiments and other activities that could be performed only in the unique conditions of weightlessness and could not be adequately accomplished by automated satellites.

Expedition 1 would also establish the basic rhythms of life aboard the ISS and create a pattern for their successors. For example, the official language of the station had been established by prior international agreement to be English.

3253

However, the practical necessities of talking with the mission controllers in Moscow would require extensive use of Russian. All three members of the Expedition 1 crew were well versed in both languages, and soon developed a habit of mixing the two in casual speech to produce a hybrid dubbed "Russlish."

Similarly, meals were a mixture of Russian and American cuisines, so that all crew members would have some familiar foods at regular intervals, as well as new and exotic ones. However, in deference to U.S. rules, there would be no alco-hol. This was a change from the protocols aboard Mir, where the occasional treat was permitted.

In the long term, the crewing of the ISS marked the creation of a stepping stone in space from which future long-range missions could be launched. Trips to the moon and to Mars would be able to use the ISS as a base, rather than having to be launched directly from Earth. Gene Rodenbury's science fictional dream of human beings at home in space has thus come one step closer to being realized.

Leigh Husband Kimmel

Pets.com Announces Withdrawal from Internet

Pets.com, the largest online retailer of pet supplies closed down because it was losing money and could not attract new investment. The closure of Pets.com was part of a general downturn in the fortunes of Internet companies.

What: Business
When: November 7, 2000
Where: San Francisco
Who:
JULIE WAINWRIGHT, chief executive officer of Pets.com

Rise and Fall

On November 7, 2000, Internet pet supply store Pets.com announced to its shareholders that it was closing down. The company explained that it was ceasing operations because it was losing money and had failed to find a purchaser or financial backer. Its losses amounted to $147 million. The closedown would be gradual, the company said, but 255 of its 320 employees would be laid off immediately. Pets.com announced it would sell the majority of its assets, including its inventory, its Web site address, and its Sock Puppet brand. This dog puppet mascot had become nationally known during the previous summer. Pets.com sold the puppet to toy stores nationwide and featured it in television commercials. The closure marked a disappointing end for a company that initially looked destined for success. Founded in 1998, Pets.com received significant investment from large Internet retailer Amazon.com, and it quickly became the biggest online retailer for pet supplies.

In September, 1999, the company was still on the rise, acquiring a much larger distribution center to cope with increased customer demand and expanding the number of items offered for sale to forty thousand. Julie Wainwright, Pets.com's chief executive officer, expressed great confidence in the venture, saying that the company was dedicated to making it fast, convenient, and cost-effective for pet owners to meet all their pet-owning needs through Pets.com.

In February, 2000, the company made an Initial Public Offering (IPO), or first sale of stock to the public. The value of Pets.com shares was $11. In March, the value rose to $14. After that, however, the company's position began to deteriorate. The value of its shares fell from $14 to just over $2 in June.

In June, the struggling company received a small boost when it acquired some of the assets of Petstore.com, another online pet supply store and a rival of Pets.com. Petstore.com was also losing money and had been forced to shut down. This was the start of what industry experts had been expecting for some months—the consolidation of the online pet supply industry, which was considered overcrowded. There were too many similar companies competing for the same limited amount of business. The acquisition failed to halt the decline in Pets.com share values, which soon plunged to $1.50.

In October, 2000, another online pet supply company, Petopia.com, reduced its operations, firing 60 percent of its staff. However, Pets.com did not benefit from this industry shakeout, as its own demise came shortly after. In the weeks before the closure, share prices fell dramatically to under $1.

Failure to Thrive

There were several reasons for the failure of Pets.com. Many of the products it offered (pet food, for example) were available in stores in every neighborhood, so for many consumers there was little incentive to shop online. Analysts pointed out that although pet owners make up a huge potential market, they have shown less inclination to

3255

shop online for their products than consumers of other retail products. The convenience of having supplies delivered directly to the purchaser's home did not seem to outweigh the disadvantage of having to wait several days for the goods to arrive.

Pets.com also made the mistake of selling its products for less than it paid for them. For example, in the third quarter of 2000, Pets.com sold goods for a total of $277,000 less than it paid for them. Economists call this a negative gross profit margin. In that third quarter, Pets.com lost approximately 19 cents on every dollar in sales. The heavily discounted merchandise was an attempt to attract a large numbers of customers, but the strategy failed, in spite of the fact that the company did attract 570,000 customers.

One problem was the high shipping costs involved in a Web-based business, which made it hard for Pets.com to make a profit. Analysts also point out that the company invested heavily in its 1Sock Puppet brand, spending millions on a thirty-second television advertisement during the Super Bowl. Although this bought Pets.com some brand recognition, it was not enough to translate into profitable sales.

Consequences

The closure of Pets.com was part of a widespread decline in the fortunes of Internet companies that began in 1999. Before that, the Internet was generating much excitement as a new way of doing business. Venture capitalists (people who provide funds for new companies) were rushing to invest in what was called the New Economy.

However, in the spring of 2000, the Internet bubble burst. Stock prices plunged, investors lost confidence, and Web businesses starting closing down. Thousands of workers were laid off. In the year ending October, 2000, more than 117 dot-coms failed, according to the Boston Consulting Group. More than 51,000 Internet workers lost their jobs over a yearlong period that began in December, 1999.

The shakeout continued to be felt in 2001. Industry experts predicted that there would be many more closures and worker layoffs for Web-based businesses as venture capital funding became more difficult to acquire.

Bryan Aubrey

AP/Wide World Photos

The Pets.com sock puppet appears next to Jeff Koons's "Puppy" in Rockefeller Plaza, New York City, in June, 2000.

Clinton Visits Vietnam

In November, 2000, Bill Clinton became the first U.S. president to visit Vietnam since President Richard Nixon visited the country during the Vietnam War in 1969.

What: International relations
When: November 18-20, 2000
Where: Vietnam
Who:
BILL CLINTON (1946-   ), president of the United States from 1993 to 2001
JOHN KING (1964-   ), senior White House correspondent for Cable News Network (CNN)

The Visit's Purpose

Many groups encouraged, supported, and arranged U.S. president Bill Clinton's trip to Vietnam. These groups included the Vietnamese Chamber of Commerce and Industry, the U.S.-Vietnam Trade Council, the American Chamber of Commerce, and the U.S.-Association of Southeast Asian Nations (ASEAN) Business Council. The purpose of Clinton's visit was to encourage Vietnam to commit itself to participating fully in the global marketplace of free trade as well as in the free exchange of ideas with other countries throughout the world. Clinton's overall desire was to complete the normalization process between Vietnam and the United States.

Clinton's visit began in Ho Chi Minh City (formerly Saigon) and continued on to Hanoi, and at each city, thousands of people lined the streets to get a glimpse of him and to hear his speeches, which promoted free trade and the free exchange of ideas among nations. Vietnam is a communist country, and Clinton's speeches about free trade irritated some Vietnamese leaders. As early as 1994, Clinton had made the decision to remove the trade barriers between the United States and Vietnam, hoping that foreign business would move in to help develop this underdeveloped nation's economy. Vietnam's population is 78 million, half of which are under the age of thirty. The average per capita income is six dollars a week. Free exchange of ideas is not encouraged; the government restricts access to the Internet to about 300,000 Vietnamese.

Throughout his three-day trip, Clinton gave positive speeches about what Vietnam had achieved during the last twenty-five years. He noted how the country had eradicated polio, brought 15 million people out of poverty, and increased personal income by almost 70 percent of what it was a decade earlier. To illustrate the possibilities of U.S.-Vietnamese business collaborations, Clinton told the success story of Nguyen Cao Thang and Truong Bich Diep, two Vietnamese American sisters. In the 1970's, these sisters founded a pharmaceutical company named Official Pharmacy of Vietnam (OPV) in Ho Chi Minh City, and by the 1990's, their success had enabled them to build a new manufacturing plant just outside that city with a loan from the U.S. government.

The Media and the Message

While in Vietnam, Clinton made a radio address that was taped at the Daewoo Hotel in Vietnam and later transmitted to the United States. In that address, Clinton outlined some of his achievements in his eight years of office, citing the family medical leave law, the expanded earned-income tax credit, and the increase in the minimum wage, which he credited to lifting millions of Americans out of poverty. He called on Congress to increase food stamp benefits in the United States. He rarely mentioned Vietnam.

On Clinton's last day in Ho Chi Minh City, he was interviewed by John King of Cable Network News (CNN). In this interview, Clinton expressed how he hoped individual, economic, religious, and political freedom might continue to accelerate in Vietnam after his visit. Clinton also expressed how his visit might bring about

U.S. President Bill Clinton (left) and Vietnamese president Tran Duc Luong sit in front of a bust of Ho Chi Minh in Hanoi.

villagers and the U.S. recovery team sift through the black mud to find any remnants of the missing captain. The experience was very moving for Evert's sons, who spoke about how they had no animosity for the Vietnamese people and that they loved them. More than 58,000 Americans and 3 million Vietnamese died in the Vietnam War.

Consequences

In Clinton's opinion, his Vietnam trip was successful. He persuaded the Vietnamese government to sign an agreement that established a $200 million line of credit to support U.S. investment in that country. In addition, the U.S. and Vietnamese governments began a dialogue on such topics as safety standards and labor issues in the workplace, the education of workers in computer skills, and the establishment of free trade with the aim of raising wages and creating more economic opportunities.

The media coverage of Clinton's visit to Vietnam met mixed reviews in the United States. About a week before the visit took place, the *Chicago Tribune* predicted that it would be "heavy on symbolism, light on substantive agreements." An editorial that ran in the *Los Angeles Times* about the same time believed the trip would not produce radical changes but that Clinton made the right decision in undertaking the trip. After the trip took place, *The New York Times* criticized Clinton for being too weak on human rights issues but praised him for taking a step toward a "faster pace for change." Vietnam veterans likewise had differing views about Clinton's trip. Some believed it would promote friendship between the two countries. Other veterans viewed Clinton's trip as an affront because he avoided military service in Vietnam.

Lloyd Johnson

closure to the Vietnam War. He told King, "We need to heal the rift within the Vietnamese community, and it divided Americans one from another. And I hope that the last eight years and the journey we've made together in moving forward with Vietnam has helped to put an end to that."

The most emotional event that occurred on Clinton's visit to Vietnam was when he and his wife, Hillary Rodham Clinton, and their daughter, Chelsea, visited an American Missing In Action (MIA) excavation site in Vin Phuc province just outside Hanoi. At this site, the Pentagon's Joint Task Force was attempting to locate the remains of Air Force captain Lawrence Evert, believed to have been shot down with his fighter jet on November 7, 1967. Evert's two sons, accompanying the Clintons, watched the

Supreme Court Resolves Presidential Election Deadlock

> *In the case of* Bush v. Gore, *the U.S. Supreme Court intervened in a presidential election for the first time in history by overturning the Florida supreme court's order for a manual recount of the vote in Florida. The controversial ruling directed that Florida's electoral college votes go to George W. Bush, who became the nation's forty-third president.*

What: Government; National politics

When: December 12, 2000

Where: Supreme Court building, Washington, D.C.

Who:

WILLIAM H. REHNQUIST (1924-   ), chief justice of the United States

GEORGE W. BUSH (1946-   ), governor of Texas and Republican candidate for president

ALBERT GORE (1948-   ), vice president of the United States and Democratic candidate for president

Florida's Disputed Results Hold Up National Elections

The U.S. Supreme Court's decision of December 12, 2000, ended a nasty and unprecedented five-week legal dispute over whether Florida's twenty-five electoral votes would count for George W. Bush or Albert Gore. Like all but two states, Florida utilized the winner-take-all principle in presidential elections, so that the candidate receiving a plurality of the state's popular vote could claim all of the state's electoral votes. Although Gore received more popular votes than Bush in the national election, the outcome of Florida's election determined who would win the majority of the votes in the electoral college and take the presidency.

The counting of Florida's popular vote was extremely messy. On November 8, the day after the election, the Florida Division of Elections reported that Bush had prevailed with a slim margin of 1,784 votes. Because the vote was so close,

Florida law required an automatic machine recount. In addition, Vice President Gore exercised his right to request manual recounts in four predominantly Democratic counties. Among various problems in the Florida election, a large number of counties used punch-out ballots that failed to record all votes. The small squares to be punched out, called chads, sometimes did not become detached from the cards, and Florida's counties had different standards for deciding which chads to count in the tally.

Florida secretary of state Katherine Harris, an active Republican, rejected Gore's request for an extension of time necessary for a manual recount. However, on November 21, the Florida supreme court, known for its liberal leanings, ordered the recounts to continue until November 26—twelve days beyond the statutory deadline. When the state's Canvassing Commission, using the revised deadline, certified Bush the winner by a 537-vote margin, Gore contested the final tally and sought additional hand recounts. On December 4, the U.S. Supreme Court, in a unanimous ruling, asked the Florida court to clarify whether its November 21 ruling was consistent with the state legislature's constitutional prerogative to decide how electors are chosen.

On December 8, the Florida supreme court, by a 4-3 margin, again ruled in Gore's favor, cutting Bush's lead to only 193 votes and ordering a hand recount of some 42,000 "undervotes" throughout the state. In addition, it instructed county officials to consider the "intent of the voters" in making the recount. Because outdated voting machines tended to be located in the Democratic precincts, most observers expected that the recount would give Gore enough additional votes to win. Republicans were outraged. Bush appealed the ruling, and the next day the nation's highest court, by a 5-4 majority, stopped the recount and scheduled oral arguments for December 11.

The *Bush v. Gore* Decision

The U.S. Supreme Court's judgment of December 12 contained two rulings. First, the Court held by a 7-2 margin that the recount, as ordered, was inconsistent with the equal protection clause of the U.S. Constitution, which required all votes to be treated equally. The justices were especially troubled by the Florida court's instructions for state officials to decide voter intent on imperfectly scored ballots without providing controls to ensure consistency and objectivity.

In the second of the rulings, the Court decided by a 5-4 vote that the recount would have to stop because there was insufficient time to complete the process according to constitutional procedures. The Court's five most conservative justices noted in a joint opinion that federal law required states to choose their electors by December 12 in order to assure a "safe harbor," meaning that Congress could not challenge the state's choice of electors when the electoral college met on December 18. Based on this federal law, the majority opinion interpreted Florida's law as requiring that the voting be completed by the safe-harbor date.

The Court four liberal dissenters argued that Florida's high court was authorized to decide that Florida law allowed the recount to continue beyond the safe-harbor date, and that the elected legislators, if necessary, could determine Florida's electors. The strong language of the dissenters demonstrated the deep division within the court. Justice John Paul Stevens, usually considered the most liberal of the justices, wrote that the majority's position would "lend credence to the most cynical appraisal of the work of the judges throughout the land."

Consequences

While most Republicans praised the Supreme Court's decision, Democrats and liberals charged that the decision had been motivated by partisan and ideological considerations. Critics noted that the five conservative justices were usually the strongest defenders of states' prerogatives and judicial restraint.

If the U.S. Supreme Court had not intervened in the controversy, it appeared likely that Gore would have prevailed in the counting of Florida's popular vote. However, the Florida legislature, controlled by Republicans, had the constitutional prerogative to decide how Florida's electors were to be chosen, and it was preparing to override the state's high court. Moreover, if Florida's electors had been challenged in Congress, the Congress also had a majority of Republicans.

The Supreme Court had never before held that a lack of objectivity and consistency in elections violated the Constitution's equal-protection requirement. Informed observers disagreed about whether this requirement might become a binding precedent for election challenges in the future. Although some experts predicted that the Court's decision could result in numerous lawsuits, others noted that the majority opinion stated that the ruling applied only to the particular circumstances of Florida's court-ordered recount.

Whatever its implications for future rulings, the case of *Bush v. Gore* convinced most Americans that electoral reform was needed. Suggestions for possible reforms included the abolition of the electoral college system, the application of statewide standards for voting, and the purchase of more reliable voting equipment. Almost everyone agreed that it would be desirable to provide the funds necessary to eliminate punch-out voting ballots.

Thomas Tandy Lewis

U.S. Submarine Sinks Japanese Fishing Vessel

When the U.S. nuclear submarine Greenville *accidentally struck and sank the Japanese fishing vessel* Ehime Maru *off Hawaii, killing nine Japanese nationals, Navy policies were questioned and U.S.-Japanese relations were strained. Ironically, another U.S. naval vessel had sunk another Japanese ship—also named* Ehime Maru—*in the Pacific in the late 1940's.*

What: Disasters; International relations; Military capability
When: February 9, 2001
Where: Hawaiian Islands
Who:
SCOTT WADDLE (1940-   ), commander of the USS *Greenville*
ROBERT L. BRANDHUBER, chief of staff of the Pacific submarine fleet,

Tragedy at Sea

On the morning of February 9, 2001, Captain Scott Waddle, commander of the submarine USS *Greenville*, got his vessel underway. Along with the crew the sub carried sixteen distinguished civilian passengers. Waddle was under orders to demonstrate the capabilities of the Los Angeles-class nuclear submarine and her crew. The U.S. military had engaged in such "incentive rides" for several years, allowing dignitaries, political contributors and public figures to fly, sail and dive in America's most advanced vessels. On this trip, however, something went terribly wrong. The submarine was beginning to surface rapidly in a main ballast blow—a maneuver designed to simulate surfacing in emergency conditions. On this occasion the *Greenville* struck the Japanese fishing vessel *Ehime Maru*, which sank within ten minutes.

One of several fishing vessels owned by the Uwajima Fisheries High School, the *Ehime Maru* carried thirty-five students, instructors and crew. It was one of more than forty similar fishing-training programs in Japan, of which more than thirty offered long-distance fishing and training excursions as part of their curricula. Students in these program were usually at sea in the waters between Japan and Hawaii for as long as two months. Often, there were as many as twenty-five student vessels in the vicinity of the Hawaiian Islands at any given time.

When the U.S. submarine struck the *Ehime Maru*, the vessel began to sink within moments. Twenty-six of the thirty-five people on board got off the sinking vessel safely and were afterward rescued; however, nine people were lost at sea and presumed dead. Of the lost, four were seventeen-year-old students, two were instructors and three were crew members. The *Greenville*'s own ability to help in the rescue was severely limited, but U.S. rescue efforts were quick. Both U.S. Coast Guard and U.S. Navy vessels and aircraft soon arrived on the scene to transport survivors to hospitals and to search for the missing.

The U.S. Navy's Response

Following the tragedy the Navy Pacific Fleet commander reassigned Waddle to an office job, pending an investigation into what had gone wrong. News soon surfaced that sixteen visitors had been aboard the sub at the time of the collision and that some of them may have been at the sub's controls when the accident occurred. Within days Navy officials announced that a court of inquiry would begin on March 5. However, many family members and political leaders in Japan suggested that both the Navy and the U.S. government were slow in coming forth with apologies and explanations.

Within three weeks additional information emerged regarding the incident, early indications were that although normal procedures were followed before the surfacing maneuver, civilians had, in fact, been at the controls of the submarine at the time of the accident. Also at issue was the suggestion that the crew of the sub had been distracted at the time of the accident and therefore not in full control of the submarine.

Also forthcoming in the weeks following the incident were letters and statements of apology by the U.S. secretary of state, the U.S. president through his personal representative, Admiral William J. Fallon. Captain Waddle—who had been born in Japan and had first-hand knowledge of Japanese culture—expressed his regret for the accident and took full responsibility for what occurred. Later, Waddle sent letters of apology to the families and the Japanese people and made plans to travel to Japan later in the year to make more personal apologies, following hearings aimed at determining causal factors and consequences for those involved.

The court of inquiry heard evidence from the submarine crew, as well as expert testimony regarding Navy policy and procedures for the operation of Los Angeles class nuclear submarines. The inquiry, which lasted more than four weeks, determined that Waddle was not guilty of any criminal wrongdoing or intent, and therefore would not face a court martial. Instead, Waddle was ordered to appear at a nonjudicial admiral's mast on April 23, 2001. These proceedings followed a review of the inquiry findings and effectively ended Waddle's military career.

In a written statement, released by Waddle, he again expressed his regrets and apologies to all of the victims and their families. Following the admiral's mast, Admiral Thomas Fargo, announced that Waddle had been given a punitive reprimand for his role in the accident and that he must resign his commission. Further, Waddle would receive an honorable discharge from his twenty-year military career. The final outcome of the admiral's mast also included a suspended forfeiture of pay for Waddle's performance as commanding officer of the *Greenville* on February 9, 2001.

Consequences

After the incident, the Navy continued to review its practice of allowing "distinguished visitors" to sail aboard its ships. The Department of Defense continued to review such visits at military installations and on military aircraft, ships and equipment. The U.S. Navy appointed divers and salvage professionals to raise the *Ehime Maru*, move it to more shallow waters, and recover the remains of the dead passengers and crew. The estimated cost of the salvage operation was estimated at $44 million.

The *Greenville* its remained in dry-dock for several weeks during the investigation and later for repairs, before heading back to sea in mid-April 2001. The cost of repairing the submarine totaled $36 million. Finally, additional reprimands were forthcoming. The fire-control technician, Patrick Seacrest was reprimanded for failing to report how close the Japanese vessel was before the submarine conducted the surfacing maneuver. Likewise, Captain Robert L. Brandhuber, chief of staff of the Pacific submarine fleet, who was escorting the sixteen civilian visitors, was faulted for failing to intervene when the procedures to prepare for the maneuver were being rushed. This event served temporarily to undermine U.S.-Japan relations, particularly in military matters.

Darlene E. Hall

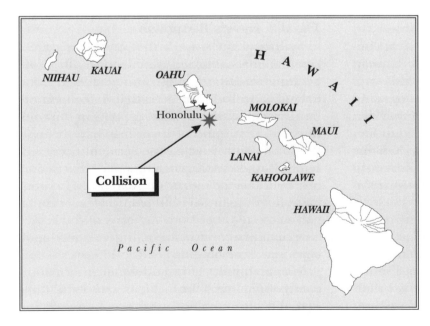

Afghanistan Has Ancient Buddhist Monuments Destroyed

The Taliban, the ruling force in Afghanistan, destroyed two massive, sixth century Buddhist statues in Bamiyan Valley, citing religious reasons.

What: Politics; Religion
When: March, 2001
Where: Bamiyan Valley, Afghanistan
Who:
TALIBAN, ruling group in Afghanistan
MOHAMMAD OMAR, leader of the Taliban

History of the Statues

Buddhism is a world religion that originated in northern India (to the east of Afghanistan) about 2,500 years ago. Eventually, it spread as far west as Afghanistan, where Buddhist kingdoms flourished. The royal capital of one of these kingdoms, the Bamiyan, was in a valley just more than 90 miles (145 kilometers) west of modern Kabul. At the height of the Bamiyan civilization, 1,500 to 1,600 years ago, thousands of Buddhist monks lived in a honeycomb of caves in a great sandstone cliff overlooking the valley. They carved two large and impressive statues of the Buddha into this cliff. The Great Buddha, almost 180 feet (55 meters) high, was decorated with polished gold and brass, and the carving of his cloak was stained a brilliant red to symbolize the power of the cosmos. A second, somewhat shorter statue (125 feet, or 38 meters, tall), was a representation of Śākyamuni, the historic Buddha. It was decorated with shining metal, but its carved clothing was stained a bright blue.

The Bamiyan kingdom was famous for its statues. People came from far away to visit the Buddhas of Bamiyan, as they came to be called. The famous Chinese pilgrim Xuanzang visited the Buddhas in the sixth century and described them in his diary. In the ninth century, many people in the Afghan region converted to Islam. As time passed, the statues lost their importance as religious icons. Their gold and brass ornaments disappeared, and the color of their cloaks faded, but they remained famous.

Genghis Khan captured the valley of Bamiyan in the thirteenth century, and a branch of his army remained in the city. They intermarried with local women, and their descendants are known as the Hazara people. Hazara people are Shia Muslims (while most Afghanis are Sunni Muslims), but they took pride in the Bamiyan Buddhas and tried to protect them over the years. In the late nineteenth century, the Hazara were attacked by other ethnic groups and became scattered across the region, making it difficult to protect the Buddhas. The Afghani king found the statues offensive and sliced off their faces. He also forced the local Hazara into slavery. Despite their difficulties, some Hazara have remained in the area and have protected the statues when they could. Throughout the twentieth century, many wars took place in Afghanistan, and the statues were further damaged by artillery fire and snipers, but they still attracted visitors.

Destruction of the Bamiyan Buddhas

The Taliban began as a group of Islamic students and academics during the Soviet occupation of Afghanistan, but they emerged as a political and religious force in 1994. They are also known as "God's students," and they often call themselves the Taliban Islamic Movement of Afghanistan (TIMA). With the support of Pakistan and Saudi Arabia, by 2000, they controlled 95 percent of the country. The Bamiyan Valley was first taken by the Taliban in 1999 but was recaptured by Hazara resistance fighters. Then, late in 2000, the Taliban took the valley again. In February of 2001, the leader of the Taliban, Mullah Mohammad Omar, declared, "Because God is

3263

Taliban soldiers stand in front of a large alcove where a statue of the Buddha once stood in Bamiyan.

one God and these statues are there to be worshiped, and that is wrong, they should be destroyed so that they are not worshiped now or in the future." He ordered his ministry of vice and virtue to destroy all Buddhist statues in Afghanistan. Many small, ancient statues of Buddha were in Kabul's museum but were already destroyed or smuggled out during the civil war. The only monumental statues still standing were the ones in Bamiyan Valley.

The day that the edict was made, an international delegation was in Kabul meeting with Taliban leaders to try to preserve Afghanistan's heritage. Therefore, Omar's announcement soon became front page news around the world. Offers to cut the statues out of the cliff and move them to a museum came from as far away as Japan and France. Omar rejected these offers.

Afghanistan had been suffering from drought the past year, and more than a million people were starving. He said that the world cared more about the statues than the Afghanis. Early in March of 2001, men from the ministry of vice and virtue used dynamite to blow up the statues.

Consequences

A number of scholars said that the destruction of the Bamiyan Buddhas served several purposes for the Taliban. First, it demoralized the Hazara people, who were actively resisting Taliban control of the area. Second, it proved to the Taliban's Saudi and Pakistani backers a willingness to follow through with an extremely strict interpretation of Islamic law and philosophy. Third, it symbolized the group's seemingly unstoppable rule and its ability to completely control any area it conquered—a message to the 5 percent of Afghanistan not yet under its rule. Finally, the destruction called the world's attention to the Taliban's moral imperative to cleanse the land of non-Islamic influences and the devastating starvation taking place within the country.

Art afficionados, historical preservationists, and Buddhists the world over were saddened by the destruction of the Bamiyan Buddhas. Although the emotions of the first two groups ranged from rage to disgust to resignation, the Buddhists responded somewhat differently. Buddhists believe in the impermanence of the physical world and try to achieve freedom from attachment. Buddhist leaders made reference to the teachings of the Buddha as expressed in the Diamond Sutra: "All phenomena are like a dream, an illusion, a bubble and a shadow, . . . like dew and lightning." Therefore, although the Bamiyan Buddhas were gone, they were, after all, no more than a part of the impermanent, illusory world. Their disappearance, therefore, did not affect the Buddhists in any significant way.

Carolyn V. Prorok

China Seizes Downed U.S. Reconnaissance Plane

A U.S. Navy EP-3E reconnaissance plane on a routine surveillance mission near China's southern coast collided with a Chinese fighter jet, seriously disrupting Sino-U.S. relations.

What: International relations; International law
When: April 1, 2001
Where: Hainan, China
Who:
GEORGE W. BUSH (1946-   ), president of the United States from 2001
JIANG ZEMIN (1926-   ), president of China from 1993 and chair of the Communist Party
DENNIS BLAIR, admiral and commander in chief of the U.S. Pacific Command

Relations Tense

On April 1, 2001, while on a routine surveillance mission near the Chinese coast off Hainan Island, a U.S. Navy EP-3E reconnaissance plane collided with a Chinese fighter jet that was closely shadowing it. The U.S. plane made an emergency landing in China, and the Chinese fighter, one of two on the scene, ditched into the ocean. Issues of blame, the fate of the crew, and the return of the high-tech aircraft created mounting tension between the Chinese and U.S. governments.

The U.S. military admitted reconnaissance flights in the region had been common for some time. Admiral Dennis Blair, commander in chief of the U.S. Pacific Command, said Chinese planes had become increasingly aggressive in tailing U.S. military aircraft in the months just before the collision, even prompting the United States to register a protest.

The incident came at a particularly sensitive moment of transition in U.S. relations with China. Within weeks, U.S. president George W. Bush was set to make a decision on whether to sell sophisticated arms and radar equipment to authorities governing Taiwan, an area China claims as a renegade province under rebel control. President Bush had taken a tougher posture toward China, one that viewed China as a military competitor first and a trade partner second.

Collision Course

The midair crash occurred about 50 miles (80 kilometers) southeast of Hainan Island. International law defines national airspace as extending only 12 miles (19 kilometers) from a nation's coastline, so the incident took place in international airspace. The EP-3E Aries II aircraft, which had taken off from a U.S. air base in Okinawa, Japan, issued a Mayday call but managed to make an emergency landing on the island. The Chinese pilot, Wang Wei, and his jet were lost at sea.

The Chinese foreign ministry claimed in an angry announcement on state television that the United States bore "total responsibility" for the incident. Chinese vice premier Qian Qichen called for the United States to take responsibility for the collision and make a full apology to the Chinese people. Qian's call was echoed by Yang Jiechi, Chinese ambassador to the United States, in meetings with the U.S. Defense Department and on U.S. television.

The twenty-four-member naval crew of the Aries II were uninjured in the emergency landing but were immediately detained by Chinese authorities. The crew remained in detention at the Chinese airbase for eleven days but was finally allowed to be flown out by civilian aircraft. However, Chinese officials refused to allow the $80 million EP-3E Aries II aircraft to be flown out of Chinese territory, although they said they would permit the plane to be removed in pieces. Before the crew's release, the U.S. government had been reluctant to force the issue of blame. However, with the crew safely at home at the Whidbey Island Naval Air Station in the American northwest, President Bush made public claims that the midair collision was the fault of the Chinese pilot.

Consequences

The release of the crew had been facilitated by a letter from President Bush and U.S. secretary of state Colin Powell stating the United States was "very sorry" about the death of the Chinese pilot, Wang. However, the letter did not completely ease tensions. The collision aroused nationalist passions in China, with some critics within the military calling for China to "shoot down the next U.S. spy plane that dares to go near [China's] coast." President Jiang Zemin reportedly placed a gag order on top military officers. He indicated that the spread of nationalistic passions could hamper efforts to repair relations with the United States, which he saw as vital to China's economic development.

Just one month after the collision, with the U.S. plane still on the ground in Hainan, former U.S. president Bill Clinton met with Jiang during a business conference in Hong Kong. Clinton told the Chinese president that he "was convinced that the [George W.] Bush administration was interested in having good relations with China" and asked the Chinese president to remember that early relations between the Clinton White House and China had been "tough."

Tensions remained high as the Bush administration took several steps viewed by the Chinese government as anything but conciliatory. On May 7, the United States resumed reconnaissance flights off the coast of China. White House spokesperson Ari Fleischer said, "It has always been the position of the United States that it is our prerogative and right to fly in international airspace, to preserve the peace by flying reconnaissance missions." The Bush administration also announced that it would allow the sale of major new weapons to Taiwan, including four naval destroyers, antisubmarine aircraft, and up to eight submarines. Another irritant loomed as the United States allowed Taiwan's president, Chen Shui-bian, to stop in New York and Houston as he traveled to and from Latin America in May and June. China has repeatedly demanded that the United States refuse to give Chen a transit permit.

Richard R. Pearce

Nepal's Crown Prince Murders His Royal Family

A member of the royal family of Nepal, the prince next in line to become king, murdered nine family members during a family gathering and then shot himself in the head. He died three days later.

What: National politics; Crime
When: June 1, 2001
Where: Kathmandu, Nepal
Who:
DIPENDRA (1971-2001), crown prince of Nepal and heir to the Nepal monarchy
BIRENDRA (1945-2001), Dipendra's father and king of Nepal
AISHWARYA (1949-2001), Dipendra's mother and queen of Nepal
NIRAJAN, Dipendra's younger brother and presumptive royal heir in his place

Explosive Cause, Shattering Effects

Nepal is a land-locked constitutional monarchy in the Himalayas. It is remote from the modern world, but its crown prince, Dipendra, was educated to his royal station. He attended an exclusive Nepali school and attended Eton, one of England's most exclusive preparatory schools. He was remembered by Eton classmates as well liked and a good student but also as someone with a sullen streak, punctuated with outbursts of physical violence.

Nearly thirty years of age in early 2001, Dipendra had been showing signs of deteriorating behavior for some time. While traditionally Nepali royals absented themselves from ordinary society, Dipendra was a regular at smart restaurants and night clubs in Kathmandu, the nation's capital. He also defied convention in other ways, such as by traveling to the Sydney Olympics in 2000 to meet Devyani Rana, a woman whom he loved and wished to marry. Devyani was not the only woman in his life. However, none of Dipendra's unorthodox behavior prepared a shocked and traumatized nation for the events of June, 2001.

The immediate cause for Dipendra's deadly fury appears to have been his family's attitude toward his intended bride. Devyani Rana was of royal lineage but—from the ruling family's point of view—from the wrong line. The family did not forbid Dipendra's marriage to Devyani outright but told him that if he married her, he would forfeit his place in the line of succession, placing his younger brother, Prince Nirajan, next in line to become king. This may or may not have been a factor in Dipendra's pumping thirty bullets into his younger brother while leaving unscathed his cousin Paras, a favorite who would be next in line for the throne.

Royal Bloodbath

On June 1, 2001, Crown Prince Dipendra shot and killed nine members of the royal family who had assembled at a Friday evening dinner gathering. Among the attendees were Dipendra's father, King Birendra; his mother, Queen Aishwarya; his younger brother, Prince Nirajan; his grandmother, assorted aunts and uncles, and his cousin, Prince Paras, whose father would become king after the slaughter.

At about seven o'clock in the evening, Dipendra began playing billiards and drinking by himself, attended by a servant. He joined the family gathering shortly afterward, pouring drinks for those assembled, leaving briefly to fetch his grandmother from her residence. Returning to the billiard room, Dipendra telephoned his lover, Devyani. He then ordered an attendant to bring him drug-laced cigarettes, which he had been smoking for at least a year.

After being joined by his cousin and his younger brother, Dipendra appeared to be drunk and had to be carried from the room. A servant heard him retching in a bathroom. Dipendra then called Devyani on his cell phone, speaking for several minutes. Shortly thereafter, aides found him on the floor struggling to undress. After they assisted him, he dismissed them and called Devyani, announcing he was going to bed.

Rather than retiring, however, Dipendra dressed himself in military attire, including a military cap, black boots and black gloves, and armed himself with four weapons, including an automatic sub-machine gun. He then returned to the royal gathering, and, according to an eyewitness, shot his father three times. After leaving the room, he soon returned, this time with a submachine gun, and shot three more persons, including his paternal uncle. Before those remaining could find safety, Dipendra returned yet again, this time with an M-16 rifle, which he used to shoot his father again, along with a number of others, including his sister. He then left the room, followed by his hysterical mother. Dipendra found his brother Nirajan in a nearby garden and killed him, then turned his gun on his mother, killing her. Finally, Dipendra shot himself in the head; however, he did not die immediately.

Crown Prince Dipendra.

Dipendra lived in a coma for three more days. Ironically, since his father was dead, and he was next in succession, he technically was Nepal's king until he died. At his death, another uncle, Prince Gyanedrea, ascended the throne. Meanwhile, Nepal was thrown into a state of shocked mourning. In the weeks that followed, a variety of conspiracy theories, none of them pointing to the actual killer, became popular. Among the suspects was the new king, implicated because he had been absent from the gathering, while his son, Paras, was present but emerged unharmed.

Consequences

Nepal had achieved an end to absolute monarchy only in 1991, when a democratic constitution took effect. The new government, however, was soon believed—with ample cause—to be corrupt. In the late 1990's a far-left revolutionary group inspired by the ideas of China's Mao Zedong began an armed insurgency that continued in Nepal's remote regions in the aftermath of the royal bloodbath. The official opposition party in the nation's legislature called itself "Marxist-Leninist," signaling its lukewarm, if not actively hostile, attitude to democracy.

Attitudes of ordinary Nepalis toward the government were also negative. Thus many Nepalis detested the nation's prime minister, widely believed to be corrupt. Added to this recipe for political uncertainty, the new king was generally distrusted and disliked. Many, including Maoist rebel leaders, subscribed to theories placing the new king at the head of a conspiracy with foreign (especially American) intelligence services, responsible for the royal killings. At a public procession the new king was greeted with silence and even jeers by assembled crowds.

Because Nepal, a poor country with only a 40 percent literacy rate, had begun to emerge from centuries of feudal monarchy only in the early 1990's, its new democratic political system had shallow roots. One of the few institutions that functioned well was the monarchy, headed by Birendra, who had been a popular king. With most of the royal family lying in ashes, political instability and attendant social insecurity seem the most likely consequences for a country so little used to responsible government.

Charles F. Bahmueller

Jeffords Quits Republican Party, Giving Democrats Control of Senate

A liberal Republican senator from Vermont, James M. Jeffords left the Republican Party to become an independent and vote with the Democrats on organizational matters, giving them formal control of the Senate with a 51-49 majority.

What: Government; National politics
When: June 6, 2001
Where: Washington, D.C.
Who:

JAMES M. JEFFORDS (1934-   ), U.S. senator from 1989

GEORGE W. BUSH (1946-   ), president of the United States from 2001

TOM DASCHLE (1947-   ), Democratic senator from 1987 who became majority leader

The Precarious Balance of Power

When Senator James M. Jeffords moved his seat to the Democratic side of the isle on June 6, 2001, the Republican Party officially turned over leadership of the Senate to the Democrats. This was the first midterm transfer of party control in the two-century history of the Senate.

Neither political party had been happy with the outcome of the congressional elections of 2000. The elections for the House of Representatives had given the Republicans a narrow 221-211 majority of its seats. In the Senate, the elections had resulted in fifty seats for each party. Senate Republicans, nevertheless, were officially in the majority because Vice President Dick Cheney, a Republican, could cast a vote if necessary to break a tie.

Because of the fifty-fifty split, the two parties were allocated an equal number of seats in the committees of the Senate. The Republicans, however, were allowed to choose the chairpersons of the committees, and they also would control a majority of votes in conference committees, which would be formed to reconcile House and Senate versions of bills.

President George W. Bush, despite the closeness of the 2000 elections, began his term with a conservative agenda that displeased progressives and moderates. His emphasis was on tax relief, calling for a tax cut of $1.6 trillion during the next decade. Senator Jeffords and other left-leaning Republicans joined most Democrats in arguing that the amount was excessive. On May 26, nevertheless, the Republicans managed to pass a bill cutting taxes by $1.35 trillion over eleven years. The bill was sent to President Bush for his signature on June 5.

Jeffords's Decision

Although Senator Jeffords had not been well known outside of Vermont, he had an impressive record of accomplishments. A lawyer with a degree from Harvard Law School, he had served as Vermont attorney general for four years, had been a member of the House of Representatives for thirteen years, and had been elected to the Senate three times.

Since 1995, Jeffords had served as chairperson of the Senate Health, Education, Labor and Pensions Committee, which had jurisdiction over about a thousand federal programs. Often when Jeffords's committee had sponsored bills with liberal spending increases for such programs, conservative Republicans managed to decrease the funding in conference committees.

In 2001, Jeffords was classified as the second-most liberal Republican in the Senate (after Lincoln Chafee). Having opposed Republican positions in 45 percent of his votes the previous year, Jeffords usually supported bills that were pro-choice, pro-welfare, pro-environment, and above all, pro-education. Years before, he had been instrumental in passing landmark legislation in which the federal government had agreed to pay 40 percent of the costs for educating the disabled, and he was upset that Congress had never authorized funds for more than 15 percent of the tab.

Until 2001, Jeffords had found enough moderate and liberal Republicans to feel comfortable in the party. However, during the early period of Bush's presidency, as never before, he felt alienated from the Republican leadership. Having agreed to support the tax cut of $1.35 trillion as long as there were additional funds for special education, he was furious when much of this funding was eliminated by the conference committee.

As the Congress was concluding the tax bill, Jeffords told his close friends that he was about ready to leave the Republican Party. Democratic senators quietly encouraged him to make the move, while President Bush and the Republican leadership did not respond to his concerns until it was too late. In Vermont on May 24, Jeffords explained why he found it unacceptable to remain a Republican, and he announced his decision to become an independent and sit with the Democrats, beginning the day that the tax bill was sent to President Bush.

Senator James M. Jeffords.

Consequences

Probably the most important result of Jeffords's defection was that it empowered the Democrats to control the agenda in the Senate. All of the chairpersons of the committees would be Democrats, and all of the committees would have one-member majorities of Democrats. In addition, the Democratic leadership would henceforth control the calendar for debates on the Senate floor.

A second important consequence of the defection was that it strengthened the ability of Democrats to prevent the confirmation of right-leaning nominations to the judicial and executive branches. This could be especially important if there were a nomination to the Supreme Court, where the justices were divided 5-4 between conservatives and liberals.

Jeffords's defection, however, did not in any way change the membership of the Senate. Democrats would find it impossible to pass bills without the support of moderate Republicans. This was because conservative Democrats could be expected to vote with the Republicans, and also because of the constant threat of a filibuster, which

would require sixty votes to overcome. In addition, the defection did not have any effect on the Republican control of the House of Representatives or President Bush's veto power.

In these circumstances, according to the new majority leader of the Senate, Tom Daschle, neither political party had the power to accomplish anything without the cooperation of the other. He pledged that the Democrats were ready to work with the Republicans in finding common ground and in finding reasonable compromises on issues such as the patient's bill of rights and campaign finance reform.

Moderate and progressive Republicans hoped that Jeffords's defection would force President Bush and the Republican Party to move toward the center and broaden the party's appeal. Otherwise, there was speculation that Senator Chafee and others might also leave the party. The most visible of the moderate Republicans, Senator John McCain of Arizona, was reported to be seriously considering the possibility of running for president as an independent in 2004.

Thomas Tandy Lewis

3271

Terrorists Attack New York City and Washington, D.C.

> *In apparently coordinated actions, radical Islamic terrorists hijacked four U.S. jetliners, crashing two of them into the towers of the World Trade Center in New York and a third into the Pentagon, while the fourth plane crashed in rural Pennsylvania.*

What: Terrorism; International politics; Disasters; Transportation
When: September 11, 2001
Where: New York City, Washington, D.C., and rural Pennsylvania
Who:
GEORGE W. BUSH (1946-), president of the United States from 2001
OSAMA BIN LADEN (1957-), head of the al-Qaeda terrorist network
RUDOLPH W. GIULIANI (1944-), mayor of New York City from 1994 to 2001
MOHAMMAD ATTA (1968-2001), apparent leader of the hijacking

Unprecedented Assaults on Peacetime America

At 8:45 A.M. on September 11, 2001, a Boeing 767 jet smashed into the north tower of the World Trade Center (WTC) in lower Manhattan, New York City. The plane was American Airlines flight 11, bound from Boston to Los Angeles, with ninety-two passengers and crew members aboard. It had been hijacked and deliberately flown into the 110-story skyscraper. Some eighteen minutes later, another jetliner crashed into the second of the Trade Center's twin towers. The second plane, also a hijacked Boeing 767—United Airlines flight 175, carrying sixty-five passengers and crew members—like the first, was en route from Boston to Los Angeles.

At 10:05 A.M., the south tower, the second to be hit, collapsed, and the massive structure turned into debris that spread over a wide area. About twenty-three minutes later, the north tower collapsed with equally catastrophic force,

sending massive amounts of steel and glass raining down on the streets below. At 5:20 P.M. the same day, a nearby building, known as 7 World Trade Center, collapsed as well. Everyone aboard the planes and thousands who were in the buildings, including hundreds of rescue workers, died. In all, more than 3,100 persons were believed to have been killed. The exact numbers remained unknown and subject to frequent revision during the months following the attack.

Meanwhile, about forty minutes after the second WTC tower was hit, at 9:43 A.M. on September 11, a third hijacked jetliner was deliberately flown into the Pentagon, the headquarters of the U.S. Department of Defense. That plane, a Boeing 757, American Airlines flight 77, bound from Dulles Airport near Washington, D.C., to Los Angeles, carried sixty-four passengers and crew members. Everyone aboard and 125 people in the building died.

Several minutes later, the White House was evacuated in case it, too, was targeted. At about 10:10 A.M. the same morning, a fourth hijacked jetliner went down in a field in Somerset County, Pennsylvania, eighty miles southeast of Pittsburgh, killing all aboard. That flight, United Airlines flight 93—a Boeing 757 aircraft flying from Newark, New Jersey, to San Francisco—carried forty-four passengers and crew members, Later it was revealed that some passengers apparently attacked their hijackers shortly before the plane crashed after hearing what happened to the other hijacked planes through cell-phone conversations with their loved ones. It was surmised that the hijackers intended to crash the plane into a Washington, D.C., landmark, such as the White House or Capitol Building.

Although no group claimed responsibility for

An airplane crashes into the second tower of the World Trade Center, creating a shower of flames and debris, as the first tower burns from an earlier plane crash.

the attacks, it soon became widely believed that they were the work of a militant Muslim terrorist network called al-Qaeda (Arabic for "the base"), headed by Saudi multimillionaire Osama bin Laden. A member of one the wealthiest families of Saudi Arabia, bin Laden lived in Afghanistan, a predominantly Muslim country then governed by the Taliban, a fundamentalist Islamic group. Bin Laden and thousands of Arab and other Muslim fighters, although not part of the Taliban, were said to be the Taliban's guests. Over the months that followed, evidence of bin Laden's role in the attacks mounted, culminating in the December discovery of a videotape of bin Laden praising Allah for the damage delivered by the attacks, which was much greater, he said, than he had expected. In describing his reaction to the attacks, bin Laden used language indicat-

ing his familiarity and involvement with the terrorists' plans, sealing virtually all public doubt about his involvement in the attacks.

Meanwhile, the nineteen hijackers directly involved in the hijackings were later identified. All Middle Eastern men, they were believed to have been led by an Egyptian, Mohammad Atta, who is also suspected to have piloted the first plane to hit the Trade Center. Fifteen of the hijackers were Saudis; the other four were from Egypt.

A Lost Sense of Security

Almost immediately after the attacks occurred, all U.S. airports were closed and incoming international flights were diverted to Canada. For several days, no international flights were allowed into the country. All major stock exchanges, led by the New York Stock Exchange,

3273

remained closed for several days. President George W. Bush, who was visiting a school in Sarasota, Florida, on the morning of the attacks, flew to Barksdale Air Force Base in Louisiana as a security precaution. In a public statement, the president said that all appropriate security measures were being taken, including placing the U.S. military on high alert worldwide. "Make no mistake," he said, "the United States will hunt down and punish those responsible for these cowardly acts." The president then traveled secretly to Offutt Air Force Base in Nebraska.

By 10:45 A.M. on September 11, all federal government offices in Washington, D.C., were ordered closed. Shortly before 4:00 P.M., it was announced that the president was conducting a National Security Council meeting by telephone. At 6:40 P.M., Secretary of Defense Donald Rumsfeld announced at a Pentagon news conference that the building was operational and would be "in business tomorrow."

At about seven o'clock that evening, Bush arrived at the White House, his flight from Nebraska escorted by jet fighters. About fifteen minutes later, Attorney General John Ashcroft announced the creation of a Federal Bureau of Investigation Web site that would receive tips on the attacks. At 8:30 P.M., the president addressed the nation, denouncing the attacks and asking for the nation's prayers for the victims. "These shameful acts shattered steel," he said, "but they cannot dent the steel of American resolve."

Three days after the attacks, a memorial service for the victims was held at the National Cathedral in Washington, D.C., where representatives of major religions spoke in a harmonious remembrance of those who died. The service was attended by President Bush and his wife, Laura Bush, as well as by major figures from the Congress and the executive departments. Expressions of sympathy for the nation poured in from many foreign leaders, including the heads of most Western European countries, Russian president Vladimir Putin, Chinese president Jiang Zemin, and Cuban leader Fidel Castro. To express solidarity with the United States, Australian legislators sang the American national anthem in their assembly in Canberra.

The following week, Bush addressed the nation at a joint session of Congress. In a forceful speech that had been several days in preparation, he sought to rally the nation, vowing to fight international terrorists throughout the world and emerge victorious. Bush's response to the attacks sent his popular approval ratings surging, from 63 percent in April to 92 percent in October. Similarly, the attacks were turning local leaders such as New York mayor Rudolph (Rudy) Giuliani into national figures. Giuliani was soon labeled "America's mayor" for his leadership throughout the early hours of the disaster as well as later in the city's efforts for recovery. Giuliani rallied New Yorkers' fighting spirit so that they could face the crisis with determination and the future with confidence. At the end of the year, *Time* magazine named him its "Man of the Year."

Consequences

The first consequence of the attacks was an instantaneous end to Americans' sense of security within a continent protected by two vast oceans. A second, almost immediate consequence was preparation for war with the Taliban regime, which was estimated to control about 95 percent of Afghanistan and believed to be harboring bin Laden. The U.S. government repeatedly stated that an Afghan campaign would be merely the beginning of a worldwide struggle against international terrorism.

The Afghanistan government refused to surrender bin Laden, so early in October, the United States, allied first with Great Britain and later with North Atlantic Treaty Organization partners and others, began a military campaign to bring down the Taliban regime and capture or kill bin Laden and other al-Qaeda leaders. In the early part of December, with the support of U.S. air power and some special forces, the fighters of an Afghan tribal coalition known as the Northern Alliance or the United Front succeeded in toppling Taliban rule and killing several key terrorist leaders.

The attacks also profoundly affected global and U.S. economies. Tourism fell sharply worldwide, and international trade decreased significantly, affecting dozens of nations. In the United States, unemployment rose to 5.4 percent in October. In November, the National Bureau of Economic Research announced that the economy was in a recession, citing the September attacks as a major factor in turning the economic downturn that had begun in March into a recession. Because of the initial airport closures and the fall in business and recreational travel caused by the public's reluctance to fly, airlines suffered steep declines in revenues that required federal financial intervention for their survival. Also hard hit were the hospitality industry—hotels, restaurants, theme parks, and convention centers—and the travel industry.

On October 26, Bush signed into law the USA Patriot Act of 2001, which gave the Federal Bureau of Investigation and other law enforcement agencies expanded powers of surveillance regarding telephone calls, e-mail messages, and financial transactions and enabled them to detain immigrants for longer periods. The bill was a compromise, reflecting concerns about constitutional rights and civil liberties. One concern involved detention of immigrants. The American Civil Liberties Union, among others, raised objections to the detention of more than one thousand foreign nationals suspected of being involved in the terrorist attack. According to the Justice Department, about five to ten of these detainees were regarded as material witnesses to the attacks.

In addition, overseas military tribunals were established to try noncitizen suspects apprehended abroad. Immigration was curtailed, and border security substantially increased, especially along the U.S.-Canadian border, where immigration controls had been lax. Greater checks were to be put into effect in issuing visas, especially for nationals of certain Middle Eastern countries. Foreign students were to be more severely scrutinized and their numbers possibly reduced. Air travel became more cumbersome, often with substantial delays for security checks. These and other changes to American life dealt significant damage to the liberal ideal of the "open society."

A further result of the attacks was dramatic confirmation of a new form of terrorism, which had been hinted at by the 1993 bombing of the World Trade Center and the 1998 bombings of the U.S. embassies in Tanzania and Kenya. Many previous terrorists, particularly hijackers, had specific, finite political goals, such as the release of prisoners or the recovery of lost territory, took limited

numbers of lives, and claimed responsibility for their crimes. The new terrorism, by contrast, was anonymous and sought simply to inflict massive, unlimited, casualties on the target country.

The entire world was affected by the September 11 attacks. One of the most significant consequences was a realignment of the relationship among various of the world's most powerful nations. The most important of these was the new cooperative relationship between Russia and the West, especially the United States. The presidents of Russia and the United States formed joint policies to fight terrorism and agreed to blunt all outstanding issues between them, including continuing differences on missile defense.

Canada was especially affected by the attacks. Evidence of al-Qaeda activity in Canada indicated that lax Canadian immigration and refugee policies made it relatively easy for terrorists to gain access to the United States through Canada. At first, Canada resisted U.S. pressure for changes in its legislation and policies but soon enacted new measures after concerns arose about terrorist activity within Canada and the economic cost of long delays at border crossings.

After the attacks, many Middle Eastern nations, including Saudi Arabia, expressed their condemnation of the acts of terrorism. However, within large segments of the Islamic world, including the educated middle class, the September attacks resulted in a heightened sense of conflict with the West, especially the United States. Although the United States had fought on the side of Muslims in conflicts in Somalia, Bosnia, and Serbia, many Muslims questioned its support for Islamic nations, particularly in light of its historical pro-Israeli stance. Some nations were also critical of the heightened scrutiny and suspicion that surrounded their citizens in the United States. Within the United States, many Middle Eastern immigrants and other Muslims (both foreign nationals and U.S. citizens), as well as anyone who even appeared to be Middle Eastern, such as Sikhs, were subjected to harassment and, in extreme cases, physical violence by people suspecting them of being involved in terrorism.

Some consequences of the attacks had a positive side. American democracy appeared to be

3275

launched toward revitalization, as those who were nominally citizens became aware of the deeper value and meaning of citizenship and its responsibilities in light of national peril. The nation's political organization gained markedly in public esteem, and support fell for the theory of capitalist anarchism, in which markets would replace the state and the government would be run like a large corporation. Millions of citizens spontaneously expressed their pride in being American through displays of flags and other patriotic symbols. Many felt a strong impulse to promote unity in a common culture of democracy among all segments of a diverse population.

Taken in their aggregate, the consequences of the September attacks spelled a turning point in both U.S. and world history. How this change in direction would affect U.S. efforts to spread the ideals of liberal democracy and human rights and the benefits of economic prosperity via the free market throughout the globe remained an imponderable question in early 2002.

Charles F. Bahmueller

TIME LINE

Date	Event	Category	Country or region
Early 1900's	Ehrlich and Metchnikoff Pioneer Field of Immunology	Medicine	Russia/France/Germany
Early 1900's	Einthoven Develops Early Electrocardiogram	Medicine/Photography	Netherlands
1900	Hilbert Presents His Twenty-three Problems	Mathematics	France
1900-1901	Landsteiner Discovers Human Blood Groups	Medicine	Austria
1900-1910	Russell and Whitehead's *Principia Mathematica* Launches Logistic Movement	Mathematics	Great Britain
1900-1913	United States Reforms Extend the Vote	Civil rights and liberties/ Political reform	United States
March 23, 1900	Evans Unearths Minoan Civilization	Archaeology	Crete/Mediterranean Sea
June-September, 1900	Boxer Rebellion Tries to Drive Foreigners from China	Civil strife	China
June, 1900-February, 1901	Reed Discovers Cause of Yellow Fever	Medicine/Health	Cuba
July 2, 1900	Zeppelin Builds Rigid Airship	Space and aviation	Germany
September 8, 1900	Hurricane Levels Galveston, Texas	Disasters	Texas/Caribbean
December 14, 1900	Planck Articulates Quantum Theory	Physics	Germany
1901, 1907	Commonwealth of Australia and Dominion of New Zealand Are Born	Political independence	Australia/New Zealand/Great Britain
1901	Booth Invents Vacuum Cleaner	Engineering	United States
1901	Dutch Researchers Discover Cause of Beriberi	Medicine/Health/Food Science	Indonesia/Netherlands
1901	Elster and Geitel Demonstrate Natural Radioactivity	Earth science	Germany
1901	First Synthetic Vat Dye Is Synthesized	Chemistry	Germany
1901	Ivanov Develops Artificial Insemination	Agriculture	Russia
1901 1904	Kipping Discovers Silicones	Chemistry	Great Britain
1901-1932	Banach Sets Stage for Abstract Spaces	Mathematics	Poland
May 27, 1901	Insular Cases Determine Status of U.S. Possessions Overseas	Law	Puerto Rico/Philippines/Virgin Islands
September 14, 1901	Theodore Roosevelt Becomes U.S. President	National politics	Washington, D.C.
December 12, 1901	Marconi Receives First Transatlantic Radio	Communications	Great Britain/Canada
1902	Kennelly and Heaviside Theorize Existence of Ionosphere	Earth science	Massachusetts/Great Britain
1902	Zsigmondy Invents Ultramicroscope	Chemistry	Germany
1902	Sutton Predicts That Chromosomes Carry Hereditary Traits	Genetics/Biology	New York
1902-1903	Pavlov Introduces Concept of Reinforcement in Learning	Biology/Education	Russia
1902-present	Mathematics Responds to Crisis of Paradox	Mathematics	Europe
January, 1902	French Archaeologists Decipher Hammurabi's Code	Archaeology/Law	Iran/France
April-June, 1902	Bayliss and Starling Discover Hormones	Biology	Great Britain
May 12, 1902	Anthracite Coal Miners Strike	Labor/National politics	Pennsylvania/Washington D.C.
1903	Tsiolkovsky Proposes Using Liquid Oxygen for Rocket	Space and aviation	Russia

Date	Event	Category	Country or region
1903-1909	Benedictus Develops Laminated Glass	Materials	France
November 18, 1903	United States Acquires Panama Canal	International relations/ Economics	Washington, D.C./Colombia/ Panama
December 17, 1903	Wright Brothers Fly First Successful Airplane	Space and aviation	North Carolina
1904	Edison Develops First Alkaline Storage Battery	Energy	New Jersey
1904	Elster and Geitel Devise First Practical Photoelectric Cell	Energy	Germany
1904	Hartmann Discovers Interstellar Matter	Astronomy	Germany
1904	Kapteyn Determines Rotation of Galaxy	Astronomy	Netherlands
1904-1905	Gorgas Controls Yellow Fever in Panama Canal Zone	Medicine/Health	Panama/Central America
February 9, 1904- September 5, 1905	Russia and Japan Battle over Territory	War	China/Russia/Japan/Korea
1905	Great Britain and France Form Entente Cordiale	International relations	Great Britain/France
November 16, 1904	Fleming Patents First Vacuum Tube	Energy	Great Britain
1905	Einstein Describes Photoelectric Effect	Physics	Switzerland
1905	Hertzsprung Notes Relationship Between Stellar Color	Astronomy/ Photography	Denmark
1905	Punnett Argues for Acceptance of Mendelism	Genetics/Biology	Great Britain
1905-1907	Baekeland Produces First Artificial Plastic	Materials	New York
1905-1907	Boltwood Uses Radioactivity to Date Rocks	Earth science	Connecticut
January 22, 1905	Russian Workers and Czarist Troops Clash on Bloody Sunday	Civil strife	Russia
August, 1905	Lowell Predicts Existence of Planet Pluto	Astronomy	Arizona/Solar system/Pluto
Fall, 1905	Einstein Articulates Special Theory of Relativity	Physics	Switzerland
October 26, 1905	Norway Becomes Independent	Political independence	Norway/Sweden
October 30, 1905	Russia's October Manifesto Promises Constitutional Government	National politics	Russia
December, 1905	Crile Performs First Human Blood Transfusion	Medicine	Ohio
1906	Anschütz-Kaempfe Perfects Gyrocompass	Transportation	Germany
1906	Nernst Develops Third Law of Thermodynamics	Physics	Germany
1906 and 1910	Oldham and Mohorovičić Determine Inner Structure of Earth	Earth science	Great Britain/Croatia
1906-1911	Coolidge Patents Tungsten Filament	Materials	New York
1906-1913	Willstätter Discovers Composition of Chlorophyll	Biology	Switzerland
May 10, 1906	First Russian Duma Convenes	National politics	Russia
August 4, 1906	Germany Launches First U-Boat	Weapons technology	Germany
December, 1906	Battleship *Dreadnought* Revolutionizes Naval Architecture	Military technology	Great Britain
December 24, 1906	Fessenden Transmits Music and Voice by Radio	Communications	Massachusetts
1907	Lumière Brothers Invent Color Photography	Photography	France
1907-1910	Thomson Confirms Existence of Isotopes	Physics	Great Britain
1907-1919	Einstein Develops General Theory of Relativity	Physics	Switzerland/Germany

Date	Event	Category	Country or region
June, 1907	Second Hague Peace Conference Convenes	International relations	Netherlands
August 31, 1907	Britain, France, and Russia Form Triple Entente	International relations	France/Russia/Great Britain
1908	Haber Converts Atmospheric Nitrogen into Ammonia	Chemistry	Germany
1908	Hardy and Weinberg Model Population Genetics	Biology/Genetics	Germany/Great Britain
1908	Hughes Revolutionizes Oil-Well Drilling	Engineering	Texas
1908	Spangler Makes First Electric Vacuum Cleaner	Engineering	Ohio
1908	Steinmetz Warns of Future Air Pollution	Earth science	New York
February 11, 1908	Geiger and Rutherford Develop Geiger Counter	Physics	Great Britain
February 24, 1908	*Muller v. Oregon* Upholds Sociological Jurisprudence	Law/Labor	Washington, D.C.
June 26, 1908	Hale Discovers Magnetic Fields in Sunspots	Astronomy	California/Solar system
July 9, 1908	Kamerlingh Onnes Liquefies Helium	Physics/Chemistry	Netherlands
October 7, 1908	Austria Annexes Bosnia and Herzegovina	International relations	Balkans/Bosnia/Austria
November 1, 1908	Belgium Annexes Congo Free State	Human rights/Political reform	Belgium/Congo (Kinshasa)/Africa
December, 1908	Boule Reconstructs Neanderthal Man Skeleton	Anthropology	France
1909	Johannsen Coins *Gene, Genotype,* and *Phenotype*	Biology/Genetics	Denmark
1909-1913	Taft Conducts Dollar Diplomacy	Economics/International relations	Washington, D.C.
January-August, 1909	Millikan Conducts Oil-Drop Experiment	Physics	Illinois
February 12, 1909	National Association for the Advancement of Colored People Is Formed	Social reform/Political reform	New York
1910's	Mexican Revolution Creates Constitutional Government	Civil war/Political reform	Mexico
1910	Crystal Sets Lead to Modern Radios	Communications	United States
1910	Electric Washing Machine Is Introduced	Engineering	United States
1910	Hale Telescope Begins Operation on Mount Wilson	Astronomy	California
1910	Rous Discovers Cancer-Causing Viruses	Medicine/Health	New York
1910-1939	Electric Refrigerators Revolutionize Home Food Preservation	Food science	Michigan/New York
April, 1910	Antisyphilis Drug Salvarsan Is Introduced	Medicine/Health	Germany
May 31, 1910	Union of South Africa Is Established	Political independence/Political reform	South Africa
October 5, 1910	Portugal Becomes Republic	Political reform/Coups	Portugal
1911	Boas Lays Foundations of Cultural Anthropology	Anthropology	New York
1911-1912	Italy Annexes Tripoli	Military conflict/Political aggression	Libya/Italy
May-December, 1911	British Parliament Passes National Insurance Act	Social reform	Great Britain
July 24, 1911	Bingham Discovers Lost Inca City of Machu Picchu	Archaeology	Peru
August 10, 1911	British Parliament Limits Power of House of Lords	Law/Political reform	Great Britain

Date	Event	Category	Country or region
November 4, 1911	France and Germany Sign Treaty on Morocco	International relations	Morocco/Germany/France
1912	First Diesel Locomotive Is Tested	Transportation	Switzerland
1912	Fischer Discovers Chemical Process for Color Films	Photography	Germany
1912	Leavitt Measures Interstellar Distances	Astronomy/ Photography	Massachusetts
1912	United States Establishes Public Health Service	Social reform	Washington, D.C.
1912-1913	Bohr Develops Theory of Atomic Structure	Physics	Denmark
1912-1913	Shuman Builds First Practical Solar Thermal Engine	Energy	Egypt
1912-1914	Abel Develops First Artificial Kidney	Medicine	Maryland
1912-1915	Braggs Invent X-Ray Crystallography	Physics/Photography/ Earth science	Great Britain
January, 1912	Wegener Articulates Theory of Continental Drift	Earth science	Germany
February 12, 1912	China's Manchu Dynasty Is Overthrown	Political reform/Coups	China
March 7, 1912	Rutherford Presents Theory of the Atom	Physics	Great Britain
April 11, 1912– September 15, 1914	Home Rule Stirs Debate in Ireland and Great Britain	Political reform	Great Britain/Ireland
April 14-15, 1912	Iceberg Sinks "Unsinkable" Ocean Liner *Titanic*	Disasters/ Transportation	Atlantic Ocean
August 3, 1912	U.S. Marines Go to Nicaragua to Quell Unrest	Political aggression	Nicaragua/Central America
August 7 and 12, 1912	Hess Determines Extraterrestrial Origins of Cosmic Rays	Physics/Astronomy	Austria
October 18, 1912	Balkan Wars Divide European Loyalties	Military conflict/War	Balkans/Bosnia/Bulgaria/Greece
November, 1912	Wilson Is Elected U.S. President	National politics	United States
1913	Burton's Thermal Cracking Process Improves Gasoline Production	Chemistry	United States
1913	Edison Develops Talking Pictures	Photography/ Entertainment	New Jersey
1913	Geothermal Power Is Produced for First Time	Energy	Italy
1913	Gutenberg Discovers Earth's Mantle-Outer Core Boundary	Earth science	Germany
1913	Salomon Develops Mammography to Detect Breast Cancer	Medicine/Photography	Germany
1913	Schick Introduces Test for Diphtheria	Medicine/Health	Austria
February 25, 1913	Sixteenth Amendment Legalizes Income Tax	Law/Economics	Washington, D.C.
March 1, 1913- January 5, 1914	Ford Develops Assembly Line	Economics	Michigan
December, 1913	Congress Passes Federal Reserve Act	Economics	Washington, D.C.
December, 1913	Russell Announces Theory of Stellar Evolution	Astronomy	New Jersey
June 28, 1914	Assassination of Archduke Ferdinand Begins World War I	War/Assassination	Europe
September 6-9, 1914	France and Germany Fight First Battle of the Marne	War	France/Germany
September 22, 1914	Submarine Warfare Begins	War/Weapons technology	North Sea/Atlantic Ocean
October 30, 1914	Spain Declares Neutrality in World War I	International relations/ War	Spain

3280

Date	Event	Category	Country or region
1915	Morgan Develops Gene-Chromosome Theory	Biology/Genetics	New York
1915-1916	Turks Massacre Armenians	Civil strife/Ethnic conflict	Turkey/Syria
January 18, 1915	Japan Presents China with Twenty-one Demands	International relations	Japan/China
February, 1915-January, 1916	Allies Fight Turkey at Gallipoli	War	Turkey
April, 1915	First Transcontinental Telephone Call Is Made	Communications	Georgia/New York/California
May, 1915	Italy Enters World War I	War	Italy
May, 1915	Corning Trademarks Pyrex Glass	Materials	New York
May, 1915	Fokker Designs First Propeller-Coordinated Machine Guns	Weapons technology	Germany
September, 1915-February, 1916	McLean Discovers Anticoagulant Heparin	Medicine	Maryland
October 21, 1915	Transatlantic Radio Communication Begins	Communications	Virginia/France
1916	Garvey Founds Universal Negro Improvement Association	Civil rights and liberties/Social reform	New York
1916	Schwarzschild Theorizes Existence of Black Holes	Astronomy	Germany
1916-1922	Ricardo Designs Modern Internal Combustion Engine	Engineering	Great Britain
February 21-December 15, 1916	Nearly One Million Die in Battle of Verdun	War	France
March 15, 1916	Pershing Leads U.S. Military Expedition into Mexico	Military conflict	Mexico/United States
April 24-30, 1916	Easter Rebellion Fails to Win Irish Independence	Civil strife	Ireland
May 31, 1916	British Navy Faces Germany in Battle of Jutland	War	North Sea/Germany/Great Britain
1917	Mexican Constitution Establishes Workers' Rights	Labor	Mexico
1917	Birdseye Develops Food Freezing	Food science	Massachusetts
1917	Langevin Develops Active Sonar	Communications/Military capability	France
1917	Insecticide Use Increases in American South	Agriculture/Environment	Louisiana/Arkansas/Mississippi
1917-1918	Congress Passes Espionage and Sedition Acts	National politics/Civil rights and liberties	Washington, D.C.
April 6, 1917	United States Enters World War I	War	Washington, D.C.
July 8, 1917	War Industries Board Is Organized	Economics	Washington, D.C.
November, 1917	Hooker Telescope Begins Operation on Mount Wilson	Astronomy	California
November 2, 1917	Balfour Declaration Supports Jewish Homeland in Palestine	Political independence	Great Britain/Israel/Palestine
November 6-7 (Old Calendar: October 24-25), 1917	Bolsheviks Seize Power During October Revolution	Political reform/Coups	Russia
1918	Influenza Epidemic Kills Millions Worldwide	Health/Disasters/Medicine	Asia/Africa/Europe
1918-1921	Soviets and Anti-Soviets Clash in Great Russian Civil War	Civil war	Soviet Union/Russia
January 8, 1918	Shapley Proves Solar System Is Near Edge of Galaxy	Astronomy	Solar system/Space

3281

Date	Event	Category	Country or region
September 26- November 11, 1918	Allies Dominate in Meuse-Argonne Offensive	War	France
November 5, 1918	Harding Is Elected U.S. President	National politics	United States
November 13, 1918	Habsburg Monarchy Ends	Political reform	Austria/Hungary
1919	Aston Builds First Mass Spectrograph	Physics/Photography	Great Britain
Spring, 1919	Frisch Learns That Bees Communicate	Biology	Germany
1919-1920	Red Scare Threatens Personal Liberties	National politics/Civil rights and liberties	United States
1919-1921	Bjerknes Discovers Atmospheric Weather Fronts	Earth science	Norway
1919-1933	Mises Develops Frequency Theory of Probability	Mathematics	Germany/Austria
May 4, 1919	Students Call for Reform in China's May Fourth Movement	Political reform/Social reform	China
June 28, 1919	League of Nations Is Created	International relations	France/Europe
June 28, 1919	Treaty of Versailles Is Concluded	International relations/War	France/Germany
August 11, 1919	Weimar Constitution Guides New German Republic	Political reform	Germany
November 6, 1919	Einstein's Theory of Gravitation Is Confirmed	Physics/Earth science	Great Britain
Early 1920's	Slipher Measures Redshifts of Spiral Nebulas	Astronomy	Arizona/Space
1920-1930	Millikan Explains Cosmic Rays	Physics	United States
1920-1940	Gandhi Leads Indian Reform Movement	Social reform/Political reform	India
January 16, 1920	Eighteenth Amendment Bans Sale of Liquor	Law/Social reform	United States
June 4, 1920	Allies and Hungary Make Peace in Treaty of Trianon	International relations	France
August 26, 1920	Nineteenth Amendment Gives Women the Vote	Political reform	United States
December 13, 1920	Michelson Measures Star's Diameter	Astronomy/Metrology	Space
1921	Calmette and Guérin Develop Tuberculosis Vaccine	Medicine/Health	France
1921	Banting and Best Discover Insulin Hormone	Biology	Quebec
1921-1922	Mussolini Formulates Fascist Doctrine	Political reform	Italy
1921-1923	Scandals Rock Harding's Administration	National politics	Washington, D.C.
1921-1928	Lenin Decrees New Economic Policy	Economics/Social reform	Soviet Union
1921-1931	Gödel Proves Incompleteness of Formal Systems	Mathematics	Germany
July 14, 1921	Sacco and Vanzetti Are Convicted of Murder	Human rights/Crime/Government	Massachusetts
November 12, 1921- February 6, 1922	Washington Disarmament Conference Meets	International relations	Washington, D.C.
December 6, 1921	Irish Free State Becomes Dominion	Political reform	Great Britain/Ireland
1922	McCollum Uses Vitamin D Against Rickets	Medicine/Health	Maryland
1922	Bocage Devises Theory of Tomography	Medicine	France
1922-1923	Compton Effect Confirms Quantum Theory of Radiation	Physics	Missouri
September, 1922	Dawes Plan Aids Germany	International relations/Economics	Washington, D.C./Great Britain/France/Germany
October 24-29, 1922	Mussolini Takes Power in Italy	Political reform	Italy

Date	Event	Category	Country or region
November 4, 1922	Carter Finds Tutankhamen's Tomb in Egypt	Archaeology	Egypt
1923	Andrews Discovers Fossil Dinosaur Eggs	Biology	Mongolia
1923	Barnay Develops Rotary Dial Telephone	Communications	France
1923	De Broglie Explains Wave-Particle Duality of Light	Physics	France
1923	Zworykin Develops Concept of Television	Communications/ Entertainment	Pennsylvania
1923 and 1951	Kahn Develops Simple Syphilis Test	Medicine/Health	Michigan
Summer, 1923	Zdansky Discovers Peking Man	Anthropology	China
1924	Hubble Measures Distance to Andromeda Nebula	Astronomy	Space
1924	Svedberg Develops Ultracentrifuge	Biology	Sweden
March, 1924	Eddington Relates Stellar Mass to Luminosity	Astronomy	Space
Summer, 1924	Dart Finds Fossil Linking Apes and Humans	Anthropology	South Africa
November, 1924	Coolidge Is Elected U.S. President	National politics/ Economics	United States
December, 1924	Hubble Shows That Universe Contains Many Galaxies	Astronomy	Space
1925	Scopes "Monkey" Trial Tests Academic Freedom	Law/Education/Social reform	Tennessee
1925	Pauli Formulates Exclusion Principle	Physics	Germany
1925	Whipple Discovers Iron's Importance in Red Blood Cells	Medicine	New York
1925-1930	Bush Builds First Differential Analyzer	Computer science	Massachusetts
1925-1977	Pahlavi Shahs Attempt to Modernize Iran	Political reform/Social reform/Religion	Iran
January 5, 1925	Widow of Wyoming's Governor Ross Becomes First Woman Governor	Gender issues/Political reform	Wyoming
April, 1925-May, 1927	*Meteor* Expedition Discovers Mid-Atlantic Ridge	Earth science	Atlantic Ocean
July 18, 1925- December 10, 1926	Hitler Publishes *Mein Kampf*	Political reform	Germany
October 5-December 1, 1925	Locarno Conference Promises Post-World War I Security	International relations	Switzerland
1926	British Workers Mount General Strike	Labor	Great Britain
1926	Schrödinger Proposes Wave-Mechanical Theory of Atom	Physics	Switzerland
1926	First Synthetic Rubber Is Invented	Materials	Germany
March 16, 1926	Goddard Launches First Liquid-Fuel-Propelled Rocket	Space and aviation	Massachusetts
August 6, 1926	Warner Bros. Puts Sound into Movie Theaters	Entertainment/ Photography	United States
1927	Bell System Demonstrates Long-Distance Television	Communications/ Entertainment	Washington, D.C.
1927	Heisenberg Proposes Uncertainty Principle	Physics	Belgium
1927	Haldane Builds First Heat Pump	Energy	Scotland
1927	Lemaître Proposes Theory of Expanding Universe	Astronomy	Space
1927	Oort Proves Spiral Structure of Milky Way Galaxy	Astronomy	Space
March, 1927-October 1, 1949	Civil War Rages in China	Civil war	China
May 20-21, 1927	Lindbergh Makes First Solo Nonstop Transatlantic Flight	Space and aviation/ Transportation	New York/France/Atlantic Ocean

3283

Date	Event	Category	Country or region
October, 1927	Kemal Institutes His Reform Program	Social reform/Political reform	Turkey
1928	Gamow Explains Radioactive Alpha Decay	Physics	Germany
1928	Papanicolaou Develops Uterine Cancer Test	Medicine/Health/Gender Issues	New York
1928	Von Neumann Publishes Minimax Theorem and Establishes Game Theory	Mathematics	Germany/New Jersey
1928-1932	Szent-Györgyi Discovers Vitamin C	Medicine	Great Britain/Hungary/Pennsylvania
1928-1933	Stalin Introduces First Soviet Five-Year Plan	Social reform/Economics	Soviet Union
August, 1928	Mead Publishes *Coming of Age in Samoa*	Anthropology	Samoa
August 27, 1928	Kellogg-Briand Pact Outlawing War Is Signed	International law/International relations	France
September, 1928	Fleming Discovers First "Wonder Drug" Penicillin	Medicine/Health	Great Britain
November 6, 1928	Smith-Hoover Campaign Revolutionizes U.S. Politics	National politics	United States
1929	Hubble Confirms That Universe Is Expanding	Astronomy	Space
1929	Marrison Fabricates Electric Clock with Quartz Resonator	Metrology	United States
1929-1936	Gericke Reveals Significance of Hydroponics	Agriculture	California
1929-1939	Great Depression Brings Economic Crisis	Economics	United States
1929-1940	France Builds Maginot Line	Military defense	France
January 19, 1929	Trotsky Is Sent into Exile	National politics	Soviet Union/Russia
February 11, 1929	Italy Recognizes Vatican Sovereignty	International relations	Italy/Vatican
April 22, 1929	Berger Devises Electroencephalogram	Medicine	Germany
July, 1929	Drinker and Shaw Invent Iron Lung	Medicine	Massachusetts
August 15, 1929	Arabs Attack Jews in Palestine	Civil strife	Palestine
October 29, 1929	Wall Street's Stock Market Crashes	Economics	New York
1930's	Practical Contact Lenses Are Developed	Materials/Health	Germany
1930	United States Adopts Good Neighbor Policy Toward Latin America	International relations	Washington, D.C.
1930	Racist Theories Give Rise to Nazi Politics	Social reform/Ethnic conflict	Germany
1930	Construction of Empire State Building Begins	Engineering	New York
1930	Midgley's Refrigerant Gas Makes Home Refrigerators Feasible	Chemistry	Ohio
1930	Zinsser Develops Typhus Vaccine	Medicine/Health	Massachusetts
1930-1931	Pauling Develops Chemical Bond Theory	Chemistry	California
1930-1932	Jansky Detects Radio Signals from Outside Solar System	Astronomy	Space
1930-1935	Armstrong Invents FM Radio	Communications	New York
February 18, 1930	Tombaugh Discovers Planet Pluto	Astronomy	Pluto/Solar system
July, 1930	First World Cup Soccer Championship Is Held in Uruguay	Sports	Uruguay
1931	Statute of Westminster Creates British Commonwealth	Political independence	Great Britain

3284

Date	Event	Category	Country or region
1931	Urey Discovers Deuterium and "Heavy Water"	Chemistry	New York
1931-1937	Scottsboro Trials Reveal American Racism	Law/Civil rights and liberties	Alabama
January 2, 1931	Lawrence Invents Particle Accelerator Cyclotron	Physics	California
March 5, 1931	India Signs Delhi Pact	Civil rights and liberties	India
April, 1931	Ruska Creates First Electron Microscope	Physics	Germany
April 14, 1931	Second Spanish Republic Is Proclaimed	Political reform	Spain
May 27, 1931	Piccard Reaches Stratosphere in Balloon	Space and aviation	Germany/Austria
September 18, 1931-February 24, 1933	Manchurian Crisis Marks Rise of Japanese Militarism	Military conflict	China/Japan
November 3, 1931	Du Pont Introduces Neoprene	Materials	Ohio
1932	Hoover-Stimson Doctrine Condemns Japanese Imperialism	International relations	United States/Japan
1932	Bacon Designs First Fuel Cell	Chemistry	Italy/France/United States
1932-1935	Domagk and Ehrlich Develop Antibacterial Drugs	Medicine	Germany
January-February, 1932	El Salvador's Army Massacres Civilians in *La Matanza*	Human rights/Civil strife	El Salvador/Central America
January 22, 1932	Reconstruction Finance Corporation Is Established	Economics	Washington, D.C.
February, 1932	Chadwick Proves Existence of Neutrons	Physics	Great Britain
February 2, 1932-June 11, 1934	Geneva Conference Seeks Disarmament	International relations	Switzerland
April, 1932	Cockcroft and Walton Split the Atom	Physics	Great Britain
June 15, 1932-June 21, 1935	Bolivia and Paraguay Battle over Chaco Wilderness	War	Bolivia/South America/Paraguay
September, 1932	Anderson Discovers Positrons	Physics	California
November 8, 1932	Franklin D. Roosevelt Is Elected U.S. President	National politics	United States
1933	United States Recognizes Soviet Union	International relations	United States/Soviet Union
1933	Fermi Predicts Existence of Neutrinos	Physics	Italy
1933	RCA Begins Electronic Television Broadcasts	Communications/Entertainment	New York
1933-1934	Joliot and Joliot-Curie Create First Artificial Radioactive Element	Physics	France
1933-1936	First Commercial Airliners Begin Operating	Space and aviation/Transportation	United States
1933-1939	Nazis Persecute Germany's Jews	Human rights	Germany
1933-1954	Flosdorf Demonstrates Freeze-Drying of Food	Food science	Pennsylvania
January 30, 1933	Hitler Takes Power in Germany	Political reform	Germany
February 24, 1933	Japan Withdraws from League of Nations	International relations	Switzerland
February 27, 1933	Germany's Reichstag Goes Up in Flames	Civil strife	Germany
March 9-June 16, 1933	Congress Passes Emergency Legislation During Roosevelt's First 100 Days	Political reform/Economics	Washington, D.C.
March 23, 1933	Germany's Enabling Act Gives Hitler Unchecked Power	Political reform	Germany
May 18, 1933	Tennessee Valley Authority Is Created	Economics/Social reform	Tennessee/Kentucky/Virginia/North Carolina/Alabama/Georgia/Mississippi
1934	Zwicky and Baade Theorize Existence of Neutron Stars	Astronomy	Space

Date	Event	Category	Country or region
1934	Benedict Publishes *Patterns of Culture*	Anthropology	New York
1934	Cherenkov Discovers Cherenkov Effect	Physics	Soviet Union
1934	Yukawa Proposes Existence of Mesons	Physics	Japan
1934-1938	Drought Creates Dust Bowl	Economics	Kansas/Oklahoma/Texas/ Colorado/New Mexico/ Illinois
February 6, 1934	Stavisky Riots Threaten French Republic	Civil strife	France
May 28, 1934	Dionne Sisters Are First Quintuplets to Survive	Health/Social reform	Canada
June 30-July 2, 1934	Hitler Purges Enemies	Assassination	Germany
July 25, 1934	Austrian Nazis Murder Chancellor Dollfuss	Assassination/Coups	Austria
August 11-15, 1934	Beebe and Barton Use Bathysphere to Explore Ocean Depths	Earth science	Atlantic Ocean
December 1, 1934	Stalin Begins Soviet Purge Trials	Human rights	Soviet Union
December 5, 1934- May 9, 1936	Italy Occupies Ethiopia	War/Political aggression	Ethiopia/Italy
1935	Chapman Measures Lunar Atmospheric Tides	Earth science	Moon
1935	Radar Is Invented	Communications	Great Britain
1935-1939	Congress Passes Neutrality Acts	International relations	Washington, D.C.
January, 1935	Richter Develops Earthquake Scale	Earth science/ Metrology	California
February 28, 1935	Carothers Synthesizes Nylon	Materials	Delaware
March 16, 1935	Germany Renounces Treaty of Versailles	International relations	Germany
April 8, 1935	Works Progress Administration Is Formed	Social reform/ Economics	Washington, D.C.
July 5, 1935	Congress Passes National Labor Relations Act	Economics/Labor	Washington, D.C.
August 14, 1935	Congress Passes Social Security Act	Social reform/ Economics	Washington, D.C.
November-December, 1935	Egas Moniz Pioneers Prefrontal Lobotomy	Medicine	Portugal
November, 1935-May, 1938	Congress of Industrial Organizations Is Formed	Economics/Labor	United States
1936	British King Edward VIII Abdicates	Political reform	Great Britain
1936	Lehmann Discovers Earth's Inner Core	Earth science	Denmark
1936-1937	Müller Invents Field Emission Microscope	Physics	Germany
1936-1937	Turing Predicts Artificial Intelligence	Computer science	Great Britain
February 17, 1936	Paraguay Introduces Corporatism to the Americas	Labor/Political reform/ Coups	Paraguay/South America
March 1, 1936	Boulder Dam Is Completed	Engineering	Arizona/Nevada
March 7, 1936	Germany Remilitarizes Rhineland	International relations	Germany/France
April-May, 1936	Popular Front Wins Power in France	Political reform	France
July 17, 1936	Civil War Erupts in Spain	Civil war	Spain
November 23, 1936	Fluorescent Lighting Is Introduced	Energy	Washington, D.C.
1937	Theiler Introduces Yellow Fever Vaccine	Medicine/Health	New York/Central America
March, 1937	Krebs Describes Body's Citric Acid Cycle	Biology	Great Britain
July 2, 1937	Aviator Earhart Disappears in Flight Over South Pacific	Gender issues/Space and aviation	New Guinea/Pacific Ocean
July 7, 1937-August, 1945	Japan Conducts Military Campaigns in China	War	China/Japan
Fall, 1937-Winter, 1938	Weidenreich Reconstructs Face of Peking Man	Anthropology	China

Date	Event	Category	Country or region
1938	HUAC Begins Investigating Suspected Communists	Civil rights and liberties	Washington, D.C.
1938	Carlson Invents Xerography	Communications	United States
1938	Hofmann Synthesizes Psychedelic Drug LSD-25	Chemistry	Switzerland
1938	Polystyrene Is Marketed	Materials	Texas
1938	Tisza Explains Superfluidity	Physics	France
1938-1941	United States Establishes Two-Ocean Navy	Military capability	United States
February 5, 1938	Hitler Takes Control of Germany's Diplomatic and Military Hierarchy	Political reform	Germany
February 12-April 10, 1938	Germany Absorbs Austria in *Anschluss*	Political aggression	Austria
April 6, 1938	Plunkett Synthesizes Teflon	Materials	Delaware
September 29-30, 1938	Chamberlain Appeases Germany at Munich	International relations	Germany/Great Britain/Czechoslovakia
October 30, 1938	Welles's Broadcast of *The War of the Worlds* Panics Nation	Communications/Entertainment	New York
December, 1938	Physicists Split Uranium Atom	Physics	Germany
1939	British White Paper Restricts Jewish Immigration to Palestine	International relations	Israel/Palestine
1939	Bourbaki Circle Publishes *Éléments de mathématique*	Mathematics	France
1939	Müller Develops Potent Insecticide DDT	Agriculture	Switzerland
1939-1941	United States Mobilizes for War	International relations	United States
1939-1942	United States Decides to Make Atomic Bomb	Technology/Military capability	United States
1939-1945	Germany Tries to Exterminate European Jews	Human rights	Europe
August 23-24, 1939	Germany and Soviet Union Form Pact	International relations	Soviet Union/Russia
September, 1939-May, 1945	Allies Win Battle of Atlantic	War	Atlantic Ocean/Germany
September 1, 1939	Germany Invades Poland	War	Germany/Poland
1940	First Color Television Signals Are Broadcast	Communications/Entertainment	New York
1940	Landsteiner Identifies Rhesus Factor	Medicine	New York
1940-1941	Royal Air Force Wins Battle of Britain	War	Great Britain/Germany
1940-1955	Spencer Discovers Principle of Microwave Cooking	Food science	Massachusetts
April 9, 1940	Germany Invades Norway	War	Norway/Germany
May, 1940	Antibiotic Penicillin Is Developed	Medicine	Great Britain
May 10, 1940	Churchill Becomes British Prime Minister	Government/War	Great Britain
June 4, 1940	British Evacuate from Dunkirk	War	France
June 22, 1940	France Surrenders to Germany	War	France/Germany
September, 1940-July, 1941	Japan Occupies Indochina	War	Japan/France/Vietnam/Laos/Cambodia
September 12, 1940	Boys Discover Prehistoric Cave Paintings at Lascaux	Archaeology	France
1941	Polyester Is Patented	Materials	Great Britain
1941	Touch-Tone Telephone Dialing Is Introduced	Communications	Maryland
February 23, 1941	Seaborg and McMillan Create Plutonium	Physics	California
April 6, 1941	Germany Invades Balkan Peninsula	War	Balkans/Yugoslavia/Germany/Italy/Bulgaria
May, 1941	Ho Chi Minh Organizes Viet Minh	Political independence	Vietnam

3287

Date	Event	Category	Country or region
May 15, 1941	Whittle's Jet Engine Is Tested	Space and aviation	Great Britain
June 22, 1941	Germany Invades Soviet Union	War	Soviet Union/Germany
December 7, 1941	Japan Bombs Pearl Harbor	War	Hawaii/Japan
December 8, 1941-August, 1945	Japan Occupies Dutch East Indies, Singapore, and Burma	War	Asia/Japan/Indonesia/Singapore/Myanmar
December 11, 1941	Germany and Italy Declare War on United States	War	Germany/Italy/United States
1942	Kodak Perfects Color Film for Infrared Photography	Photography	New York
1942-1947	Reber Makes First Radio Maps of Universe	Astronomy	Space
1942-1950	Orlon Fibers Are Introduced	Materials	Delaware/Germany
February 19, 1942	Roosevelt Approves Internment of Japanese Americans	Civil rights and liberties/Ethnic conflict	Washington, D.C./California/Oregon/Washington State/Arizona
May 30-31, 1942	British Mount First Thousand-Bomber Raid on Germany	War	Germany
June 3-6, 1942	Battle of Midway Turns Tide in Pacific	War	Pacific Ocean/Japan
August, 1942-February, 1943	Allies Win Battle of Guadalcanal	War	Solomon Islands/Pacific Ocean
October 23-November 3, 1942	Battle of Alamein Heralds German Defeat	War	Egypt
November 7-8, 1942	Allied Forces Invade North Africa	War	North Africa/Germany
November 19, 1942-January 31, 1943	Germans Besiege Stalingrad	War	Soviet Union
November 28, 1942	Fire in Boston's Coconut Grove Nightclub Kills 492 People	Disasters	Massachusetts
December 2, 1942	Fermi Creates Controlled Nuclear Fission Chain Reaction	Physics	Illinois
1943-1944	DNA Is Identified as Genetic Key	Biology/Genetics	New York
1943-1944	Weizsäcker Completes Theory of Planetary Formation	Astronomy	Solar system
1943-1946	Eckert and Mauchly Create ENIAC Computer	Computer science	Pennsylvania
1943-1955	Frederick Sanger Determines Structure of Insulin	Chemistry	Great Britain
January 14-24, 1943	Allied Leaders Meet at Casablanca	International relations	Morocco
Spring, 1943	Cousteau and Gagnan Develop Aqualung	Earth science/Sports	France
September, 1943-March, 1944	Waksman Discovers Streptomycin Antibiotic	Medicine	New Jersey
September 3-9, 1943	Western Allies Invade Italy	War	Italy
November 4, 1943	First Nuclear Reactor Is Activated	Energy	Tennessee
December, 1943	British Scientists Create Colossus Computer	Computer science	Great Britain
1944	German V-2 Rockets Enter Production	War/Military technology/Space and aviation	Germany
1944	IBM Completes Development of Mark I Calculator	Computer science	New York
1944-1949	Hodgkin Uses Computer to Find Structure of Penicillin	Chemistry/Computer science	Great Britain
1944-1952	Ryle Locates First Known Radio Galaxy	Astronomy	Space
April, 1944 to mid-1945	Soviets Invade Eastern Europe	War	Europe
June 6, 1944	Allies Invade France	War	France/Europe
June 15, 1944	U.S. B-29's Bomb Japan	Weapons technology/War	China/Japan

Date	Event	Category	Country or region
June 22, 1944	Congress Passes G.I. Bill	Education/Social reform	United States
September 25, 1944-March 22, 1945	Arab League Is Formed	International relations	Egypt
October 19-25, 1944	Allies Gain Control of Pacific in Battle for Leyte Gulf	War	Philippines/Pacific Ocean/Japan
November 29, 1944	Blalock Performs First "Blue Baby" Operation on Heart	Medicine	Maryland
December 16, 1944-January 28, 1945	Allies Defeat Germany at Battle of the Bulge	War	Belgium/Germany
1945	Duggar Discovers First Tetracycline Antibiotic, Aureomycin	Medicine	New York
1945-1946	Americans Return to Peace After World War II	Economics/Social change	United States
1945-1947	Marshall Mission Attempts to Unify China	International relations	China
January, 1945	Municipal Water Supplies Are Fluoridated	Medicine	United States
January-May, 1945	Battle of Germany Ends World War II in Europe	War	Germany/Europe
February 4-11, 1945	Churchill, Roosevelt, and Stalin Meet at Yalta	International relations	Soviet Union
April 26-June 26, 1945	United Nations Is Created	International relations	California
May 8, 1945	France Quells Algerian Nationalist Revolt	Civil strife/Political independence	Algeria
May 8, 1945	Germany Surrenders on V-E Day	War	France/Germany
July 16, 1945	First Atom Bomb Is Detonated	Weapons technology	New Mexico
July 17-August 2, 1945	Potsdam Conference Decides Germany's Fate	International relations	Germany
July 26, 1945	Labour Party Leads Postwar Britain	Political reform	Great Britain
August 6, 1945	United States Drops Atomic Bomb on Hiroshima	War/Weapons technology	Japan
September 2, 1945-April 28, 1952	General MacArthur Oversees Japan's Reconstruction	Political reform/Economics	Japan
November 20, 1945-August 31, 1946	Nuremberg Trials Convict War Criminals	International law	Germany
February 20, 1946	Congress Passes Employment Act of 1946	Economics/Labor	Washington, D.C.
February 24, 1946	Perón Creates Populist Alliance in Argentina	National politics/Labor	Argentina
March 5, 1946	Churchill Delivers "Iron Curtain" Speech	International relations	Missouri
July 4, 1946	Philippines Republic Regains Its Independence	Political independence	Philippines
July 12, 1946	Schaefer Seeds Clouds with Dry Ice	Earth science	New York
July 29, 1946-February 10, 1947	Paris Peace Conference Makes Treaties with Axis Powers	International relations	France/Italy/Romania/Finland/Hungary/Bulgaria
November, 1946	Physicists Build Synchrocyclotron	Physics	California
1947	Gabor Invents Holography	Physics/Photography	Great Britain
1947-1949	Libby Uses Carbon 14 to Date Ancient Objects	Archaeology	Illinois
March 12, 1947	Truman Presents His Doctrine of Containment	International relations	Washington, D.C.
March 12, 1947	Marshall Plan Helps Rebuild Europe	Economics/International relations	Washington, D.C./Europe
Spring, 1947	Dead Sea Scrolls Are Found	Archaeology	Palestine
July 26, 1947	Congress Passes National Security Act	National politics/Military	Washington, D.C.

3289

Date	Event	Category	Country or region
August 15, 1947	India Is Partitioned	Political reform	India
October 14, 1947	Bell's X-1 Rocket Plane Breaks Sound Barrier	Space and aviation	California
November-December, 1947	Shockley, Bardeen, and Brattain Invent Transistors	Energy	New Jersey
1948	Nationalists Take Power in South Africa	Political reform/Civil rights/Human rights	South Africa
1948	Gamow Proposes "Big Bang" Theory of Universe	Astronomy	Space
1948	Lyons Constructs First Atomic Clock	Metrology	Washington, D.C./Europe
1948	Hoyle Advances Steady-State Theory of Universe	Astronomy	Great Britain/Space
1948-1960	Piccard Invents Bathyscaphe	Earth science	Belgium/Italy
February 20-25, 1948	Communists Take Over Czechoslovakia	Political reform/Coups	Czechoslovakia
March-May, 1948	Organization of American States Is Formed	International relations	Colombia/South America/Central America
April 18, 1948	Communists Lose Italian Elections	Political reform	Italy
May 14, 1948	Jewish State of Israel Is Born	Political independence	Palestine
June 3, 1948	Hale Telescope Begins Operation at Mount Palomar	Astronomy	California
June 24, 1948-May 12, 1949	Western Allies Mount Berlin Air Lift	International relations/Economics	Germany
June 28, 1948	Cominform Expels Yugoslavia	International relations	Europe/Soviet Union/Yugoslavia
July 26, 1948	Truman Orders End of Segregation in U.S. Military	Civil rights and liberties/Military	United States
August 23, 1948	World Council of Churches Is Formed	Religion	Netherlands
November 2, 1948	Truman Is Elected U.S. President	National politics	United States
November 26, 1948	Land Invents Instant Photography	Photography	Massachusetts
February 24, 1949	First Multistage Rocket Is Launched	Space and aviation	New Mexico
April 4, 1949	Twelve Nations Sign North Atlantic Treaty	International Relations	Europe/Canada/United States
August, 1949	BINAC Computer Uses Stored Programs	Computer science	Pennsylvania
August 29, 1949	Soviet Union Explodes Its First Nuclear Bomb	Technology/Military	Soviet Union
September 21, and October 7, 1949	Germany Is Split into Two Countries	International relations	Germany
October 1, 1949	People's Republic of China Is Formed	National politics	China
1950's	Li Isolates Human Growth Hormone	Biology	California
1950	McCarthy Conducts Red-Baiting Campaign	National politics	West Virginia/Washington, D.C.
1950	Abbott Labs Introduces Artificial Sweetener Cyclamate	Food science	Illinois
1950	Oort Postulates Origin of Comets	Astronomy	Netherlands/Solar system/Space
1950-1964	Morel Multiplies Plants In Vitro, Revolutionizing Agriculture	Agriculture	France
June, 1950	United States Enters Korean War	War	Korea
July 8, 1950-April 11, 1951	Truman and MacArthur Clash over Korea Policy	International relations/National politics	Korea/Washington, D.C.
September 15, 1950	United Nations Forces Land at Inchon	War	Korea
October 7, 1950	China Occupies Tibet	Human rights/Political aggression	Tibet/China
November 1, 1950	President Truman Escapes Assassination Attempt	Assassination/National politics	Washington, D.C.
Early 1951	United States Explodes Its First Hydrogen Bomb	Technology/Military capability	Siberia/Pacific Ocean
1951	Lipmann Discovers Acetyl Coenzyme A	Biology	Massachusetts

Date	Event	Category	Country or region
1951	UNIVAC I Computer Is First to Use Magnetic Tape	Computer science	Pennsylvania
1951-1953	Watson and Crick Develop Double-Helix Model of DNA	Biology/Genetics	Great Britain
December 20, 1951	First Breeder Reactor Begins Operation	Energy	Idaho
1952	Bell Labs Introduces Electronically Amplified Hearing Aids	Communications	United States
1952	Wilkins Introduces Reserpine to Treat High Blood Pressure	Medicine	Massachusetts
1952	Aserinsky Discovers REM Sleep	Medicine	Illinois
1952-1956	Müller Invents Field Ion Microscope	Physics	Pennsylvania
February 23, 1952	Bevis Shows How Amniocentesis Can Reveal Fetal Problems	Medicine	Great Britain
July 2, 1952	Salk Develops First Polio Vaccine	Medicine	Pennsylvania
October 20, 1952	Kenya's Mau Mau Uprising Creates Havoc	Civil strife	Kenya
November 4, 1952	Eisenhower Is Elected U.S. President	National politics	United States
1953	De Vaucouleurs Identifies Local Supercluster of Galaxies	Astronomy	Space
1953	Du Vigneaud Synthesizes First Artificial Hormone	Medicine	New York
1953	Miller Synthesizes Amino Acids	Biology	Illinois
1953	Ziegler Develops Low-Pressure Process for Making Polyethylene	Materials	Germany
1953-1959	Cockerell Invents Hovercraft	Transportation	Great Britain
March 5, 1953	Stalin's Death Ends Oppressive Soviet Regime	Political reform	Soviet Union/Russia
May 6, 1953	Gibbon Develops Heart-Lung Machine	Medicine	Massachusetts
May 29, 1953	Hillary and Tenzing Are First to Reach Top of Mount Everest	Sports/International relations	Nepal
June 19, 1953	Rosenbergs Are First Americans Executed for Espionage During Peace Time	Human rights/Crime	New York
November 20, 1953- May 7, 1954	Battle of Dien Bien Phu Ends European Domination in Southeast Asia	War	Vietnam/France
Mid-1950's	Ochoa Creates Synthetic RNA Molecules	Biology/Genetics	New York
January 21, 1954	Atomic-Powered Submarine *Nautilus* Is Launched	Engineering/Military/ Technology	Connecticut
April 30, 1954	Barghoorn and Tyler Discover Two-Billion-Year-Old Microfossils	Earth science	Wisconsin/Massachusetts
May, 1954	Bell Labs Develops Photovoltaic Cell	Energy	New Jersey
May 17, 1954	U.S. Supreme Court Orders Desegregation of Public Schools	Social reform/ Education/Law	Washington, D.C.
September 8, 1954- February 19, 1955	Southeast Asia Treaty Organization Is Formed	International relations/ Military defense	Philippines/Australia/New Zealand/Great Britain/ Pakistan/Thailand/France/ Asia
1955	Ryle Constructs First Radio Interferometer	Astronomy	Great Britain
1955-1957	Backus Invents FORTRAN Computer Language	Computer science	New York
January 24-29, 1955	Congress Passes Formosa Resolution	International relations	Washington, D.C./Taiwan
May 14, 1955	East European Nations Sign Warsaw Pact	International relations	Eastern Europe
July 18-23, 1955	Geneva Summit Conference Opens East-West Communications	International relations	Switzerland
August 18-30, 1955	Sudanese Civil War Erupts	Civil war	Sudan
December 5, 1955	American Federation of Labor and Congress of Industrial Organizations Merge	Labor/Economics	New York

Date	Event	Category	Country or region
December 5, 1955- December 21, 1956	African Americans Boycott Montgomery, Alabama, Buses	Civil rights and liberties/ Social reform	Alabama
1956	First Transatlantic Telephone Cable Starts Operation	Communications	United States/Atlantic Ocean/ Europe
1956	Heezen and Ewing Discover Midoceanic Ridge	Earth science	Atlantic Ocean
1956-1961	Pottery Suggests Early Contact Between Asia and South America	Anthropology/ Archaeology	Ecuador/Asia/South America
February, 1956- October, 1964	Soviet-Chinese Dispute Creates Communist Rift	International relations/ Military capability	China/Soviet Union
February 24-25, 1956	Soviet Premier Khrushchev Denounces Stalin	National politics	Soviet Union
April-December, 1956	First Birth Control Pills Are Tested	Medicine	Puerto Rico
May, 1956	Mao Delivers His "Speech of One Hundred Flowers"	Civil rights and liberties	China
July 26, 1956	Egypt Seizes Suez Canal	International relations	Egypt
October-November, 1956	Soviets Crush Hungarian Uprising	Civil strife/Political reform	Hungary/Soviet Union
December 2, 1956- January 1, 1959	Castro Takes Power in Cuba	Military conflict/Coups/ Political reform	Cuba/Caribbean
1957	Calvin Explains Photosynthesis	Biology	California
1957	Isaacs and Lindenmann Discover Interferons	Medicine	Great Britain
1957	Sabin Develops Oral Polio Vaccine	Medicine	Ohio
1957	Esaki Demonstrates Electron Tunneling in Semiconductors	Physics	Japan
1957	Sony Develops Transistor Radio	Communications	Japan
1957	Velcro Is Patented	Materials	Switzerland
1957-1958	Eisenhower Articulates His Middle East Doctrine	International relations	Middle East/Lebanon
1957-1972	Senning Invents Implanted Heart Pacemaker	Medicine	Sweden
February-July, 1957	Bardeen, Cooper, and Schrieffer Explain Superconductivity	Physics	Illinois
March 6, 1957	Ghana Is First Black African Colony to Win Independence	Political independence	Ghana/Africa
March 25, 1957	Western European Nations Form Common Market	Economics/ International relations	Italy/Europe
August, 1957	Jodrell Bank Radio Telescope Is Completed	Astronomy	Great Britain
September, 1957	Federal Troops Help Integrate Little Rock's Central High School	Civil rights and liberties/ Education	Arkansas
October 4, 1957	Soviet Union Launches First Artificial Satellite, Sputnik	Space and aviation	Soviet Union/Earth orbit
October 22, 1957	Duvalier Takes Control of Haiti	Human rights/Coups/ Political reform	Haiti/Caribbean
December 2, 1957	United States Opens First Commercial Nuclear Power Plant	Energy	Pennsylvania
1958	Ultrasound Permits Safer Examination of Unborn Children	Medicine	Scotland
1958	Van Allen Discovers Earth's Radiation Belts	Earth science	Iowa
January 2, 1958	Parker Predicts Existence of Solar Wind	Astronomy	Illinois/Solar system
January 31, 1958	United States Launches Its First Satellite, Explorer 1	Space and aviation	Florida/Earth orbit
February 1, 1958	Egypt and Syria Join to Form United Arab Republic	International relations/ Government	Egypt

Date	Event	Category	Country or region
Spring, 1958	China's Great Leap Forward Causes Famine and Social Dislocation	Social reform	China
May 13-June 1, 1958	France's Fourth Republic Collapses	Political reform	France/Algeria
Summer, 1958	Cohen Begins Research on Neutron Bomb	Weapons technology	California
1959	Hopper Invents COBOL Computer Language	Computer science	Washington, D.C.
1959	IBM Model 1401 Computers Initiate Personal Computer Age	Computer science	United States
1959	Price Investigates Ancient Greek Astronomical Computer	Archaeology	Greece
1959	X-Ray Image Intensifier Makes Daylight Fluoroscopy Possible	Medicine	France
1959-1965	Congo Crisis Ends in Independence	Political independence/ Political reform	Congo (Kinshasa)
January 3, and August 21, 1959	Alaska and Hawaii Gain Statehood	Law/National politics	Alaska/Hawaii
June 26, 1959	St. Lawrence Seaway Opens	Engineering	Canada/United States
July 17, 1959	Leakeys Find 1.75-Million-Year-Old Hominid Fossil	Anthropology	Tanzania
July 21, 1959	First Commercial Atomic-Powered Ship, *Savannah*, Is Launched	Transportation	New Jersey
September 13, 1959	Luna 2 Is First Human-Made Object to Land on Moon	Space and aviation	Soviet Union/Moon/Space
October, 1959	Luna 3 Captures First Views of Far Side of Moon	Space and aviation	Earth orbit/Moon
Early 1960's	Sperry Conducts Split-Brain Experiments	Biology	California
1960's	American Society Becomes Politically Polarized	Social change	United States
1960's	Congress Passes Civil Rights Acts	Civil rights and liberties/ Law/Social reform	Washington, D.C.
1960's	Consumers Call for Safer Products	Social reform/ Economics/Business	United States
1960's	Drug Use Rises	Social change/Crime	United States
1960's	Iraq Tries to Exterminate Kurds	Human rights/Ethnic conflict	Iraq
1960's	Poverty Haunts Affluent U.S. Society	Economics/Social reform	United States
1960's	IBM 370 Computer Is First Virtual Machine	Computer science	Massachusetts
1960	Buehler Develops First Memory Metal, Nitinol	Materials	Maryland
1960	First Electronic Telephone Switching System Begins Operation	Communications	Illinois
1960	Friedman and Smith Discover How to Date Ancient Obsidian	Archaeology	Washington, D.C.
1960	International System of Units Is Adopted	Metrology/Mathematics	France
1960	Mössbauer Effect Helps Detect Gravitational Redshifting	Physics	Massachusetts
1960	Many New Nations Emerge During Africa's Year of Independence	Political independence	Africa
1960-1962	Hess Attributes Continental Drift to Seafloor Spreading	Earth science	New Jersey
1960-1969	German Measles Vaccine Is Developed	Medicine	Maryland/Pennsylvania
1960's-1970's	Chávez Organizes Farm Workers	Labor	California
1960-1976	Separatist Movement Grows in Quebec	National politics	Quebec

Date	Event	Category	Country or region
Spring, 1960	Oró Detects Chemical Reaction Pointing to Origin of Life	Biology	Texas
April 1, 1960	First Weather Satellite, Tiros 1, Is Launched	Space and aviation	Florida/Earth orbit
May, 1960	Soviets Capture American U-2 Spy Plane	International relations	Soviet Union/Washington, D.C.
July, 1960	Maiman Develops First Laser	Physics	California
August 12, 1960	First Passive Communications Satellite, Echo, Is Launched	Communications	Florida/Earth orbit
August 16, 1960	Cyprus Becomes Independent	Political independence	Cyprus
September 10-14, 1960	Organization of Petroleum Exporting Countries Is Formed	Economics	Iraq
November, 1960	Kennedy Is Elected U.S. President	National politics	United States
1961	Peace Corps Is Formed	Education/Economics	United States/Asia/Africa/South America
1961	Horsfall Discovers Cause of Cancer	Medicine	New York
1961	Nirenberg's Experimental Technique Cracks Genetic Code	Biology	Maryland
1961	Slagle Invents First Expert System, SAINT	Computer science/ Mathematics	Massachusetts
April 12, 1961	Gagarin Is First Human to Orbit Earth	Space and aviation	Soviet Union/Earth orbit
April 17-19, 1961	Cuba Repels Bay of Pigs Invasion	Military conflict/Coups	Cuba/Caribbean
May 5, 1961	Shepard Is First American in Space	Space and aviation	Florida/Earth orbit
August 13, 1961	Soviets Build Berlin Wall	International relations/ Political reform	Germany/Soviet Union
1962	Atlas-Centaur Rocket Is Developed	Space and aviation	Ohio
1962	Unimates, First Industrial Robots, Are Mass-Produced	Engineering	United States
1962	Giacconi Discovers X-Ray Sources Outside Solar System	Astronomy	Space
1962	Lasers Are Used in Eye Surgery	Medicine	New York
1962-1964	Supreme Court Rules on Reapportionment Cases	Law/National politics	Washington, D.C.
1962-1967	Researchers Reconstruct Ancient Near Eastern Trade Routes	Archaeology	Great Britain
February 20, 1962	Glenn Is First American to Orbit Earth	Space and aviation	Florida/Earth orbit
July 3, 1962	Algeria Wins Its Independence	Political independence	Algeria
July 10, 1962	Telstar Relays Live Transatlantic Television Pictures	Communications/ Entertainment	Maine/France/Earth orbit
August, 1962-January, 1963	Mariner 2 Is First Spacecraft to Study Venus	Space and aviation	Space/Venus
September 27, 1962	Carson's *Silent Spring* Warns of Pesticides	Biology	Maryland
October, 1962	Cuban Missile Crisis Fuels Fears of War	Military capability/ International relations	Washington, D.C.
October, 1962- December, 1965	Second Vatican Council Meets	Religion	Italy/Vatican
October 1, 1962	Meredith Integrates University of Mississippi	Civil rights and liberties	Mississippi
1963	Audiocassette Recording Is Introduced	Communications/ Entertainment	Netherlands
1963	Cohen Proves Independence of Continuum Hypothesis	Mathematics	California
1963	Schmidt Discovers Quasars	Astronomy	Space
1963	Lorenz Demonstrates Chaotic Dynamics in Weather Systems	Earth science	Massachusetts

Date	Event	Category	Country or region
1963-1965	Penzias and Wilson Discover Cosmic Microwave Background Radiation	Astronomy	Space
March 18, 1963	*Gideon v. Wainwright* Defines Rights of Accused Criminals	Law	Washington, D.C.
April-May, 1963	U.S. Civil Rights Protesters Attract World Attention	Civil rights and liberties/ Social reform	Alabama
May 25, 1963	Organization of African Unity Is Founded	International relations	Ethiopia
June 17, 1963	Supreme Court Rules Against Bible Reading in Public Schools	Education/Law	Washington, D.C.
August 28, 1963	Martin Luther King, Jr., Delivers "I Have a Dream" Speech	Civil rights and liberties/ Social reform	Washington, D.C.
September 24, 1963	Nuclear Test Ban Treaty Is Signed	International relations/ Military defense	Washington, D.C./Switzerland/ Great Britain
November 22, 1963	President Kennedy Is Assassinated	Assassination	Texas
December, 1963	Ethnic Conflicts Split Republic of Cyprus	Civil strife/Ethnic Conflict	Cyprus
1964	United States Enters Vietnam War	War/Civil war	Vietnam/United States
1964	Gell-Mann and Zweig Postulate Existence of Quarks	Physics	California/Switzerland
1964	Moog and Others Produce Portable Electronic Synthesizers	Communications/ Entertainment	United States
1964-1965	Kemeny and Kurtz Develop BASIC Computer Language	Computer science	New Hampshire
May, 1964	Refugees Form Palestine Liberation Organization	Human rights/Political independence	Middle East/Israel/Palestine
Summer, 1964- Summer, 1965	Doell and Dalrymple Discover Magnetic Reversals of Earth's Poles	Earth science	California
June 21, 1964	Three Civil Rights Workers Are Murdered in Mississippi	Civil rights and liberties	Mississippi
September-December, 1964	Students Revolt in Berkeley, California	Social change/Civil strife	California
October 1, 1964	First Bullet Train Begins Service in Japan	Transportation/ Engineering	Japan
October 13-14, 1964	Soviet Hard-Liners Oust Khrushchev	National politics	Soviet Union
November 3, 1964	Johnson Is Elected U.S. President	National politics	United States
November 21, 1964	New York's Verrazano-Narrows Bridge Opens	Engineering	New York
1965-1980	Zimbabwe Nationalist Movements Arise in Rhodesia	Civil war	Rhodesia/Zimbabwe
1965-1980's	Civil War Ravages Chad	Civil war	Chad
February 21, 1965	Black Nationalist Leader Malcolm X Is Assassinated	Assassination/Civil rights and liberties	New York
March 18, 1965	Soviet Cosmonauts Make First Space Walk	Space and aviation	Soviet Union/Earth orbit
April 28, 1965	U.S. Troops Occupy Dominican Republic	International relations/ Civil war	Dominican Republic/Caribbean
August-October, 1965	Sealab 2 Expedition Concludes	Earth science	California
August 11, 1965	Race Riots Erupt in Los Angeles's Watts District	Civil rights and liberties/ Civil strife	California
November 11, 1965	Rhodesia Unilaterally Declares Independence	Political independence	Rhodesia/Zimbabwe
December, 1965	Gemini 6 and 7 Rendezvous in Orbit	Space and aviation	Florida/Earth orbit
1966	First Tidal Power Plant Begins Operation	Energy	France
1966-1976	China's Cultural Revolution Starts Wave of Repression	Civil strife/Social reform	China

3295

Date	Event	Category	Country or region
1966-1980	Shock Waves Are Used to Pulverize Kidney Stones	Medicine	Germany
1966-1991	Soviet Intellectuals Disagree with Party Policy	Human rights	Soviet Union
January, 1966	Simons Identifies Thirty-Million-Year-Old Primate Skull	Anthropology	Egypt
February-March, 1966	France Withdraws from NATO	International relations	France/Europe
February 3, 1966	Luna 9 Soft-Lands on Moon	Space and aviation	Soviet Union/Moon/Space
March 1, 1966	Venera 3 Lands on Venus	Space and aviation	Venus/Space
June, 1966	National Organization for Women Is Formed	Social reform/Civil rights and liberties	Washington, D.C.
July 1, 1966	U.S. Government Begins Medicare Program	Health/Social reform	United States
August 10-October 29, 1966	Lunar Orbiter Sends Images from Moon's Surface	Space and aviation	Moon/Space
1967	Davis Constructs Giant Neutrino Detector	Astronomy	South Dakota
1967	Favaloro Develops Coronary Artery Bypass Surgery	Medicine	Ohio
1967	Dolby Invents Noise Reducer for Music Recording	Communications/Entertainment	United States
1967	Manabe and Wetherald Warn of Global Warming	Earth science	New Jersey
1967-1968	Amino Acids Are Found in Three-Billion-Year-Old Rocks	Earth science	Massachusetts/California
1967-1970	Ethnic Conflict Leads to Nigerian Civil War	Ethnic Conflict/Civil war	Nigeria
April 21, 1967	Greek Coup Leads to Military Dictatorship	Political reform/Coups	Greece
June 5-10, 1967	Israel Fights Arab Nations in Six-Day War	War	Middle East/Israel/Egypt/Jordan/Syria
August-September, 1967	Kornberg Discovers How to Synthesize DNA	Biology/Genetics	California
October 2, 1967	Marshall Becomes First African American Supreme Court Justice	Government/Law/National politics	Washington, D.C.
November, 1967-February, 1968	Bell Discovers Pulsars	Astronomy	Space
November 9, 1967	Saturn 5 Booster Rocket Is Tested at Cape Kennedy	Space and aviation	Florida/Earth orbit
December, 1967	Barnard Performs First Human Heart Transplant	Medicine	South Africa
1968	*Glomar Challenger* Explores Ocean Floor	Earth science	Atlantic Ocean/Pacific Ocean
1968	International Practical Temperature Scale Is Established	Metrology	France
1968	Friedman, Kendall, and Taylor Discover Quarks	Physics	California
1968	Wheeler Coins Term "Black Hole"	Astronomy	New Jersey/Space
1968 to the present	International Terrorism Increases	Terrorism	World
1968-1977	Alaska Pipeline Is Built	Economics	Alaska
1968-1989	Brezhnev Doctrine Limits Soviet Satellites' Independence	Political aggression	Soviet Union
January 23, 1968	North Korea Seizes U.S. Navy Ship *Pueblo*	International relations/War/Military	North Korea
January 31-March 31, 1968	Viet Cong Lead Tet Offensive in Vietnam	War	Vietnam
April 4, and June 4, 1968	Martin Luther King, Jr., and Robert F. Kennedy Are Assassinated	Assassination	Tennessee/California

Date	Event	Category	Country or region
July 25, 1968	Pope Paul VI Denounces Contraception	Religion/Social change	Vatican
August 20-21, 1968	Soviet Union Invades Czechoslovakia	Political reform/ Political aggression/ Military conflict	Czechoslovakia/Soviet Union
August 24-30, 1968	Chicago Riots Mar Democratic Party Convention	Civil strife	Illinois
September 27, 1968	Caetano Becomes Premier of Portugal	Political reform	Portugal
October, 1968	First Cameron Press Begins Operation	Engineering/ Communications	Tennessee
November 5, 1968	Nixon Is Elected U.S. President	National politics	United States
December 13, 1968	Brazil Begins Period of Intense Repression	Political reform	Brazil
1969	Bubble Memory Is Touted as Ideal Information Storage Device	Computer science	New Jersey
1969	Internetting Project Begins	Computer science	California
1969	Soyuz 4 and Soyuz 5 Dock in Orbit	Space and aviation	Earth orbit
1969-1970	Jumbo Jet Passenger Service Is Introduced	Space and aviation/ Transportation	United States/Europe
1969-1974	Very Long Baseline Interferometry Is Developed	Astronomy	Great Britain
1969-1979	Congress Launches Environmental Legislation	Law/Environment	Washington, D.C.
1969-1979	United States and Soviet Union Conduct SALT Talks	International relations/ Military	United States/Soviet Union/ Europe
1969-1981	United States and Soviet Union Approach Détente	International relations	United States/Soviet Union
1969-1983	Optical Disks Revolutionize Computer Data Storage	Computer science/ Photography	Netherlands
March, 1969	Soviet Union and China Wage Border War	Political aggression/ War	China/Soviet Union
April 28, 1969	France's President De Gaulle Steps Down	Political reform	France
June 28, 1969	Stonewall Rebellion Launches Gay Rights Movement	Civil rights and liberties/Gender issues	New York
July 20, 1969	Humans Land on Moon	Space and aviation	Space/Moon
August 15-17, 1969	Woodstock Festival Marks Cultural Turning Point	Entertainment	New York
October, 1969	Brandt Wins West German Elections	National politics	Germany
1970	U.S. Voting Age Is Lowered to Eighteen	Political reform	Washington, D.C.
1970	IBM Introduces Computer Floppy Disk	Computer science	California
1970	Philips Introduces Home Videocassette Recorder	Communications/ Entertainment	Great Britain
1970-1975	Hybrid Mouse-Human Cells Make Pure Antibodies	Biology	Great Britain
1970-1980	Virtual Reality Becomes Reality	Computer science	Utah
April-June, 1970	United States Invades Cambodia	Political aggression/ War	United States/Cambodia/ Vietnam
May 4, 1970	National Guard Kills Demonstrators at Kent State University	National politics/Social reform/War	Ohio
November 17, 1970- October 1, 1971	Soviet Union Lands Self-Propelled Lunokhod on Moon	Space and aviation	Soviet Union/Moon
1971	Computer Chips Are Introduced	Computer science	California
1971	Direct Transoceanic Telephone Dialing Begins	Communications	United States/Europe

Date	Event	Category	Country or region
1971	Oral Rehydration Therapy Reduces Diarrhea Deaths	Medicine/Health	Bangladesh
1971-1972	Mariner 9 Orbits Mars	Space and aviation	Mars/Space
January-December, 1971	India-Pakistan War Leads to Creation of Bangladesh	War/Political independence	Pakistan/India
January 25, 1971	Amin's Regime Terrorizes Uganda	Political reform/Coups/ Human rights	Uganda
April 20, 1971	Supreme Court Approves Busing to Desegregate Schools	Civil rights and liberties/ Law/National politics	Washington, D.C.
August 15, 1971	United States Lowers Value of Dollar	Economics	Washington, D.C.
August 15, 1971- June, 1974	International Monetary System Is Reformed	Economics	Washington, D.C.
October 25, 1971	United Nations Admits People's Republic of China	International relations	New York
November 27, 1971	Mars 2 Lands on Mars	Space and aviation	Mars/Space
1972	Congress Passes Equal Employment Opportunity Act	Civil rights and liberties/ Labor	Washington, D.C.
1972	Gell-Mann Formulates Theory of Quantum Chromodynamics	Physics	California
1972	Hounsfield Develops CAT Scanner	Medicine	Great Britain
1972	Janowsky Explains Manic Depression	Medicine	Tennessee
1972-1982	Equal Rights Amendment Passes Congress but Is Not Ratified	Civil rights and liberties	United States
January 30, 1972	Irish and British Clash in Ulster's Bloody Sunday	Civil strife/Political reform	Ireland
February 21, 1972	United States and China Reopen Relations	International relations	China
March 2, 1972-1983	Pioneer 10 Explores Outer Planets	Space and aviation	Solar system
September, 1972	Philippines President Marcos Declares Martial Law	Political reform/Human rights	Philippines
September, 1972	Texas Instruments Introduces Pocket Calculators	Computer science	Texas
September 5, 1972	Arab Terrorists Kill Eleven Israeli Athletes at Munich Olympics	Terrorism/Crime/ International relations	Germany
December 31, 1972	U.S. Government Bans Insecticide DDT	Earth science	United States
May 27, 1905	Cohen and Boyer Develop Recombinant DNA Technology	Biology/Genetics	California
January 1, 1973	Great Britain Joins Common Market	Economics	Great Britain/Europe
January 22, 1973	Supreme Court's *Roe v. Wade* Decision Legalizes Abortion	Law/Social change	Washington, D.C.
January 27, 1973	U.S. Military Forces Withdraw from Vietnam	War/International relations	Vietnam/France
February 27-May 8, 1973	Native Americans Take Stand at Wounded Knee	Civil strife/Civil rights and liberties	South Dakota
May 14, 1973	Skylab Space Station Is Launched	Space and aviation	United States/Earth orbit
June 21, 1973	East and West Germany Establish Relations	International relations	Germany
June 27, 1973	Oppressive Military Rule Comes to Democratic Uruguay	Political reform	Uruguay
September 11, 1973	Chile's President Allende Is Overthrown	Civil strife/Coups/ Political reform	Chile
October, 1973	Arabs and Israelis Clash in Yom Kippur War	War	Israel/Egypt/Syria
October, 1973-March, 1974	Arab Oil Embargo Precipitates Energy Crisis	Economics/Energy	Middle East/United States
October 10, 1973	Vice President Agnew Resigns After Being Charged with Tax Evasion	National politics/Crime	Washington, D.C.

Date	Event	Category	Country or region
November 3, 1973- March 24, 1975	Mariner 10 Uses Gravity of Venus to Reach Mercury	Space and aviation	Venus/Mercury/Space
February, 1974	Ethiopian Emperor Haile Selassie Is Overthrown	Political reform/Civil war/Coups	Ethiopia
February, 1974	Georgi and Glashow Propose First Grand Unified Theory	Physics	Massachusetts
February-March, 1974	Organic Molecules Are Found in Comet Kohoutek	Biology	Germany
February 4, 1974	Symbionese Liberation Army Kidnaps Patty Hearst	Crime/Terrorism	California
June, 1974	Rowland and Molina Warn of CFC Gas Threat to Ozone Layer	Earth science/Environment	California
July 15-August 16, 1974	Greeks Stage Coup on Cyprus	Political aggression/Coups	Cyprus
August 9, 1974	Watergate Affair Forces President Nixon to Resign	National politics	Washington, D.C.
November 30, 1974	Johanson Discovers "Lucy," Three-Million-Year-Old Hominid Skeleton	Anthropology	Ethiopia
December 12, 1974	United Nations Adopts Economic Rights Charter	International relations/Economics	New York
1975	Cambodia Falls to Khmer Rouge	Civil war/Political reform	Cambodia
1975-1976	Cuba Intervenes in Angola's Civil War	Civil war/International relations	Angola/Cuba
1975-1979	Philips Corporation Develops Laser-Diode Recording Process	Communications	Netherlands
1975-1990	Fax Machines Become Common Office Equipment	Communications	United States
January 7-April 30, 1975	South Vietnam Falls to North Vietnam	Civil war/Political reform	Vietnam
April, 1975-September, 1990	Lebanese Civil War Redistributes Sectarian Power	Civil war	Lebanon
May 12, 1975	Cambodia Seizes U.S. Merchant Ship *Mayaguez*	War/International relations	Cambodia
July 17, 1975	Apollo-Soyuz Rendezvous Inaugurates International Cooperation in Space	Space and aviation	Earth orbit
August 1, 1975	Helsinki Agreement Seeks to Reduce East-West Tensions	International relations	Finland
September, 1975	President Ford Survives Two Assassination Attempts	Assassination/Politics	California
October, 1975	Venera Space Probes Transmit Pictures from Venus	Space and aviation	Soviet Union/Space/Venus
November 20, 1975	Franco's Death Returns Spain to Democracy	Political reform	Spain
1976	Kibble Proposes Theory of Cosmic Strings	Astronomy	Great Britain/Space
1976	Seymour Cray Creates First Supercomputer	Computer science	Wisconsin
1976-1979	Argentine Military Conducts "Dirty War" Against Leftists	Human rights	Argentina
1976-1988	Radar-Invisible Stealth Aircraft Are Introduced	Weapons technology	California
January 8, and September 9, 1976	Deaths of Mao and Zhou Rock China's Government	Political reform	China
June 16, 1976	South Africa's Black Nationalists Mobilize	Civil strife	South Africa
July 2, 1976	Supreme Court Reverses Itself on Death Penalty	Law	Washington, D.C.
July 4, 1976	Israeli Commandos Rescue Hijack Hostages in Entebbe, Uganda	Terrorism/International relations	Uganda/Israel

Date	Event	Category	Country or region
July 20, 1976, and September 3, 1976	Viking Spacecraft Transmit Pictures from Mars	Space and aviation	Mars/Space
November 4, 1976	Carter Is Elected U.S. President	National politics	United States
1977	Mandelbrot Introduces Fractals	Mathematics	New York
1977	Oceanographers Discover Deep-Sea Vents and Strange Life-Forms	Earth science	Pacific Ocean
1977-1978	Communists Struggle for Africa's Horn	International relations	Africa/Ethiopia/Somalia/Cuba
1977-1980	Radar Images Reveal Ancient Mayan Canals	Archaeology	Guatemala/Central America
1977-1985	Modern Cruise Missiles Are Introduced	Weapons technology	United States
January, 1977	United States Pardons Vietnam War Draft Evaders	Civil rights and liberties	Illinois
January 1, 1977	Episcopal Church Ordains Its First Woman Priest	Gender issues/Religion	Indiana
April, 1977	Preassembled Apple II Computer Is Introduced	Computer science/ Business	California
May 11, 1977	AT&T Tests Fiber-Optic Telecommunications	Communications	Illinois
August 20, 1977-1989	Voyagers 1 and 2 Explore Solar System	Space and aviation	Jupiter/Saturn/Solar system
September 16, 1977	Gruentzig Uses Angioplasty to Unclog Arteries	Medicine/Health	Switzerland
November 4, 1977	United Nations Imposes Mandatory Arms Embargo on South Africa	International relations	South Africa
November 19-21, 1977	Egyptian President Sadat Is First Arab Leader to Visit Israel	International relations	Israel/Middle East/Egypt
1978	Toxic Waste Is Discovered at New York's Love Canal	Environment	New York
1978	Nuclear Magnetic Resonance Produces Brain Images	Medicine	Great Britain
1978	First Cellular Telephone System Is Tested	Communications	Illinois/New Jersey
1978-1979	Islamic Revolutionaries Overthrow Iran's Shah	Coups/Religion/ Political reform	Iran
1978-1981	Rohrer and Binnig Invent Scanning Tunneling Microscope	Physics	Switzerland
1978-1985	Guatemalan Death Squads Target Indigenous Indians	Civil strife/Human rights	Guatemala/Central America
March 16, and April 18, 1978	Panama Canal Treaties Are Ratified	International relations	Washington, D.C.
July 25, 1978	English Woman Delivers First Test-Tube Baby	Medicine	Great Britain
August 12, 1978	China and Japan Sign Ten-Year Peace Treaty	International relations	China
October 16, 1978	Poland's Wojtyła Becomes First Non-Italian Pope Since 1522	Religion	Vatican/Poland
November 18, 1978	People's Temple Cultists Commit Mass Suicide in Guyana	Religion	Guyana
December 8, 1978	Compressed-Air-Accumulating Power Plant Opens in Germany	Energy	Germany
1979	Vietnam Conquers Cambodia	Political aggression	Cambodia
1979	Philips Corporation Proposes Compact Disc	Communications	Netherlands
1979-1980	World Meteorological Organization Starts World Climate Programme	Earth science	Switzerland
1979-1989	Ayatollah Khomeini Rules Iran	Human rights/Political reform	Iran
1979-1990	Scientists Implicate P53 Gene in Many Cancers	Biology/Genetics/ Medicine	United States

3300

Date	Event	Category	Country or region
January 1, 1979	United States and China Establish Full Diplomatic Relations	International relations	United States/China
February 17, 1979	China Invades Vietnam	War	Vietnam
March 4-7, 1979	Voyager 1 Discovers Jupiter's Ring	Astronomy	Space/Solar system/Jupiter
March 26, 1979	Egypt and Israel Sign Peace Treaty	International relations	Egypt/Israel/Middle East
March 28, 1979	Nuclear Accident Occurs at Three Mile Island	Environment/Social reform	Pennsylvania
May 4, 1979	Thatcher Becomes Great Britain's First Woman Prime Minister	National politics	Great Britain
June 7-10, 1979	Direct Election to European Parliament Begins	International relations	Europe
June 18, 1979	United States and Soviet Union Sign SALT II Agreement	International relations	Switzerland/United States/Soviet Union
July 17, 1979	Sandinistas Overthrow Nicaragua's President Somoza	Civil war/Coups/Political reform	Nicaragua/Central America
August, 1979	Archaeologists Find Ancient Sanctuary in Spain	Archaeology	Spain
October 15, 1979	Salvadoran Coup Topples Romero Dictatorship	Political reform/Coups	El Salvador/Central America
November 4, 1979	Iranian Revolutionaries Take Americans Hostage	Civil strife/International relations	Iran
December 9, 1979	World Health Organization Announces Eradication of Smallpox	Health/Medicine/Biology	Switzerland/Somalia
December 24-27, 1979	Soviet Union Invades Afghanistan	Political aggression	Afghanistan
1980's	Tamil Separatist Violence Erupts in Sri Lanka	Civil war	Sri Lanka
1980	Spain Grants Home Rule to Basques	Political reform/Civil strife	Spain
1980	South Africa Invades Angola	Political aggression	Angola
1980	Gilbert and Sanger Develop DNA Sequencing Techniques	Biology/Genetics	United States/Great Britain
1980	Guth Proposes Inflationary Model of Universe	Astronomy	California/Space
1980	Paleontologists Postulate Asteroid That Caused Extinction of Dinosaurs	Earth science	Italy
January 9, 1980	Saudi Arabia Executes Radical Muslims for Attack on Mecca	Religion/Civil strife	Saudi Arabia
April 12, 1980	Liberian Soldiers Overthrow Americo-Liberian Regime	Political reform/Coups	Liberia
April 18, 1980	Rhodesia Becomes Independent as Zimbabwe	Political independence/Political reform	Zimbabwe
May, 1980	International Monetary Fund Predicts Worldwide Recession	Economics	Washington, D.C.
May, 1980	Pluto's Atmosphere Is Discovered	Astronomy	Pluto/Solar system/Space
August 30, 1980	Abscam Trial Ends in Convictions	National politics/Political reform	New York
September 17, 1980	Solidarity Is Founded in Poland	Labor/Political reform	Poland
September 22, 1980	Iran-Iraq Conflict Escalates into War	War	Iran/Iraq/Middle East
November 4, 1980	Reagan Is Elected U.S. President	National politics	Washington, D.C.
1981	U.S. Centers for Disease Control Recognizes Existence of AIDS	Medicine	United States/Africa/Congo (Kinshasa)
January 25, 1981	China Sentences "Gang of Four"	Political reform	China
March 30, 1981	Would-be Assassin Shoots President Reagan	Assassination/National politics	Washington, D.C.
May 10, 1981	Mitterrand Is Elected France's President	National politics	France

Date	Event	Category	Country or region
June 7, 1981	Israel Destroys Iraqi Nuclear Reactor	Military/International relations	Iraq
August 12, 1981	IBM Introduces DOS-Based Personal Computer	Computer science	Florida
August 13, 1981	Reagan's Budget and Tax Plans Become Law	Economics	Washington, D.C.
September 25, 1981	O'Connor Becomes First Woman on Supreme Court	Civil rights and liberties/ Law	Washington, D.C.
October 6, 1981	Egypt's President Sadat Is Assassinated	Assassination	Egypt
October 22, 1981	Professional Air Traffic Controllers Are Decertified	Labor	Washington, D.C.
November 12-14, 1981	*Columbia* Flight Proves Practicality of Space Shuttle	Space and aviation	Florida/Earth orbit
1982	Great Britain Wins Falkland Islands War	War	Falkland Islands/Argentina/Great Britain
1982	Antarctic Survey Team Confirms Ozone Depletion	Earth science/ Environment	Antarctica
1982	DeVries Implants First Jarvik-7 Artificial Heart	Medicine	Utah
1982	Exploratory Oil Well Is Drilled at Record Depth	Engineering	France
1982	First Household Robot, Hero 1, Is Invented	Engineering	United States
1982	Baulieu Develops Abortion Pill	Medicine/Health	France
1982 and 1989	Astronomers Discover Neptune's Rings	Astronomy	New Zealand/Solar system/ Neptune
1982-1983	Compact Disc Players Are Introduced	Communications	Japan/Europe
1982-1986	Congress Bans Covert Aid to Nicaraguan Contras	Military/International relations	Washington, D.C./Nicaragua/ Central America
March 15, 1982	Nicaragua Suspends Its Constitution	National politics/Civil right and liberties	Nicaragua/Central America
March 28, 1982	El Salvador Holds Elections	National politics	El Salvador/Central America
April, 1982	Solar One Power Plant Begins Operation	Energy	California
April 17, 1982	Canadian Charter of Rights and Freedoms Is Enacted	Civil rights and liberties	Canada
April 30, 1982	Law of the Sea Treaty Is Completed	International relations/ International law	New York
May 14, 1982	Eli Lilly Markets Genetically Engineered Human Insulin	Medicine/Genetics	Indiana
September 6, 1982	Salvadoran Troops Are Accused of Killing Civilians	Civil strife	El Salvador/Central America
September 16-18, 1982	Lebanese Phalangists Massacre Palestinians in West Beirut	Human rights	Lebanon/Palestine
1983	Peru Faces State of Emergency	National politics/ Economics	Peru
1983	FDA Approves Artificial Sweetener Aspartame	Food science	Washington, D.C.
1983	Rubbia and Van Der Meer Discover Three New Particles	Physics	Switzerland
1983-1984	Woman Bears Transplanted Child	Biology/Medicine	California
March 8, 1983	IBM Introduces Personal Computers with Hard Drives	Computer science	Florida
April 4, 1983	NASA Places First TDRS in Orbit	Communications	United States/Earth orbit
June 18, 1983	Ride Becomes First American Woman in Space	Space and aviation/ Gender issues	Florida/Earth orbit
July 21, 1983	Martial Law Ends in Poland	Political reform/Civil rights and liberties	Poland

Date	Event	Category	Country or region
August 21, 1983	Philippine Opposition Leader Aquino Is Assassinated	Assassination	Philippines
September, 1983	Murray and Szostak Create First Artificial Chromosome	Biology/Genetics	Massachusetts
September 1, 1983	Soviet Union Downs South Korean Airliner	International relations/ Transportation/ Disasters	Soviet Union/Russia
October, 1983	Laser Vaporization Is Used to Treat Blocked Arteries	Medicine/Health	France
October 25, 1983	United States Invades Grenada	Political aggression	Grenada/United States/ Caribbean
November 3, 1983	Jackson Is First Major Black Presidential Candidate	National politics	Washington, D.C.
November 28, 1983	Spacelab 1 Is Launched	Space and aviation	Europe/Earth orbit
December 31, 1983	Nigerian Government Is Overthrown	Political reform/Coups/ Civil strife	Nigeria
1984	Sikhs Revolt in India's Punjab State	Civil strife/Ethnic conflict	India
1984	Iran-Iraq War Escalates	War/International relations	Iran/Iraq/Middle East
1984	Humans and Chimpanzees Are Found to Be Genetically Linked	Anthropology/Genetics	Connecticut
1984	Willadsen Splits Sheep Embryos	Biology/Genetics	Great Britain
1984-1987	Twelve European Community States Resolve to Create True Common Market	Economics	Europe
January 1, 1984	AT&T Divestiture Takes Effect	Communications/ Business	United States
May 6, 1984	El Salvador Has First Military-Free Election in Fifty Years	National politics	El Salvador/Central America
May 10, 1984	World Court Orders U.S. to Stop Blockading Nicaragua	International relations	Central America/Nicaragua
September 26, 1984	Great Britain Promises to Leave Hong Kong in 1997	International relations	China/Hong Kong
October 20, 1984	China Adopts New Economic System	Economics	China
October 31, 1984	India's Prime Minister Gandhi Is Assassinated	Assassination	India
November 22, 1984	Ethiopian Famine Gains International Attention	Economics/ International relations	Ethiopia/Africa
December 2 3, 1984	Union Carbide Gas Leak Kills Hundreds in Bhopal, India	Environment/Disasters/ Medicine	India
1985	Buckminsterfullerene Is Discovered	Chemistry	Texas
1985	Construction of Keck Telescope Begins on Hawaii	Astronomy	Hawaii
1985	Norway Begins First Reactor to Use Wave Energy	Energy	Norway
1985	Scientists Discover Hole in Ozone Layer	Earth Science/ environment	Antarctica
1985-1987	Iran-Contra Scandal Taints Reagan Administration	National politics/ International relations	Iran/Nicaragua
March 6, 1985	Jeffreys Produces Genetic "Fingerprints"	Biology/Genetics	Great Britain
March 11, 1985	Gorbachev Becomes General Secretary of Soviet Communist Party	National politics	Soviet Union
March 16, 1985	Japan's Seikan Tunnel Is Completed	Engineering/ Transportation	Japan

Date	Event	Category	Country or region
July 10, 1985	French Agents Sink Greenpeace Ship *Rainbow Warrior*	Environment/ International relations	New Zealand
October, 1985	Tevatron Accelerator Begins Operation	Physics	Illinois
October 7, 1985	Palestinians Seize *Achille Lauro* Cruise Ship	International relations	Mediterranean Sea/Palestine
November 15, 1985	Great Britain Approves Ulster Pact	International relations/ Terrorism	Ireland/Northern Ireland/Great Britain
November 19, 1985	U.S.-Soviet Summit Begins in Geneva	International relations	Switzerland
December 9, 1985	Argentine Leaders Are Convicted of Human Rights Violations	Human rights/Military	Argentina
1986-1987	Tully Discovers Pisces-Cetus Supercluster Complex	Astronomy	Space
January 27, 1986	Müller and Bednorz Discover Superconductivity at High Temperatures	Physics	Switzerland
January 28, 1986	Space Shuttle *Challenger* Explodes on Launch	Technology/Disasters/ Economics	Florida
February 7, 1986	Haiti's Duvalier Era Ends	Political reform/Coups	Haiti/Caribbean
February 20, 1986	Mir Space Station Is Launched	Space and aviation	Soviet Union/Earth orbit
February 25, 1986	President Ferdinand Marcos Flees Philippines	Political reform	Philippines
March 24-25, 1986	U.S. and Libyan Forces Clash over Gulf of Sidra	International relations	Mediterranean Sea
April 26, 1986	Accident at Ukraine's Chernobyl Nuclear Plant Endangers Europe	Environment/Disasters	Soviet Union/Ukraine
June 1, 1986	United Nations Drafts Plan for African Economic Recovery	Economics	Africa
September 26, 1986	Rehnquist Becomes Chief Justice of the United States	Law	Washington, D.C.
December 14-23, 1986	Rutan and Yeager Fly *Voyager* Nonstop Around World	Space and aviation	California
December 17-19, 1986	Muslims Riot Against Russians in Kazakhstan	Civil strife	Soviet Union/Kazakhstan/Russia
1987	Black Workers Strike in South Africa	Civil rights and liberties/ Labor	South Africa
1987-1988	Anthropologists Find Earliest Evidence of Modern Humans	Anthropology	Israel
February 10, 1987	Surgeon General Koop Urges AIDS Education	Medicine/Social change	Washington, D.C.
February 23, 1987	Supernova 1987A Supports Theories of Star Formation	Astronomy	Space
March 31, 1987	Gallo and Montagnier Are Recognized as Codiscoverers of HIV	Medicine	Maryland/France
May 17, 1987	Iraqi Missiles Strike U.S. Navy Frigate *Stark*	Military	Persian Gulf/Iraq/United States
July 31, 1987	Hundreds Die in Riots at Mecca's Great Mosque	International relations	Saudi Arabia/Middle East
September, 1987	Miller Discovers Oldest Known Embryo in Dinosaur Egg	Earth science	Utah
December, 1987	Palestinian Intifada Begins	Civil strife	Israel/Palestine
December 8, 1987	United States and Soviet Union Sign Nuclear Forces Agreement	International relations	Washington, D.C.
1988	Ethnic Riots Erupt in Armenia	Civil strife/Ethnic conflict	Soviet Union/Armenia/ Azerbaijan
1988	Hunger Becomes Weapon in Sudan's Civil War	Civil war/Human rights	Sudan
1988	"Peace" Virus Attacks Mac Computers	Computer science/ Business	United States
1988-1989	Recruit Scandal Rocks Japan's Government	National politics	Japan

3304

Date	Event	Category	Country or region
1988-1991	Pakistani Bank Is Charged with Money Laundering	International relations/ Economics	United States/Great Britain/ Pakistan
January 2, 1988	Canada and United States Sign Free Trade Agreement	Economics/ International relations	United States/Canada
July 3, 1988	U.S. Cruiser *Vincennes* Shoots Down Iranian Airliner	Military Conflict/ International relations	Persian Gulf/Iran/United States
July 20, 1988	Ayatollah Khomeini Calls Halt to Iran-Iraq War	War	Iran/Iraq/Middle East
July 31, 1988	Jordan Abandons Its West Bank	International relations	Jordan/Israel/Palestine
August 10, 1988	Congress Formally Apologizes to Wartime Japanese Internees	Civil rights and liberties	Washington, D.C.
Late August, 1988	Iraq Uses Poison Gas Against Kurds	Human rights/Ethnic conflict	Iraq
October 1, 1988	Gorbachev Becomes President of Supreme Soviet	National politics	Soviet Union
October 22, 1988	Congress Passes Comprehensive Drug Bill	Law/Crime/Social change	Washington, D.C.
November, 1988	Bhutto Is First Woman Elected Leader of Muslim Country	National politics/Civil rights and liberties	Pakistan
November 8, 1988	George Bush Is Elected U.S. President	National politics	United States
December 22, 1988	Namibia Is Liberated from South African Control	Political independence	Namibia/Africa/South Africa
1989	Vietnamese Troops Withdraw from Cambodia	Political independence	Cambodia/Vietnam
1989	Soviet Troops Leave Afghanistan	Political aggression	Afghanistan/Soviet Union
1989	Kashmir Separatists Demand End to Indian Rule	Ethnic conflict/Political independence	India
1989	Hungary Adopts Multiparty System	Political reform	Hungary
1989-1990	Three Colombian Presidential Candidates Are Assassinated	Assassination	Colombia
1989-1991	Kenya Cracks Down on Dissent	Civil strife/Human rights	Kenya
1989-1992	Green and Burnley Solve Mystery of Deep Earthquakes	Earth science	California
January 4, 1989	U.S. Fighter Planes Down Two Libyan Jets	International relations/ Military conflict	Mediterranean Sea/Libya/United States
January 7, 1989	Japan's Emperor Hirohito Dies and Is Succeeded by Akihito	Government	Japan
March 24, 1989	*Exxon Valdez* Spills Crude Oil Off Alaska	Environment/Disasters	Alaska
April, 1989	Solidarity Regains Legal Status in Poland	Labor/Political reform	Poland
April-June, 1989	China Crushes Prodemocracy Demonstrators in Tiananmen Square	Civil rights and liberties	China
May 7, 1989	Fraud Taints Panama's Presidential Election	National politics	Panama/Central America
June 3, 1989	Iran's Ayatollah Khomeini Dies	National politics	Iran
October, 1989-January, 1990	Panamanian President Noriega Survives Bungled Coup	Civil strife/Coups/ National politics	Panama/Central America
November 9, 1989	Berlin Wall Falls	Political reform	Germany
December 14, 1989	Chileans Vote Pinochet Out of Power	Political reform	Chile/South America
December 23, 1989	Romanian President Ceausescu Is Overthrown	Political reform/Coups	Romania
December 29, 1989	Havel Is Elected President of Czechoslovakia	Political reform	Czechoslovakia
1990	Political Reforms Come to Albania	Political reform	Albania

3305

Date	Event	Category	Country or region
February 11, 1990	South African Government Frees Mandela	Civil rights and liberties/ Political reform	South Africa
February 25, 1990	Sandinistas Lose Nicaraguan Elections	National politics	Nicaragua/Central America
February 26, 1990	Soviet Troops Withdraw from Czechoslovakia	International relations	Czechoslovakia
June 11, 1990	Supreme Court Overturns Law Banning Flag Burning	Civil rights and liberties/ Law	Washington, D.C.
June 12, 1990	Algeria Holds First Free Multiparty Elections	Political reform	Algeria
August 2, 1990	Iraq Occupies Kuwait	War/Political aggression	Iraq/Kuwait
September 10, 1990	Cambodians Accept U.N. Peace Plan	Civil war/International relations	Asia/Indonesia/Cambodia
September 14, 1990	First Patient Undergoes Gene Therapy	Medicine/Genetics/ Health	Maryland
October 1, 1990	Biologists Initiate Human Genome Project	Biology/Genetics	United States
October 3, 1990	East and West Germany Reunite	Political reform	Germany
November 22, 1990	Thatcher Resigns as British Prime Minister	Government	Great Britain
December 9, 1990	Wałęsa Is Elected President of Poland	National politics/ Political reform	Poland
1991	Ethnic Strife Plagues Soviet Union	Civil strife/Ethnic conflict	Soviet Union
1991	Hostages Are Freed from Confinement in Lebanon	International relations	Lebanon
1991	Scientists Identify Crater Associated with Dinosaur Extinction	Earth science	Mexico
1991-1992	Civil War Rages in Yugoslavia	Civil war/Political independence	Yugoslavia
January 11 and 20, 1991	Soviet Union Tries to Suppress Baltic Separatism	Civil strife	Lithuania/Estonia/Soviet Union/ Latvia
February 1, 1991	South Africa Begins Dismantling Apartheid	Civil rights and liberties	South Africa
July 1, 1991	East European Nations Dissolve Warsaw Pact	International relations/ Military defense	Eastern Europe
August-December, 1991	Soviet Union Disintegrates	Political reform/Social reform	Soviet Union
September 27, 1991	President George Bush Announces Nuclear Arms Reduction	International relations/ Weapons technology	Washington, D.C.
October 15, 1991	Senate Confirms Thomas's Appointment to Supreme Court	National politics/Law	Washington, D.C.
October 30-November 4, 1991	Middle East Peace Talks Are Held in Madrid	International relations	Spain/Middle East
1992	European Economic Community Tries to Create Single Market	Economics/ International relations	Belgium
1992	Scientists Find Strong Evidence for "Big Bang"	Astronomy	Space
1992	Astronomers Detect Planets Orbiting Another Star	Astronomy	Space
January 15, 1992	Yugoslav Federation Dissolves	Political independence	Yugoslavia
January 16, 1992	El Salvador's Civil War Ends	Civil war	El Salvador/Central America
January 18, 1992	Kenyans Demonstrate Against Moi's Government	Civil rights and liberties/ National politics	Kenya
February-April, 1992	Muslim Refugees Flee Persecution in Myanmar	Human rights/Religion	Myanmar/Bangladesh
February 1, 1992	Russia and United States End Cold War	International relations	Maryland
February 1, 1992	United States Resumes Repatriating Haitian Boat People	International relations/ Law	Caribbean/Haiti

3306

Date	Event	Category	Country or region
February 6, 1992	International Agencies Pledge Funds for Ethiopia	Social reform/ International relations	Ethiopia
February 16, 1992	Israelis Assassinate Hezbollah Leader	Assassination	Lebanon/Israel
March 17, 1992	Civil War Intensifies in Sudan	Civil war	Sudan
March 17, 1992	South Africans Vote to End Apartheid	National politics/Civil rights and liberties	South Africa
March 25, 1992	Kurds Battle Turkey's Government	Military conflict/Ethnic conflict	Turkey/Iraq
Spring, 1992	Commonwealth of Independent States Faces Economic Woes	Government/Politics	Russia/Soviet Union/Belarus/ Ukraine/Georgia/Moldova/ Tajikistan
April 6, 1992	Nepalese Police Fire on Kathmandu Protesters	Civil strife/Social reform/Economics	Nepal
April 10, 1992	Keating Is Sentenced for Securities Fraud	Law/Business	California
April 15, 1992	United Nations Allows Sanctions Against Libya	International relations/ Terrorism	Libya
April 29, 1992	Acquittals of Police Who Beat King Trigger Riots	Civil strife/Law	California
May 4, 1992	Muslim Fundamentalists Attack Egyptian Coptic Christians	Civil strife/Religion	Egypt
May 6, 1992	Lebanon's Prime Minister Karami Resigns	National politics	Lebanon
May 18, 1992	Thailand Declares National Emergency	National Politics/Civil strife	Thailand
May 21, 1992	China Detonates Powerful Nuclear Device	Military technology/ International relations	China
June 3-14, 1992	United Nations Hold Earth Summit in Brazil	International relations/ Social change	Brazil
June 26, 1992	Navy Secretary Resigns in Wake of Tailhook Scandal	Social reform/ Economics	United States/Nevada
July 6-8, 1992	Group of Seven Holds Summit	Economics/ International relations	Germany
July 19-24, 1992	AIDS Scientists and Activists Meet in Amsterdam	Medicine/Health	Netherlands
July 26, 1992	Americans with Disabilities Act Goes into Effect	Social reform/Civil rights and liberties	Washington, D.C.
August, 1992-May, 1993	Violence Against Immigrants Accelerates in Germany	Civil rights and liberties/ Ethnic conflict/Law	Germany
August 24, 1992	South Korea Establishes Diplomatic Ties with China	International relations	South Korea/China
August 27, 1992	Canadians Discuss Constitutional Reform	National politics/ Political reform	Canada
August 27, 1992	Iraq is Warned to Avoid "No Fly" Zone	International relations/ Military conflict	Iraq
September 1, 1992	Non-Aligned Movement Meets in Indonesia	International relations	Indonesia
September 11, 1992	World Court Resolves Honduras— El Salvador Border Dispute	International relations/ International law	Netherlands
October 4, 1992	Peace Accord Ends Mozambique Civil War	Civil war	Mozambique
November 20, 1992	British Queen Elizabeth Begins Paying Taxes After Castle Fire	National politics	Great Britain
November 24, 1992	United Nations Asks for End to Cuban Embargo	International relations	Cuba/Caribbean
December 6, 1992	Militant Hindus Destroy Mosque in India	Religion/Civil strife/ Ethnic conflict	India

Date	Event	Category	Country or region
December 9, 1992	International Troops Occupy Somalia	International relations/ Human rights	Somalia
December 17, 1992	Israel Deports 415 Palestinians	International relations	Israel/Palestine
December 18, 1992	Kim Young Sam Is Elected South Korean President	National politics	South Korea
1993	Wiles Offers Proof of Fermat's Last Theorem	Mathematics	Great Britain/New Jersey
January 1, 1993	"Velvet Divorce" Splits Czechoslovakia	International relations/ Political independence	Czech Republic/Czechoslovakia/ Slovakia
January 3, 1993	Russia and United States Sign START II	International relations/ Military	Russia/United States
January 20, 1993	Clinton Becomes U.S. President	National politics	Washington, D.C.
February 26, 1993	Terrorists Bomb World Trade Center	International relations/ Terrorism	New York
March 5, 1993	World Meteorological Organization Reports Ozone Loss	Environment	Switzerland
March 7, 1993	Afghan Peace Pact Fails to End Bloodshed	Civil war	Pakistan/Afghanistan
March 10, 1993	Egyptian Government Troops Attack Islamists	Civil strife/Religion	Egypt
March 12, 1993	North Korea Withdraws from Nuclear Non-Proliferation Treaty	Military/International relations	North Korea
March 31, 1993	Elections Increase Speculation About China's Future	National politics	China
April 19, 1993	Branch Davidian Cult Members Die in Waco Fire	Civil rights and liberties/ Religion	Texas
April 24, 1993	IRA Bomb Explodes in London	Civil strife/Terrorism	Great Britain/Northern Ireland
May 1, 1993	Tamil Rebels Assassinate Sri Lanka's President	Assassination/National politics	Sri Lanka
May 9, 1993	Paraguay Holds Democratic Presidential Election	Civil rights and liberties/ National politics	Paraguay/South America
May 23, 1993	Demonstrations Publicize China's Occupation of Tibet	Human rights/Civil rights and liberties	Tibet/China
May 26, 1993	Armenia and Azerbaijan Sign Cease-Fire Agreement	Military conflict	Armenia/Azerbaijan
May 27, 1993	Bomb Damages Florence's Uffizi Gallery	Law/Crime	Italy
June 6, 1993	Chinese Immigrants Are Detained in New York	International relations	New York/China
August 25, 1993	United States Orders Trade Sanctions Against China	International relations/ Military	Washington, D.C./China
August 26, 1993	Nigerian President Babangida Resigns	National politics	Nigeria
August 30, 1993	Brazilian Police Massacre Rio Slum Dwellers	Civil rights and liberties/ Crime	Brazil
September 13, 1993	Israel and PLO Sign Peace Agreement	International relations	Washington, D.C.
September 24, 1993	Sihanouk Returns to Power in Cambodia	National politics/Civil war	Cambodia
September 27, 1993	Georgia's Abkhaz Rebels Capture Sukhumi	Civil war	Georgia (nation)
October 4, 1993	Yeltsin Battles Russian Parliament	National politics	Russia
October 6-19, 1993	Bhutto's Pakistan People's Party Returns to Power	National politics	Pakistan
October 25, 1993	Liberal Party's Chrétien Becomes Canada's Prime Minister	Government/National politics	Canada
November 1, 1993	Maastricht Treaty Formally Creates European Union	International relations	Netherlands
November 2, 1993	Algeria Cracks Down on Islamic Militants	Civil war/Religion	Algeria

3308

Date	Event	Category	Country or region
November 18-20, 1993	APEC Nations Endorse Global Free Trade	Economics/ International relations	Washington State
November 30, 1993	Clinton Signs Brady Gun Control Law	Law/Crime	Washington, D.C.
December 15, 1993	General Agreement on Tariffs and Trade Talks Conclude	Economics/ International relations	Switzerland
1994	Astronomers Detect Black Hole	Astronomy	Space
January, 1994	Hubble Space Telescope Transmits Images to Earth	Astronomy	Earth orbit
January 1, 1994	Indian Peasants Revolt in Mexico's Chiapas State	Civil rights and liberties/ Civil strife	Mexico
January 1, 1994	North American Free Trade Agreement Goes into Effect	Economics/ International relations	Washington, D.C.
January 10, 1994, and March 29, 1994	Guatemala Takes Steps Toward Peace	Civil war	Mexico/Guatemala/Central America
February 3, 1994	United States and Vietnam Improve Relations	International relations	Washington, D.C.
March 12, 1994	Church of England Ordains Women	Religion	Great Britain
March 23, 1994	Mexican Presidential Candidate Colosio Is Assassinated	Assassination/National politics	Mexico
March 28, 1994	Ugandans Prepare to Draft Constitution	National politics/Politics	Uganda/Africa
April-October, 1994	Commonwealth of Independent States Strengthens Economic Ties	Economics	Russia/Belarus
April 10, 1994	U.S. Military Enters Bosnian War	Civil war/International relations	Bosnia
April 11, 1994	Florida Sues for Reimbursement of Costs of Illegal Immigrants	Law/National politics	Florida
April 24, 1994	Calderón Sol Is Elected President of El Salvador	National politics	El Salvador/Central America
April 26-May 9, 1994	African National Congress Wins South African Elections	Politics/Political reform	South Africa
May 6, 1994	Channel Tunnel Links Great Britain and France	Technology/ Transportation	Great Britain/France/Europe
May 6, 1994	Colombia Legalizes Some Private Use of Drugs	Law/Social reform	Colombia
May 30, 1994	Pope John Paul II Forbids Ordination of Women	Religion	Vatican
June 6, 1994	Veterans Observe Fiftieth Anniversary of D-Day	War/International relations	France
June 15, 1994	Vatican and Israel Establish Full Diplomatic Ties	International ties	Israel
June 21, 1994	Indonesian Government Suppresses Three Magazines	National politics/Civil rights and liberties	Indonesia
June 30, 1994	Hong Kong Legislature Votes to Expand Democracy	Politics	Hong Kong
July, 1994	Rwandan Refugees Flood Zaire	Civil war/International relations	Rwanda/Zaire/Congo (Kinshasa)
July 28, 1994	U.S. Congress Advances Crime Bill	Law/Crime	Washington, D.C.
August 10, 1994	German Officials Seize Plutonium	International relations	Germany
August 17, 1994	Lesotho's King Moshoeshoe II Dissolves Government	National politics	Lesotho
August 19, 1994	United States Changes Policy on Cuban Refugees	International relations/ Human rights	Washington, D.C.
August 31, 1994	Provisional Wing of IRA Declares Cease-Fire	Civil strife	Northern Ireland/Ireland

3309

Date	Event	Category	Country or region
September 5-13, 1994	World Delegates Discuss Population Control	Social reform/ Environment	Egypt
September 18, 1994	Haiti's Military Junta Relinquishes Power	National politics	Haiti/Caribbean
September 26, 1994	Congress Rejects President Clinton's Health Plan	Medicine/Social reform	Washington, D.C.
October 19, 1994	*The Bell Curve* Raises Charges of Renewed Racism	Education	United States
October 21, 1994	North Korea Agrees to Dismantle Nuclear Weapons Program	Military/International relations	Switzerland
October 26, 1994	Israel and Jordan Sign Peace Treaty	International relations	Israel/Jordan/Middle East
November 8, 1994	Californians Vote to Limit Aid to Undocumented Immigrants	Social reform/ Economics/Politics	California
November 8, 1994	Republicans Become Congressional Majority	National politics	United States
November 10, 1994	Iraq Recognizes Kuwait's Sovereignty	International relations	Iraq/Middle East/Kuwait
November 16, 1994	Ukraine Agrees to Give Up Nuclear Weapons	International relations/ Military capability	Ukraine
November 20, 1994	Angolan Government Shares Power with UNITA Rebels	National politics/Civil war	Angola
December 9-11, 1994	Western Hemisphere Nations Approve Free Trade Area of the Americas	Economics/ International relations	Florida/South America/Central America/Caribbean
December 10, 1994	Reformers Realign Japan's Political Parties	National politics	Japan
December 11, 1994	Russian Army Invades Chechnya	Political aggression	Chechnya/Russia
December 22, 1994	Berlusconi Resigns as Prime Minister of Italy	National politics	Italy
January-June, 1995	Ebola Epidemic Breaks Out in Zaire	Medicine/Health	Zaire/Congo (Kinshasa)
January 13, 1995	Gray Wolves Are Returned to Yellowstone National Park	Biology/Environment	Wyoming
January 17, 1995	Powerful Earthquake Hits Kobe, Japan	Disasters/Earth science	Japan
January 31, 1995	Clinton Announces U.S. Bailout of Mexico	International relations/ Economics	Washington, D.C./Mexico
February 1, 1995	U.S. State Department Criticizes Human Rights in China	International relations/ Human rights	Washington, D.C.
February 17, 1995	Peru and Ecuador Agree to a Cease-Fire in Their Border Dispute	International relations/ Military/War	South America/Brazil
February 26, 1995	Illegal Trading Destroys Barings, Britain's Oldest Bank	Business/Economics	Singapore
March 2, 1995	Top Quark Is Tracked at Fermilab	Physics	Illinois
March 20, 1995	Terrorists Release Toxic Gas in Tokyo Subway	Civil strife/Terrorism/ Crime/Religion	Japan
April 19, 1995	Terrorists Bomb Oklahoma City's Federal Building	Terrorism/Crime	Oklahoma
May 15, 1995	Dow Corning Seeks Chapter 11 Bankruptcy	Business/Law	Michigan
May 22, 1995	Supreme Court Voids State Term-Limit Laws	Government/Law/ Political reform	Washington, D.C.
June 12, 1995	Supreme Court Limits Affirmative Action	Civil rights and liberties/ Law	Washington, D.C.
June 13, 1995	French Plans for Nuclear Tests Draw Protests	Weapons technology/ Environment	France/Pacific Ocean
June 27-July 7, 1995	Shuttle *Atlantis* Docks with Space Station Mir	Technology/ International relations	Earth orbit/United States/Russia
July 11, 1995	United States Establishes Full Diplomatic Relations with Vietnam	International relations	Vietnam/United States

Date	Event	Category	Country or region
July 31, 1995	Disney Company Announces Plans to Acquire Capital Cities/ABC	Business/Entertainment	California
August 2, 1995	INS Frees Thai Workers Held Captive in Clothes Factory	Human rights/Law	California
August 10, 1995	Principal *Roe v. Wade* Figure Takes Stand Against Abortion	Gender issues/Human rights/Social reform	Texas
August 24, 1995	Microsoft Releases Windows 95	Business/Computer science/Technology	Washington State
September 4-15, 1995	Beijing Hosts U.N. Fourth World Conference on Women	Human rights/International relations	China
September 19, 1995	*Washington Post* Prints Unabomber Manifesto	Social reform/Terrorism/Crime	Washington, D.C.
September 20, 1995	Turner Broadcasting System Agrees to Merge with Time Warner	Business/Communications	New York
September 24, 1995	Israel Agrees to Transfer Land to Palestinian National Authority	International relations/Political independence	Israel/Palestine/Middle East
October 3, 1995	O. J. Simpson Is Acquitted	Law/Crime	California
October 16, 1995	Million Man March Draws African American Men to Washington, D.C.	Civil rights and liberties/Gender issues	Washington, D.C.
October 30, 1995	Quebec Voters Narrowly Reject Independence from Canada	National politics/Political independence	Canada/Quebec
November 4, 1995	Israeli Premier Rabin Is Assassinated	Assassination	Israel
November 10, 1995	Nigeria Hangs Writer Saro-Wiwa and Other Rights Advocates	Civil strife/International relations	Nigeria
November 27-29, 1995	International Panel Warns of Global Warming	Environment/International relations	Spain
November 29, 1995	Chinese Government Selects Tibet's Panchen Lama	International relations/Religion	Tibet/China
December 7, 1995	Galileo Spacecraft Reaches Jupiter	Space and aviation	Space/Jupiter
December 14, 1995	NATO Troops Enter Balkans	Military/International relations	Bosnia/Balkans
January 29, 1996	France Announces End to Nuclear Bomb Tests	International relations/Weapons technology	France/Pacific Ocean
March 13, 1996	Liggett Group Agrees to Help Fund Antismoking Program	Business/Health/Law	Louisiana
March 20, 1996	Britain Announces Mad Cow Disease Has Infected Humans	Medicine/Health/Food science	Great Britain
April 3, 1996	FBI Arrests Suspected Unabomber	Crime/Terrorism	Montana
April 4, 1996	Congress Reduces Federal Farm Subsidies and Price Supports	Agriculture/Government	Washington, D.C.
April 9, 1996	Clinton Signs Line-Item Veto Bill	Government/National politics	Washington, D.C.
April 10, 1996	Clinton Blocks Ban on Late-Term Abortions	Gender issues/Health/Medicine/Religion	Washington, D.C.
May 8, 1996	South Africa Adopts New Constitution	National politics/Social reform	South Africa
June 6, 1996	Arsonists Burn Thirtieth Black Church in Eighteen Months	Civil rights and liberties/Religion/Crime	North Carolina
June 13, 1996	Last Members of Freemen Surrender After Long Standoff	Civil strife/Crime/Law/Social reform	Montana
June 25, 1996	Truck Bomb Explodes at U.S. Base in Saudi Arabia Killing Nineteen	Terrorism/War	Saudi Arabia

Date	Event	Category	Country or region
July 3, 1996	Yeltsin Wins Runoff Election for Russian Presidency	Government/Political independence	Russia
July 12, 1996	House of Representatives Passes Legislation Barring Same-Sex Marriages	Human rights/Social reform	Washington, D.C.
July 12, 1996	Prince Charles and Princess Diana Agree to Divorce	Government	Great Britain
July 27, 1996	Pipe Bomb Explodes in Park During Summer Olympics in Atlanta, Georgia	Terrorism/Crime/Sports	Georgia
August 2, 1996	Clinton Signs Bill to Raise Minimum Wage	Business/Labor/ Economics	Washington, D.C.
August 7, 1996	NASA Announces Possibility of Life on Mars	Astronomy/Biology	Mars/Solar system
September 27, 1996	Taliban Leaders Capture Capital of Afghanistan	Civil war/Coups/ National politics	Afghanistan
October 21, 1996	Supreme Court Declines to Hear Challenge to Military's Gay Policy	Law/National politics	Washington, D.C.
November 5, 1996	Clinton Is Reelected	National politics	United States
November 15, 1996	Texaco Settles Racial Discrimination Lawsuit	Law/Civil rights and liberties/Labor	New York
December 13, 1996	Ghana's Annan Is Named U.N. Secretary-General	International relations	New York/Ghana
December 19, 1996	Television Industry Panel Proposes Program Rating System	Communications/ Entertainment	Washington, D.C.
December 29, 1996	Guatemalan Government and Rebels Sign Accord	Civil war/Human rights	Guatemala/Central America
January 27, 1997	Researchers Announce Development of Atom Laser	Physics/Technology	Massachusetts
February 19, 1997	Chinese Leader Deng Xiaoping Dies	Economic reform/ Government	China
February 20, 1997	Galileo Makes Closest Pass to Jupiter's Moon Europa	Space and aviation/ Astronomy	Jupiter/Solar system
February 22, 1997	Scottish Researchers Announce Cloning Breakthrough	Biology/Genetics	Scotland
March 3, 1997	CIA Official Admits Espionage Activity	International relations/ Crime	Washington, D.C.
March 22, 1997	Hale-Bopp Comet's Appearance Prompts Scientific and Spiritual Responses	Astronomy/Religion	Space/Solar system
April-May, 1997	Seven Climber Deaths on Mount Everest Echo Previous Year's Tragedy	Disasters/Environment	Nepal
April 1, 1997	Presbyterian Church Groups Adopt Law Against Ordaining Homosexuals	Religion/Gender issues/ Social reform	United States
April 8, 1997	U.S. Appeals Court Upholds California's Ban on Affirmative Action Programs	Human rights/Law	California
April 22, 1997	Peruvian Government Commandos Free Japanese Embassy Hostages	Terrorism/International relations	Peru/Japan
May 1, 1997	Labour Party's Blair Becomes British Prime Minister	Government/National politics	Great Britain
May 16, 1997	Zaire President Sese Seko Flees Capital as Rebels Take Over	Civil strife/Coups/ National politics	Zaire/Congo (Kinshasa)
May 30, 1997	Last Common Ancestor of Humans and Neanderthals Found in Spain	Anthropology/ Archaeology	Spain
June 25, 1997	Caribbean Volcano Erupts, Destroying Much of Island of Montserrat	Disasters/Earth science	Montserrat/Caribbean
June 26, 1997	Supreme Court Overturns Communications Decency Act	Law/Computer science	Washington, D.C.
July 1, 1997	Hong Kong Reverts to Chinese Sovereignty	International relations/ National politics	Hong Kong/China/Great Britain

3312

Date	Event	Category	Country or region
July 4, 1997	U.S. Spacecraft Mars Pathfinder Lands on Mars	Space and aviation	Mars/Space
July 25, 1997	Khmer Rouge Try Longtime Cambodian Leader Pol Pot	Civil war/Human rights/ International law	Cambodia
July 30, 1997	Nuclear Submarine Helps Find Ancient Roman Shipwrecks in Mediterranean	Archaeology	Mediterranean Sea
August 3, 1997	Microsoft and Apple Form Partnership	Computer science/ Business	California
August 14, 1997	117,000-Year-Old Human Footprints Are Found Near South African Lagoon	Anthropology/ Archaeology	South Africa
August 31, 1997	Great Britain's Princess Diana Dies in Paris Car Crash	National politics	France/Great Britain
September 11, 1997	Scots Vote for First Separate Legislature Since 1707	Government/Political independence	Scotland/Great Britain
September 15, 1997	Diet Drugs Are Pulled from Market	Medicine/Health	United States
September 18, 1997	Welsh Vote to Establish Own Assembly	Political independence/ Political reform	Wales/Great Britain
October 25, 1997	Million Woman March Draws Many Thousands to Philadelphia	Gender issues/Social reform	Pennsylvania
December 11, 1997	World Nations Approve Kyoto Plan for Reducing Harmful Emissions	Environment/Earth science	Japan
January 21, 1998	Pope John Paul II Visits Cuba	Religion/International relations	Cuba/Caribbean/Vatican
February 3, 1998	Texas Executes First Woman Since Civil War	Crime/Law	Texas
February 3, 1998	U.S. Military Plane Cuts Italian Ski Cable, Killing Twenty People	Disasters/Military	Italy
February 11, 1998	Disabled Golfer Wins Right to Use Golf Cart in Tournaments	Civil rights and liberties/ Law/Sports	Oregon
February 23, 1998	U.S. Supreme Court Rejects Challenge to New Jersey's Megan's Law	Human rights/Law	New Jersey/Washington, D.C.
March 5, 1998	U.S. Scientists Report Discovery of Ice Crystals on Moon	Space and aviation/ Astronomy	Moon
March 13, 1998	Astronomers See Light from Object 12.2 Billion Light-Years Away	Astronomy	Space
March 16, 1998	Vatican Issues Document on Jewish Holocaust and Condemns Racism	Religion	Vatican
March 16, 1998	Democratic Fund-Raiser Pleads Guilty to Making Illegal Contributions	Political reform	United States
March 22, 1998	Meteorite Is First Extraterrestrial Object Found to Contain Water	Astronomy	Space/Solar system
March 26, 1998	Food and Drug Administration Approves Sale of Viagra	Biology/Gender issues/ Health/Medicine	Washington, D.C.
April 1, 1998	Judge Dismisses Sexual Harassment Lawsuit Against President Clinton	Gender issues/Law	Arkansas
May, 1998	India and Pakistan Conduct Nuclear Weapon Tests	Military capability/ International relations	India/Pakistan
May 3, 1998	Europeans Agree on Euro Currency	Economics/ International relations	Belgium/Europe
May 21, 1998	Indonesian Dictator Suharto Steps Down	Government	Indonesia
June, 1998	California Voters Reject Bilingual Education	Education/Ethnic conflict/Politics	California
June 4, 1998	School Prayer Amendment Fails in House of Representatives	Government/Religion/ National politics	Washington, D.C.
June 11, 1998	Mitsubishi Motors Pays Record Settlement in Sexual Harassment Suit	Gender issues/Law	Illinois

3313

Date	Event	Category	Country or region
July 8, 1998	Dow Corning Agrees to Pay Women with Faulty Breast Implants	Consumer issues/ Gender issues/ Health/Medicine	Michigan
July 9, 1998	Congress Approves Overhaul of Internal Revenue Service	Business/Government	Washington, D.C.
July 16, 1998	Food and Drug Administration Approves Thalidomide to Fight Leprosy	Biology/Chemistry/ Health/Medicine	Washington, D.C.
July 20, 1998	International Monetary Fund Approves Rescue Plan for Russia	International relations/ Economics	Russia/Washington, D.C.
July 28, 1998	General Motors and Union End Production-Stopping Strike	Economics/Labor	Michigan
August 7, 1998	Terrorists Attack U.S. Embassies in East Africa	Terrorism/International relations	Kenya/Africa/Tanzania
August 12, 1998	Swiss Banks Agree on Compensation for Holocaust Victims	Business/International relations	Switzerland
August 14, 1998	Court Rules That Food and Drug Administration Cannot Regulate Tobacco	Law/Government/ Health	West Virginia
August 20, 1998	U.S. Missiles Hit Suspected Terrorist Bases in Afghanistan and Sudan	International relations/ Military/Terrorism	Afghanistan/Sudan
August 21, 1998	Former Ku Klux Klan Member Found Guilty of Murder in 1966 Firebombing	Civil rights and liberties/ Crime	Mississippi
August 31, 1998	North Korea Fires Missile Across Japan	International relations/ Military capability	North Korea/Japan/United States
September 4, 1998	United Nations Tribunal Convicts Rwandan Leader of Genocide	International law/Ethnic conflict/Human rights	Rwanda/Netherlands
September 19, 1998	Scientists Report Record-Size Hole in Antarctic Ozone Layer	Environment/ International relations	Antarctica/Earth orbit
September 24, 1998	Iran Lifts Death Threat Against Author Rushdie	International relations/ Religion	Iran/Great Britain
September 27, 1998	German Chancellor Kohl Loses Election to Schröder	National politics	Germany
October 11, 1998	Vatican Grants Sainthood to Formerly Jewish Nun	Religion/Gender issues	Vatican
October 16, 1998	Former Chilean Dictator Pinochet Is Arrested in London	Human rights/ International law	Great Britain/Chile
October 29 to November 7, 1998	Former Astronaut Glenn Returns to Space	Space and aviation/ Health/Technology	Earth orbit
October 31, 1998	First Digital High-Definition Television Signals Are Broadcast	Communications/ Entertainment	New York
November 5, 1998	DNA Tests Indicate That Jefferson Fathered Slave's Child	Genetics/Technology	Virginia
November 6, 1998	Speaker of the House Gingrich Steps Down	Government/Politics/ Law	Washington, D.C.
November 13, 1998	Brazil's Ailing Economy Gets $41.5 Billion in International Aid	Economics/ International relations	Brazil/South America
November 30, 1998	Hun Sen and Norodom Ranariddh Form Cambodian Coalition	International relations/ Government	Cambodia
December, 1998	Scandal Erupts over 2002 Winter Olympic Games in Utah	Sports/Crime	Utah
December 1, 1998	Exxon and Mobil Merge to Form World's Largest Oil Company	Business/Energy	Texas
December 13, 1998	Puerto Ricans Vote Against Statehood	International relations/ National politics	Puerto Rico/Caribbean

Date	Event	Category	Country or region
December 16, 1998	United States and Great Britain Bomb Military Sites in Iraq	War/National politics	Iraq/Washington, D.C.
December 19, 1998	House of Representatives Impeaches Clinton	Government/Law	Washington, D.C.
January 12, 1999	Haiti's President Préval Dissolves Parliament	National politics	Haiti/Caribbean
January 24, 1999	First Hand Transplant Is Performed	Medicine	Kentucky
January 31, 1999	Scientists Trace HIV to Chimpanzees	Health/Medicine/ Genetics/Biology	Africa/Illinois
February, 1999, to January, 2000	Researchers Generate Nuclear Energy on Tabletop	Physics/Energy	California
February 7, 1999	King Hussein of Jordan Dies	Government	Jordan
February 18, 1999	Scientists Slow Speed of Light	Physics	Massachusetts
February 23, 1999	White Supremacist Is Convicted in Racially Motivated Murder	Civil rights and liberties/ Crime	Texas
March 2, 1999	Plan to Save California Redwoods Is Approved	Environment/Business	California
March 5, 1999	New Fault Is Discovered Under Los Angeles	Earth science/Disasters	California
March 12, 1999	Czech Republic, Poland, and Hungary Join NATO	International relations	Europe/Czech Republic/Poland/ Hungary
March 20, 1999	First Nonstop Around-the-World Balloon Trip Is Completed	Space and aviation/ Transportation	Switzerland/Africa
March 24-June 10, 1999	North Atlantic Treaty Organization Wars on Yugoslavia	War	Yugoslavia
March 26, 1999	Assisted-Suicide Specialist Kevorkian Is Found Guilty of Murder	Crime/Human rights	Michigan
March 29, 1999	Dow Jones Industrial Average Tops 10,000	Business/Economics	New York
March 30, 1999	Jury Awards Smoker's Family $81 Million	Law/Health	Oregon
April 1, 1999	Canada Creates Nunavut Territory	Government	Canada/Nunavut
April 15, 1999	Astronomers Discover Solar System with Three Planets	Astronomy/Physics	Space
April 15, 1999	Former Pakistani Prime Minister Bhutto Is Sentenced for Corruption	Politics/Crime	Pakistan
April 17, 1999	Vajpayee's Government Collapses in India	Government	India
April 20, 1999	Littleton, Colorado, High School Students Massacre Classmates	Crime	Colorado
April 22, 1999	United States to Retain Smallpox Virus Sample	Medicine/Biology/ Terrorism	Washington, D.C.
May 7, 1999	NATO Bombs Hit Chinese Embassy in Belgrade	International relations/ War	Yugoslavia/China
May 8, 1999	Citadel Graduates First Woman Cadet	Gender issues/ Education	South Carolina
May 25, 1999	U.S. House Report Describes Chinese Spy Activity	International relations	Washington, D.C.
June 7, 1999	Two New Elements Are Added to Periodic Table	Chemistry/Physics	California
June 9, 1999	Clinton Denounces Racial Profiling	Law/Civil rights and liberties	Washington, D.C.
July 10, 1999	Women's World Cup Soccer Draws Unprecedented Attention	Sports/Gender issues/ Education	California
July 22, 1999	China Detains Thousands in Falun Gong	Religion/Human rights	China
July 23, 1999	Morocco's King Hassan II Dies	National politics	Morocco
August 14, 1999	Kansas Outlaws Teaching of Evolution in Schools	Biology/Education/ Religion	Kansas

3315

Date	Event	Category	Country or region
August 27, 1999	Full Crew Leaves Russian Space Station Mir	Space and aviation	Earth orbit/Russia
August 30, 1999	Decline in U.S. AIDS Death Rate Slows	Health/Medicine	Georgia
August 31, 1999	East Timor Votes for Independence	Civil war/Political independence	East Timor/Indonesia
September 7, 1999	Viacom Buys Columbia Broadcasting System	Business/ Communications	New York
September 10, 1999	School Busing Ends in North Carolina	Education/Civil rights and liberties	North Carolina
September 12, 1999	Scientists Produce Most Proton-Rich Nucleus	Physics	France
September 23, 1999	Mars-Orbiting Craft Is Destroyed	Space and aviation/ Astronomy	Mars/Solar system
September 30, 1999	Japan Has Its Worst Nuclear Accident	Energy/Disasters	Japan
October 7, 1999	Clarkson Becomes Governor-General of Canada	Government	Canada
October 12, 1999	World's Six Billionth Inhabitant Is Born	Environment	Bosnia
October 12, 1999	Pakistani Government Is Ousted by Coup	Government/Coups/ Coups/Military	Pakistan
October 13, 1999	Philip Morris Company Admits Smoking Is Harmful	Business/Health	Florida
October 21, 1999	Enzyme Is Linked to Alzheimer's Disease	Health/Medicine/ Genetics	California
November 5, 1999	Federal Court Rules Microsoft a Monopoly	Business/Computer science	Washington, D.C.
November 6, 1999	Australia Votes Against Becoming a Republic	Politics/Government	Australia
November 12, 1999	Hereditary Peers Are Removed from British House of Lords	Political reform/ Government	Great Britain
November 14, 1999	Ukraine Votes Down Return to Communism	National politics	Ukraine
November 15, 1999	United States and China Reach Trade Accord	Business/International relations	China
November 25, 1999	Coast Guard Rescues Cuban Refugee Elián González	International relations/ Human rights/Law	Florida/Cuba/Caribbean
November 29, 1999	Protesters Disrupt Seattle Trade Conference	International relations/ Economics	Washington State
December 2, 1999	Northern Ireland Gains Home Rule	Political independence	Northern Ireland/Great Britain
December 2, 1999	Ireland Amends Its Constitution	International relations/ Political reform	Ireland/Northern Ireland/Great Britain
December 9, 1999	Evidence of Martian Ocean Is Reported	Astronomy/Space and aviation/Technology	Solar system/Mars
December 11, 1999	European Union Accepts Turkey as Candidate for Membership	International relations/ Political reform	Finland
December 24, 1999	Astronauts Renovate Hubble Telescope	Astronomy/Space and aviation	Earth orbit
December 31, 1999	Panama Takes Full Control of Panama Canal	International relations/ Transportation	Panama/Central America/ Caribbean
December 31, 1999	Russian President Yeltsin Resigns	Government/Political reform	Russia
January, 2000	Hybrid Gas-Electric Cars Enter U.S. Market	Energy/Engineering/ Transportation	United States
January, 2000	U.S. Mint Introduces Sacagawea Dollar Coin	Economics/Social reform	United States
January 1, 2000	Computers Meet Y2K with Few Glitches	Technology/Computer science	United States

Date	Event	Category	Country or region
January 10, 2000	America Online Agrees to Buy Time Warner	Communications/ Business	New York
January 11, 2000- January 19, 2001	Clinton Creates Eighteen New National Monuments	Environment/National politics	Washington, D.C.
January 14, 2000	Hague Court Convicts Bosnian Croats for 1993 Massacre	Civil war/Human rights/ International law	Netherlands/Bosnia
January 21, 2000	Gene Therapy Test at University of Pennsylvania Halted After Man Dies	Medicine/Genetics	Pennsylvania
February 5, 2000	Intel Unveils 1.5 Gigahertz Microprocessor	Business/Computer science/Technology	California
February 6, 2000	Hillary Clinton Announces Candidacy for U.S. Senate	Politics	New York
February 7, 2000	Hacker Cripples Major Web Sites and Reveals Internet Vulnerability	Computer science/ Business	California/Quebec
February 9, 2000	Kurdish Rebels Lay Down Arms and Seek Self-Rule Without Violence	Political independence/ Civil war/Ethnic conflict	Turkey
February 11-22, 2000	Shuttle *Endeavour* Gathers Radar Images for Three-Dimensional Maps of Earth	Earth science/Space and aviation	Earth orbit
February 14, 2000	NEAR Is First Satellite to Orbit an Asteroid	Astronomy/Space and aviation	Space/Solar system
March 1, 2000	Investigation Finds Misconduct in Los Angeles Police Department	Civil rights and liberties/ Crime/Law	California
March 12, 2000	Pope Apologizes for Catholic Church's Historical Errors	Religion	Vatican
March 13, 2000	Tribune Company and Times Mirror Company to Merge	Communications/ Business	California/Illinois
March 17, 2000	Smith and Wesson to Change Handguns to Avoid Lawsuits	Social reform/National politics	Washington, D.C.
March 18, 2000	Taiwan's Nationalist Party Is Voted Out	National politics	Taiwan
March 30, 2000	Highest-Ranking Army Woman Charges Sexual Harassment	Military/Gender issues	Washington, D.C.
April 4, 2000	Controversy Arises over Estrogen and Heart Attacks	Health/Medicine/ Gender issues	Washington, D.C.
April 20, 2000	Petrified Dinosaur Heart Is Found in South Dakota	Biology	South Dakota
April 26, 2000	Vermont Approves Same-Sex Civil Unions	Civil rights and liberties/ Social reform	Vermont
May 4, 2000	ILOVEYOU Virus Destroys Data on Many Thousands of Computers	Computer science/ Crime/Technology	Philippines
May 4, 2000	Unemployment Rate Reaches Lowest Level in Thirty Years	Business/Economics/ Labor	Washington, D.C.
May 19, 2000	Fijian Rebels Attempt Coup	Civil strife/Coups/ National politics	Fiji
May 24, 2000	Israeli Troops Leave Southern Lebanon	Military conflict/ International relations	Israel/Lebanon/Palestine/Middle East
June 15, 2000	South and North Korea Sign Peace and Unity Agreement	International relations/ War	South Korea/North Korea/Asia
June 8, 2000	Environmental Protection Agency Bans Pesticide Chlorpyrifos	Environment/Health	Washington, D.C.
June 26, 2000	Scientists Announce Deciphering of Human Genome	Biology/Genetics/ Medicine	Washington, D.C.
July-November, 2000	Gore and Bush Compete for U.S. Presidency	Government/National politics	United States
July 2, 2000	Fox's Election Ends Mexico's Institutional Revolutionary Party Rule	Political reform/ National politics	Mexico

Date	Event	Category	Country or region
July 8, 2000	Defense Missile Fails Test and Clinton Shelves System Development	Military defense/Military technology	Washington, D.C.
July 14, 2000	Florida Jury Orders Tobacco Companies to Pay Record Damages	Business/Law/Health	Florida
August 5, 2000	Clinton Vetoes Bill Reversing Taxpayer "Marriage Penalty"	Economics/ Government	Washington, D.C.
August 7, 2000	Nine Extrasolar Planets Are Found	Astronomy/Physics/ Space and aviation	Space
September 6, 2000	Bridgestone Apologizes for Tires That Caused Numerous Deaths	Business/Transportation	Washington, D.C.
September 13, 2000	Unjustly Accused Scientist Lee Is Released	Human rights/Law	New Mexico
September 20, 2000	Whitewater Inquiry Clears Clintons of Wrongdoing	Law/National politics	Washington, D.C.
September 25, 2000	Senate Approves Plan to Restore Everglades Ecosystem	Environment/Earth science	Florida
September 25, 2000	Scientists Find Interferon, Drug That Fights Multiple Sclerosis	Biology/Medicine	New York
September 28, 2000	FDA Approves Sale of Mifepristone for Nonsurgical Abortion	Gender issues/Health/ Medicine	Washington, D.C.
October 5, 2000	Protesters Topple Yugoslavian President Milošević	Civil strife/Coups/ Political reform	Yugoslavia
October 12, 2000	Explosion Tears Hole in U.S. Destroyer *Cole* Killing Seventeen	International relations/ Terrorism	Yemen
October 26, 2000	Post-Election Violence Rocks Ivory Coast	Civil strife/National politics	Ivory Coast/Africa
October 31, 2000	Napster Agrees to Charge Fees for Downloading Music over Internet	Business/Computer science/ Entertainment	California
November, 2000	Peruvian President Fujimori Resigns While in Japan	National politics	Japan
November 2, 2000	International Space Station Welcomes First Crew	International relations/ Space and aviation	Russia/United States/Earth orbit/Space
November 7, 2000	Pets.com Announces Withdrawal from Internet	Business	California
November 18-20, 2000	Clinton Visits Vietnam	International relations	Vietnam
December 12, 2000	Supreme Court Resolves Presidential Election Deadlock	Government/National politics	Washington, D.C.
February 9, 2001	U.S. Submarine Sinks Japanese Fishing Vessel	Military capability/ International relations	Hawaii/Pacific Ocean/Japan/ United States
March, 2001	Afghanistan Has Ancient Buddhist Monuments Destroyed	Politics/Religion	Afghanistan
April 1, 2001	China Seizes Downed U.S. Reconnaissance Plane	International relations/ International law	China
June 1, 2001	Nepal's Crown Prince Murders His Royal Family	Crime/National politics	Nepal
June 6, 2001	Jeffords Quits Republican Party Giving Democrats Control of Senate	Government/National politics	Washington, D.C.
September 11, 2001	Terrorists Attack New York City and Washington, D.C.	Terrorism/Disasters	New York/Washington, D.C./ Pennsylvania

GREAT EVENTS

1900-2001

CATEGORY INDEX

LIST OF CATEGORIES

AGRICULTURE

Congress Reduces Federal Farm Subsidies and Price Supports, **7**-2760

Gericke Reveals Significance of Hydroponics, **2**-569

Insecticide Use Increases in American South, **1**-359

Ivanov Develops Artificial Insemination, **1**-44

Morel Multiplies Plants in Vitro, Revolutionizing Agriculture, **3**-1114

Müller Develops Potent Insecticide DDT, **2**-824

ANTHROPOLOGY

Anthropologists Find Earliest Evidence of Modern Humans, **6**-2220

Benedict Publishes *Patterns of Culture*, **2**-698

Boas Lays Foundations of Cultural Anthropology, **1**-216

Boule Reconstructs Neanderthal Man Skeleton, **1**-183

Dart Finds Fossil Linking Apes and Humans, **2**-484

Humans and Chimpanzees Are Found to Be Genetically Linked, **5**-2129

Johanson Discovers "Lucy," Three-Million-Year-Old Hominid Skeleton, **5**-1808

Last Common Ancestor of Humans and Neanderthals Found in Spain, **7**-2841

Leakeys Find 1.75-Million-Year-Old Hominid Fossil, **4**-1328

Mead Publishes *Coming of Age in Samoa*, **2**-556

117,000-Year-Old Human Footprints Are Found Near South African Lagoon, **7**-2860

Pottery Suggests Early Contact Between Asia and South America, **3**-1230

Simons Identifies Thirty-Million-Year-Old Primate Skull, **4**-1553

Weidenreich Reconstructs Face of Peking Man, **2**-786

Zdansky Discovers Peking Man, **2**-476

ARCHAEOLOGY

ASSASSINATION

ASTRONOMY

BIOLOGY

BUSINESS

COMMUNICATIONS

COMPUTER SCIENCE

EDUCATION

ENERGY

ENGINEERING

INTERNATIONAL LAW

INTERNATIONAL RELATIONS

LABOR

XIX

XXI

PHOTOGRAPHY

PHYSICS

XXIII

POLITICS

XXIV

SPACE AND AVIATION

SPORTS

WEAPONS TECHNOLOGY

GEOGRAPHICAL INDEX

XXXII

XXXV

XXXVI

XXXVIII

XL

XLVI

XLVIII

XLIX

L

WASHINGTON STATE

WEST INDIES. *See* CARIBBEAN

WEST VIRGINIA

WISCONSIN

PERSONAGES INDEX

SUBJECT INDEX

This Subject Index should be used in conjunction with the other indexes in the back of this volume. Particular attention should be paid to the **Geographical Index** (beginning on page XXIX), which lists states, countries, continents, planets, and other broad regions that are *not* listed here.

G.I. Bill, **3:** 963-964
Giacconi, Riccardo, **4:** 1429
Giaever, Ivar, **3:** 1260
Gibbon, John H., Jr., **3:** 1177
Gibbon, Mary Hopkinson, **3:** 1177
Gibbons, John J., **7:** 2888
Gibbs, Lois Marie, **5:** 1899
Giberson, Walker E. (Gene), **5:** 1787
Gibson, Everett, **7:** 2901
Gibson, Robert, **7:** 2700
Gideon, Clarence E., **4:** 1470
Gideon v. Wainwright, **4:** 1470-1471
Gidzenko, Yuri, **8:** 3252
Gierek, Edward, **5:** 2009, 2100
Gilbert, Walter, **5:** 1988, **6:** 2347
Gilruth, Robert R., **4:** 1419
Gingrich, Newt, **7:** 2650, 2801, 2918, 2971
Giolitti, Giovanni, **1:** 218, 322
Giraud, Henri-Honoré, **3:** 930
Giscard d'Estaing, Valéry, **5:** 1830, 2024
Giuliani, Rudolph, **8:** 3155, 3272
Glashow, Sheldon L., **5:** 1793, 2089, **7:** 2685
Glass, **1:** 77; laminated, **1:** 77-78; Pyrex, **1:** 324-325
Glass, Carter, **1:** 296
Glenn, Annie, **7:** 2963
Glenn, John H., Jr., **4:** 1437, **7:** 2963
Global Atmospheric Research Program, **5:** 1935
Global warming, **4:** 1575-1576, **7:** 2740-2742
Glomar Challenger, **4:** 1602-1604
Glushko, Valentin P., **4:** 1557, 1698, **5:** 1724
Goddard, Robert H., **1:** 74, **2:** 520, **4:** 1284
Gödel, Kurt, **1:** 6-7, 13, 66, **2:** 443, **4:** 1460
Godke, R. A., **5:** 2131
Goebbels, Joseph, **2:** 679, 684, 689, 713
Gold, Thomas, **3:** 1059
Golden Venture grounding, **6:** 2530
Goldin, Daniel S., **7:** 2793, 2963
Goldmark, Josephine, **1:** 172
Goldmark, Peter Carl, **3:** 842
Goldstein, Eugen, **1:** 150
Goldstine, Herman Heine, **3:** 926, 1205
Goldwater, Barry M., **4:** 1518
Golombek, Matthew, **7:** 2851
Goncz, Arpad, **6:** 2287
Gongadze, Georgiy, **8:** 3107
González, Elián, **8:** 3112
González, Juan Miguel, **8:** 3112
González, Lázaro, **8:** 3112
Good Friday Agreement, **8:** 3117, 3120
Good Neighbor Policy, **2:** 591-592; and Dominican invasion, **4:** 1534
Goodacre, Glenna, **8:** 3137
Goodman, Andrew, **4:** 1507
Goodyear, Charles, **2:** 518, 773
Gorbachev, Mikhail, **4:** 1551, 1616, **5:** 1976, **6:** 2168, 2181, 2215, 2240, 2243, 2266, 2281, 2302, 2314, 2322, 2356, 2366, 2371, 2373, 2376, 2403, 2421, 2441, 2493, **8:** 3132
Gorbunovs, Anatolijs, **6:** 2366
Gordon-Reed, Annette, **7:** 2968

Gore, Al, **7:** 2898, **8:** 3206, 3228, 3259
Goremykin, Ivan Loginovich, **1:** 138
Gorgas, William Crawford, **1:** 21, 93, 377
Gorie, Dom, **8:** 3161
Göring, Hermann, **2:** 674, 684, 713, 803, 831, **3:** 847, 859, 1012
Gosslau, Ing Fritz, **3:** 945
Gottwald, Klement, **3:** 1064
Goudsmit, Samuel, **2:** 493
Gould, R. Gordon, **5:** 2110
Gould, Stephen Jay, **7:** 2640
Goulian, Mehran, **4:** 1589
Governor, first woman, **2:** 502-504
Gowon, Yakubu, **4:** 1580
Granados, Hector Rosada, **7:** 2577
Grand unified theory, **5:** 1793-1795
Granger, Walter Willis, **2:** 466
Graves, Bill, **8:** 3064
Gravitation, **1:** 153; theory of, **1:** 411-413
Gravitational redshifting, **4:** 1367-1369
Gray, Donald M., **5:** 1888
Gray, Elisha, **2:** 468
Gray v. Sanders, **4:** 1434
Gray wolves, **7:** 2670-2672
Great Depression, **2:** 571-572; Dust Bowl, **2:** 705-707; and stock market crash, **2:** 587-588
Great Leap Forward (China), **4:** 1304-1305
Greaves, Ronald I. N., **2:** 677
Green, Harry W., II, **6:** 2294
Green, William, **2:** 748
Green v. County School Board of New Kent County, **5:** 1714
Greenglass, David, **3:** 1181
Greenhouse effect, **4:** 1575
Greenough, William Bates, III, **4:** 1704
Greenpeace; and nuclear tests, **7:** 2750; and *Rainbow Warrior*, **6:** 2172-2174, **7:** 2750; and *Rainbow Warrior II*, **7:** 2698
Greenspan, Alan, **8:** 3024
Greenville, USS, **8:** 3261-3262
Gregg, Troy L., **5:** 1854
Gregg v. Georgia, **5:** 1854-1855
Gregorich, K. E., **8:** 3051
Greiff, Gustavo de, **7:** 2605
Grey, Sir Edward, **1:** 155, 158, 179, 271, 300, 314
Griffith, Arthur, **2:** 450
Griffith, Frederick, **3:** 921
Grijns, Gerrit, **1:** 37
Gringauz, K. I., **4:** 1297
Grissom, Virgil I. (Gus), **4:** 1437
Gromyko, Andrei, **4:** 1657, **6:** 2240, 2266
Grosz, Karoly, **6:** 2287
Grotefend, Georg Friedrich, **1:** 68
Group of Seven, **6:** 2448
Group of Seven Summit (1992), **6:** 2448-2449
Grove, Andy, **8:** 3153
Grove, Sir William Robert, **2:** 641
Groves, Leslie Richard, **2:** 810, **3:** 998
Gruening, Ernest, **4:** 1322
Gruentzig, Andreas, **5:** 1890

XCVII

C

CIV

CXVI